AF704561

HUSBANDRY:

THE SUREST, CHEAPEST WAY TO LEISURE, PLENTY PROSPERITY AND CONTENTMENT PLAINLY ILLUSTRATED

◊

NATHAN GRIFFITH

COBBLEMEAD
PUBLICATIONS
TROUT VALLEY ◊ WEST VIRGINIA
1998

For Mišelka & Suzanka,

To lighten heavy days; Proverbs 8:21

©1998 by Nathan Griffith

FIRST EDITION

Published by
COBBLEMEAD PUBLICATIONS
HCR 68 Box 185
Trout, WV 24991

ISBN 0-9665103-0-5
Library of Congress Card Catalog Number: 98-93022

Manufactured in the United States by:
Morris Publishing ◊ 3212 East Highway 30 ◊ Kearney, NE 68847
1-800-650-7888

CONTENTS

1.	PARTING	1
2.	AWAY AT LAST	2
3.	GETTING STARTED	4
4.	FARMING FOR MONEY	6
5.	LAND & AUTONOMY	9
6.	BUYING LAND	13
7.	HUSBANDING LABOR	14
8.	WATER	19
9.	PREPARING LAND	27
10.	HENS	33
11.	GOATS	45
12.	COWS	47
13.	HOGS	58
14.	SHEEP	76
15.	RABBITS, ETC.	93
16.	DRAFT BEASTS	94
17.	OXEN	117

CONTENTS

18.	DAIRY	121
19.	POULTRY	127
20.	PORK	130
21.	CHARCUTERIE	136
22.	MUTTON	146
23.	PASTURE	151
24.	MEADOW	157
25.	POTATOES	164
26	PULSES	168
27.	SMALL GRAINS	174
28.	MAIZE	181
29.	ROOT CROPS	189
30.	GARDENING	193
31.	FRUITS AND NUTS	206
32.	HUSBANDING THE HARVEST	221
33.	SWEETS	229
34.	DUNG	236
35	ENERGY	239
36.	SMITHING	243
37.	ROUGH CONSTRUCTION	247

CONTENTS

38. VERMIN 253

39. ODDS & ENDS 258

40. EPILOGUE 266

APPENDIX 1--HUNGRY GAP 269

APPENDIX 2--DONKEYS AND DOGS 270

BIBLIOGRAPHY 271

INDEX 277

USEFUL ADDRESSES 289

1
PARTING

For where your treasure is, there will your heart be also. Matthew 6:21

As a musician I found out early in life what it meant to go long periods without "cash flow." Those times were commonly punctuated by sudden bursts of very good pay, but I had to spend many hours per week practicing music, whether or not the playing jobs came along. I couldn't seem to make much headway without immediately slipping backwards with every business doldrum.

While I was a child, my parents taught me about horticulture in our suburban lot. Mine was not a wealthy family; what little money I received was by labor. I helped out by gardening in the back yard and assisting my mother on occasion with her beehives.

Once on my own, I started the way everybody else did: I rented a room, and hunted for work. There is always at least *some* work for musicians; I was a music teacher for 6 or 7 years. I also performed, wrote and arranged music. I found I worked about 12 hours per day and still wasn't getting ahead. To top it off, I didn't really like the indoor life of commerce.

All my life I have enjoyed being outdoors. As an 8-year old boy I would march to the woods near town almost every day, just to be there. I chased rabbits, made little forts, and sometimes "played hooky" from school. I started my career as a scholastic shirker in the first grade, skipping school for two weeks while pursuing crawdads in the creek of a nearby park. As a ten-year-old I kept tame rabbits in the back yard. I began trapping muskrats for fur and meat in creeks right there in suburbia. By the time I was in high school I tended a garden, did yard work for many neighbors, grew ginseng under mom's backyard rose bushes, worked at blacksmithing using a make-shift outdoor forge, hunted, trapped, fished, and continued my musical education. I loved doing it all.

I missed the outdoors when money-making became the engrossing feature of life. Though the place where I lived was near the city, it was truly rural. Alas, it was rented, so all my improvements profited only others when I moved.

Earning money left precious little time to enjoy life. I discovered I was frittering away what little spending money there was. I constantly needed a change of pace, but couldn't afford to take a vacation. The weekends were times to "get away." I went to historical parks, woodland retreats, anywhere I could forget "life" for awhile and unwind.

Right there was my answer. If the daily business of making a living needed to be "gotten away from" then how happy was making a living for me?

It began to dawn on me that making money and making a living are not the same thing. Don't animals "make a living?" Yet they earn not a cent. I too, would learn to "make a living."

Mere animals also had more spare time than I; my goal was to reverse this. I would live, like all nature does, by making a living, and relegate money-making and other such pursuits to my leisure time. Like Noah of old, I would "begin to be an husbandman," enjoying the benefits of commerce, but not living by it.

I began looking for land I could afford. Wherever it was, my place needed to be able to support me, just in case I couldn't find paying work. It needed clean water, living quarters, a source of firewood, and cleared ground.

Being from among the urban/suburban masses, I thought what I wanted was privacy. Little did I realize at the time how much more privacy there is in an apartment or even suburban tract house than a rural home of any description.

Family and friends all vowed I had gone crazy. Vainly they tried to dissuade me from abandoning urbanity. Frightening fairy tales were invented to change my mind. Invitations to less rural locations were suggested, even offered! (This from the same folks who constantly muse how delightful it would be to live in the country!)

"You can't earn much money in the country!" they insisted.

"I can grow my basic needs, how much money would I need?" I asked.

"There's no culture, no museums, no fine dining."

"I'll make pilgrimages back when I miss 'em enough."

"You'll sacrifice convenience!"

"Working all day treading water isn't convenient."

And on and on....

2
AWAY AT LAST

I will lift my eyes unto the hills, from whence cometh my help. Psalm 121:1

Thinking I needed as much land as I could get, I eschewed more expensive areas for places where land was cheapest. For me, that meant buying a place near Clarksburg, West Virginia. There was an old cabin, a pond, two springs, a couple acres level, an acre of clear ground, and over 75 acres of land. It was at the back of a "holler" and was reached only by a poor dirt road. I found that the owner didn't need any credit check. He would finance the whole thing! Only 20 minutes from a good-sized town with all the amenities. I signed the deal.

Providence determined I would never move there. I slept in the place quite a few times. I moved furniture up there, and tableware, and a small backpacker's gasoline stove I had received as a gift years before. Before the end of a few weeks the door was kicked in during my absence, the contents of the house stolen or vandalized. I repeatedly came and found trash strewn about the pond. Mostly the locals resembled all the bad stories one hears about people in rural Appalachia. Even timber pirates worked hard stealing valuable trees. I soon put the place back on the market, and continued making payment after bitter payment. Then one day, another buyer appeared. I sold the place on the same kind of Land Sales Contract the former seller had used with me. The Land Sales Contract allows the transfer of title only after the final payment has been made. It contains many pitfalls for the buyer, which I learned about only after my own purchase. By the time I had sold it, I had it paid for, so I at least now had collateral enough to try again.

Now I was experienced. I had come to realize that large holdings didn't insulate me from intruders, guaranteed no privacy, drove up the taxes, and supplied too many tasks for just one man (low on capital) to carry out. My mistake was thinking I would earn enough money to pay for food, taxes, tools, and the land itself, all from the land.

It occurred to me much later that if I only bought as much land as I absolutely needed, I could afford it without debt, tend it without tears, and protect it from intrusion without angering interlopers. All the world loves, or at least respects, the smallholder. Two unsatisfactory land purchases later, I found better environs in the incomparable Greenbrier Valley of West Virginia. I found ten acres for sale, which though terribly rocky, had room for pasture, gardens and trees. It was mostly treeless, but completely covered with very heavy brush. There was a large trout stream flowing through, which the governmental authorities had altered so the valley didn't flood any more (quite a problem some decades ago). There was a hovel of a house, a good garage, a cellar, and other outbuildings. It cost less than the other place, and this one was right on a paved road, utilities already in place. There was a gravity springwater system and a hand dug well. I signed the contract. Within a year I was moved in, never to return to suburban Maryland except on rare visits.

Along the way, I found the people in primarily agricultural districts to be much friendlier, more law-abiding, and more helpful than those in coal/gas regions. This may be true only for West Virginia. It may not be true at all, except in my own experience. Regardless, one local person told me then, and I think it is probably a faithful saying: "God made mountains so decent poor people could afford land."

This place is approximately all flat, by mountain standards. The soil is varies from 5% to 75% boulders, with some soil and some gravel between them. Farming it commercially is out of the question. The same could be said of steep ground, forest, swamp, or sand desert. All of these conditions can be ameliorated to some extent, but neither government, nor big landowners, nor even angry mobs will covet such places. They wag their heads in pity for you, that's all. The taxes don't rise fast; on this kind of land taxes aren't much to begin with! A few acres of almost any land will make one's living, and even provide a little money from surpluses now and then. Our "unfarmable" land does so through not much more than 3 hours' daily husbandry, and the land just keeps improving.

3
GETTING STARTED

And it shall come to pass in that day that *a man shall nourish a young cow and 2 sheep; and it shall come to pass, for the abundance of milk* that *they shall give he shall eat butter. Isaiah 7:21-2*

The land was full of briars, burrs, brushy trees, and old bottles and other trash. Little by little it all fell to the scythe, chainsaw and brush pile fire. Inside the little cottage, books, tools and other possessions took up so much floor space there was hardly room to sit. The old stove in the kitchen had been burned out by former residents having used coal in it. It was March, and having been from the Atlantic coast I was unused to the cooler temperatures.

After picking out all the burr plants by hand, I turned loose a bunch of sheep I had acquired. I kept hens and a rooster on the back porch for a while--a big mistake. Aside from their droppings, which stank, I was summoned to wake long before daylight. I quickly made them a little house.

I bought mail-order ducks and geese. The ducks were a mistake, as I found out. They messed all over the place, turned the whole yard into mud puddles with their digging, laid eggs "on the run" without a thought to nesting, and when they did nest, they were the poorest mothers of all poultry. Rarely did a duckling survive his mother's leading them all over the countryside, and into the fast-moving creek. They needed lots grain. And if not butchered promptly at a specific time, they're full of "bristles," so the skin isn't edible.

The geese were much better. A few ate some bees, and died. Besides that, they were ready for limited grazing at less than a week old. The mothers have been superb. The young are hardier than any other poultry. Geese don't even like being indoors, except when the roof is high and at least one side of the building is fully open. Unlike any other poultry, geese respect most fences, at least our white Embden geese do. They are much better to pluck than ducks and their eggs, which come at Easter time are very salable as Easter eggs, being about fist sized.

I put sheep out on the grass, such as it was. They did well.

Before long I found out where local livestock auctions were held, and picked up three young pigs. These were not my first swine. As a boy I helped my uncle tend his hogs in South Carolina. I had raised hogs two years in a row on my semi-rural rental place in Maryland. It was there I found out how unreliable would-be butchering helpers were. No one who promised even showed up to help with the big job! The plus side of this was that I learned how to manage without 'em.

I put the swine, two young females (properly called "gilts") and a castrated male (barrow) into the erstwhile garden space to sort of clean it up.

Shortly afterwards I bought a cow. The old folks said "The cow is the center of farm life." Without it, there is little dung for the garden, and little milk (unless you have goats, but that's another story). Without milk, there's no butter, no cheese, no cream, no yogurt, no ice cream.

My cow had a birth defect, which lowered her selling price. She had a sway-back. I found out later that she ate too much for the amount of milk she gave, though she was still more efficient than many other locals' cows. I was so excited when I brought her home. I sent a photo of her to Eva, who was still living in Maryland at that time. Eva was later to join me in this venture, and the cow surprised and intimidated her, I think. Me too. I learned to milk her the day I bought her. I enjoyed good milk from that time forth, except a couple months a year before calving. But that's another story too.

By May of that year (1986) I had a garden started. I didn't follow the methods I had grown up with. Those old ways included 8-inch thick mulching with grass clippings, planting only by pulling aside just enough mulch to put a seed or two in the soil. Grass clippings were rare in the country, common as they had been in suburbia. There, I could just drive the car through town on "trash day," and pick up all I needed for free. Here all there was was hay, most of which had copious seed matter. This sprouted, making the weeds worse than at first!

The other method was that of cultivation. The cardinal rule of cultivation is to keep the soil stirred constantly, both to check evaporation, and to discourage the sprouting of weeds. I learned my lesson when I planted about a sixth of an acre of corn back in Maryland. I cultivated only once, and had about a bushel and a half yield, all nubbins. A good yield there would have been 20 to 30 bushels. Naturally, dung would have helped, but I was informed by a kindly farming neighbor that even without it, I should have gotten seven to ten bushels. Had I access to a few tons of grass clippings, that sixth of an acre should have yielded at least 15 bushels.

The leaky little pony shed here afforded a place for sheep to escape the rain. The cow stayed out that summer on grass. Much of my food was bought, because the garden wasn't yet productive and only the hens and the cow were giving regular produce. I was buying feed for the pigs and cow. The hens ate bugs, grass and grain from the other animals' droppings.

My plan, a poor one in retrospect, was to make enough money selling lambs, goslings, ducklings, eggs, piglets, wool, and surplus milk to pay for any other expenses I might have. But the grubstake I arrived with continued to dwindle with the passing of the year, especially so when I bought another pig at the auction, a boar, and thus another (big) mouth to feed.

While at the market buying that boar, I met a man who was "real big" in the pig business. He had about 200 hogs, and was expanding to about 1000. (I found out with time that modern officials consider these numbers about minimal and call this a "family farm.") He heard I wanted to sell piglets, and proposed a deal with me for him to buy my pigs at an average price, regardless of market fluctuations. I was in the pig business.

4
FARMING FOR MONEY

Ho, every one that thirsteth, come ye to the waters, and he that hath no money; come ye, buy, and eat; yea, come, buy wine and milk without money and without price. Wherefore do ye spend money for that which is *not bread? and your labor for* that which *satisfieth not? hearken diligently unto me, and eat ye* that which is *good, and let your soul delight itself in fatness.*

Isaiah 55:1,2

I had the barrow slaughtered at a local abattoir, and cut up the carcass at home.

I still kept the two sows and the boar together, and the sows were pregnant. I was keeping these hogs behind two strands of electric fence, but they still got out almost daily. I was feeding them plenty, but that didn't satisfy them. Here's a lesson: hogs with pricked-up ears can see the wire. They run through it, getting stung on the way out. At least they killed the brush in that enclosure, and dug up lots of rocks.

I had no idea when the sows were bred, because I wasn't paying enough attention to them. I had my new "business associate," the local hog expert, examine them and he said they would farrow (give birth) in two weeks.

There was no time to lose. I quickly slapped up some plywood farrowing houses according to a design (I later found it highly inadequate) I got from a "home-steader" build-it-yourself type book. In two weeks the sows had their pigs. One had 12, the other, 8. They soon stomped some piglets to death, but I managed to end up with 17, which I sold for $30 apiece. I knew I didn't have $510 invested, and felt satisfied.

Shortly afterward, a relative died and left me a small sum. I spent it all on more fence supplies, more hogs, better farrowing houses, and, (ugh!) feed. I sold dozens of pigs at $30 each, even though that year the market climbed to over $70! We were about breaking even, because I had to buy more swine tools, lots of feed, more fences.

I don't have to tell prudent readers the pig prices soon crashed. I was happy as a clam. I was still getting $30, tho' the market had dropped to between $10 and $15. I was gloating over the fact that I had sold pigs at "half price" for a year, because now I was getting double, even triple!

Then my business "partner" quit buying. He said he couldn't afford it. We had no contract, just a gentleman's handshake. But only one of us proved a gentleman. Even if we had had a contract, he could have simply declared bank-ruptcy, and I would have great difficulty receiving equity in the case. I had dozens of pigs, many pregnant sows, no money for feed, and the market was very bad. My mother had just passed away, and I again received a small inheritance.

I used it buying an old tractor, and more feed.

I did what I could for a while. I eventually even borrowed money (which took years to pay back) to feed the remaining hogs until they could be sold. But the market stayed very bad, and feed prices began creeping up. Unable to sleep for weeks, very early one wintry morning, I resolved to sell all but two small pigs. After paying some of the debt, I was totally broke.

I needed money just to pay for the telephone (needed for emergencies) and property taxes. There was little to eat, because with all the work of a hog farrowing operation, I couldn't tend the garden. Eva found it too difficult work for her alone. We had no money. We were "ashamed to beg; ashamed to let the neighbors know."

A friend of mine needed some part-time help with his small business. His house (he worked at home) was three miles away. I was paid very little, but I was very grateful, and so was he.

He sold old and rare books. I got to read them during lunch hour. With all the work I still had at home, and with his meager budget, I only worked 2 days a week, at most. But it got me through. I remember some days walking there through 2 feet of snow, setting out before dawn, and working until very late at night. I also fed his sheep when he went to fairs to sell books.

Through the efforts of his very kind and quick-witted wife, I got started back in music. She got me my first playing job in the area. I played at a very prestigious hotel, in a borrowed tux. I played again a few days later. The people greatly appreciated "civilized" music "way out here" and I have enjoyed their patronage on a limited basis every year since then.

I climbed back up out of the financial hole I had dug being a "money farmer." Looking back it is very good that I turned down a government loan to fund my hog business. I would have lost my home when I couldn't make payments. The loan I did have was from a very understanding relative; I paid it back as quickly as I could.

I learned during the whole ordeal that I didn't need money as much as I had thought I did. Having very, very little of it (I lived about two years on a grand total of $100/month budget) I found myself taking few trips to town (an hour's drive away). The garden became more of a priority. I soon dismantled the farrowing buildings, making a new hen-house, some fences, and a "new, improved" pig house.

I sold the tractor and its implements and bought a small horse and harness tack. I have never for a moment regretted that decision. I sold the roto-tiller I had, too. It was a top of the line model, a BCS, but I found it created about as much toil as it saved, and there was no end to its noise and fumes. I sold a power shredder/grinder I had. I had only been using it to grind grain for all those hogs.

Each of the items we sold brought very little money, but it was all we needed to live on for a year. In fact, we discovered we were beginning to get ahead. This, without ever rising up to the "official" poverty level in income.

Readers might find this a bit scary if they're contemplating a move to the country. They well ought to. The thing to remember is America grows about a third more food than it consumes. It will always be a buyer's market until that changes. Farming for money tends to be successful when items are grown which no one *truly needs*.

Perverse as it sounds, agricultural enterprises are very, very commonly initiated just to (on purpose) lose money. The wealthy find it a wonderful way to cut their unfairly high tax rates while sheltering their capital in the land. Capital is not taxed if it is reinvested quickly, so they sell the improved land and acquire profitable enterprises with the net gain. Voila! No tax, and the cycle begins again. Competing with people trying to lose money at farming is no way to make a profit.

Very few big investors bother with such things as market gardening. Nor do they "lose" money on floriculture. Nor on more-or-less ornamental things like gourds, Indian "rainbow" corn, walking stick cabbage, bedding plants, etc. Nor on craft supplies such as handspinning wool, goose down & Easter eggs, ornamental (colored) cotton, luffa sponges, etc. Strictly local markets exist for some items, like maple syrup, or sorghum molasses. The honey of commerce is rarely sorted according to its origin, i.e., basswood, tulip poplar, sourwood, buckwheat, etc.; health-food outlets and country stores will often retail it for you.

My own experience is that most people who pursue a single money-making scheme often begin wasting too much of their time trying to earn money to make a living. Thus they give up their self-sufficiency, their independence, and start back on the road to urbanity. I came very close to it myself with the hog business.

I think the best course of action is to find enjoyable work. Take it for granted the cash flow will be considerably less than it would be in the city. But don't forget to tally up the money saved in taxes, housing, food, water, the costs of attending a daily job, and the usual crime, disease and ill will of crowded places.

But I have digressed.

I neglected to mention that my antiquarian-bookselling friend handled mainly religious books, but he also specialized in old agricultural books. In the year I worked for him I read, borrowed, and bought dozens of books. I learned a great deal, things which made life easier in the days before America became petroleum-based. Finding much scientific lore too, I began experimenting with low-cost ways of making a good living with less money, less labor. Little by little, our tiny farm began taking shape. It continues to change as improvements are carried out, and "new" old ways are implemented. I suppose my ultimate goal is that the place will practically operate itself, so that a cranky, senile old invalid can live here without help. That eccentric geezer will be me.

5
LAND & AUTONOMY

For the LORD thy God bringeth thee into a good land, a land of brooks of water, of fountains and depths that spring out of valleys and hills; a land of wheat, and barley, and vines and fig trees, and pomegranates; a land of oil olive, and honey; a land wherein thou shalt eat bread without scarceness, thou shalt not lack any thing *in it. Deuteronomy 8:7-9*

Human Needs

"Making a living" consists of procuring those things we cannot do without. My list of human needs looks like this: Air, water, food, shelter, sanitation, energy, clothing. All these needs can be satisfied with a small parcel of land, a bit of labor, and common sense. Using these is learned, not inborn. No one yet knows how much food, building materials or energy an acre may provide us. Human ingenuity seems almost miraculous in its ability to improve the land's productivity.

AIR

Reliably clean air is where one finds it. There's no swamp or marsh gas here in the mountains. I turned away from buying an otherwise nice seaside place one time because it was downwind of a nuclear power plant; a trapper friend had showed me numerous photos of animals he caught in the vicinity. I saw animals with extra-short legs or tails, and with fur curled up like wool; one possum had no ears, just little holes; a fox had a lower jaw too short to match the upper one. Despite official promises of safety, I guess my superstitions just got the best of me. Find a place with air you like.

WATER

After air, the foremost need is water. Clean water must be abundant and reliable. Cisterns (rainwater storage) wells, and springs are the most common clean water resources. In former days, rainwater was for drinking, while more abundant sources, like streams and ponds, were for other uses.

Well water is usually safe to drink, but often not good tasting. Deep wells are the most likely to contain iron, calcium or sulfur. These all have undesirable tastes and interfere with sudsing ability. Shallow wells can be contaminated by nearby hazards like septic tank effluent, or livestock feedlot runoff. Such things ought to be downhill from the shallow well, especially if they're within 150 feet. Ask the seller for a drink of water.

Someone else's well may get pumped so much that the water table falls below the level of your shallower one. Then yours must be abandoned or deepened.

There are two kinds of springs. The first is a well-defined hole from which water flows. The other is a wet place in the soil in some seasons, and a visibly flowing trickle in others. The former may be running through caves and rock fissures for miles before surfacing, and is sometimes contaminated from sink hole debris and other stuff. The latter is simply a place where the topsoil and subsoil are so shallow they expose a sandy or gravelly layer carrying water; i.e., the aquifer. If this type of spring occurs high enough a gravity flow system may allow running water without any pump at all! It's the land's best asset.

If they're pure, streams and brooks are very good sources of water. If not, they still warrant considerable merit, because they can supply water to animals, irrigation, and power generation. Over 90% of the water we use need not be drinking quality. Streams and brooks identify the level of the water table; if at any point it is higher than your homesite an "Indian well" may supply pure water despite upstream hazards. More about this anon. But speaking of hazards, many streams overflow their banks, and can be a nuisance and even a threat to life.

FOOD GROWING CAPABILITY

The Bible says "All Flesh is Grass." The grass family not only includes grazing land but also wheat, corn and other grains. These grasses feed us and our animals. But most require workable ground for a successful crop.

Obstructions to tillage include stumps & trees, stones, poor drainage, and steep slopes. All these can be fixed, but it costs. If the land is so cheap that you have money left over to bulldoze or blast out rocks & stumps, then go for it. If the soil was once productive cropland, but is now forested, it may pay to sell the timber and use the proceeds to pay for the land's improvement. Poor drainage can be improved. The Romans often used open ditches with grass on them to prevent erosion. Water exits by the path of least resistance. Land drainage is expensive. Besides, official busybodies may hound you for it with their red tape and fines. Of course, what they don't know won't hurt them.

Steep slopes are often made tillable with terraces. These aren't suitable for tractor tillage unless exceptionally wide. But large machines can be hired to make terraces. Terraces tend to be the most productive of all farmland. This is because drainage is good, including air drainage. Cold air can sink away from young springtime plantings; the growing season is longer. Terraces are cultivated by hand or horse. Crops and dung must be hauled up and down hill a lot. Drought often occurs unless irrigation is possible.

Assessing Fertility

Existing vegetation says much about the land's potential. Some trees, like beech, hickory, sugar maple, white oaks, and walnut, especially if they are large and healthy show the land is very good. White and Virginia Pines, scrub oaks and rough, weakling trees grow on poor, or at least abused, land. Healthy willows, elderberry bushes, alders or silver maples, show poor drainage.

Weeds grow mainly where the land has been farmed before. Sheep sorrel, blackberry brambles, blueberry bushes and some kinds of green moss indicate acidity, which can often be corrected by liming the land. Mullein, Queen Anne's lace, ox-eye daisy, broom sedge and hawkweed show a lack of humus. The latter two often show the ground is untillable, since they can't take tillage. Healthy dark green weeds mean the ground is rich, but neglected.

If the land has good fences, so much the better. Putting up fences is a whole lot of work, but is necessary where grazing animals or poultry are kept. Even vegetarian people need them to protect gardens from vegetarian animals. Contact farm stores to get an idea of their cost, and you'll see that aside from the labor, they're infernally expensive. Making fences from the land's bounty, like split rails or stone, is perhaps cheaper, but the time they take is something rising husbandmen may not have.

SHELTER

It is a huge advantage to have livable shelter already on the land. Some "back-to-nature" faddists of past decades slapped up tents and shanties, finding when their supply of narcotics petered out they were no longer comfortable and not very secure against wild and domestic pests. Better an unsightly trailer than that. In all fairness though, the Pilgrims started with tents and a dugout; their hard work and faith in God saw their plans succeed.

Structures

Don't forget barns and outbuildings. I went for 7 years with no hay barn, storing hay here and there in a garage, in leaky little sheds, under tarpaulins, sometimes just buying it as I used it. What an affliction it all was! Every building I have added here has decreased my daily labor, while adding to the land value (if I ever want to sell out). Buildings you'll need sooner than later are tool sheds, animal housing, a cellar for food storage, and a place to keep animal feeds like hay and grain.

SANITATION

Septic tanks and dry wells are the two main sanitation devices in rural places. The former isn't really all that sanitary, and many aren't working right anyway. The latter simply allows soapy "gray water" to soak into the ground via a pit filled with rocks. Be sure neither one affects the well, if there is one. Some places just drain effluent right into nearby water courses. This is not only criminal but not very bright.

ENERGY

For husbandmen, firewood is used for heating the house, cooking, hot water, and less well-known things like blacksmithing, soil sterilization, smoking meats, etc. In fact, there is very little use for forms of energy besides firewood. Here in the mountains summertime cooling isn't really a problem. Woodlots usually grow about a half-cord of new wood per acre per year. With some care, they can probably grow a good bit more. So reckon about two acres for every cord of wood you'll need to heat the house in winter. This is hard for beginners. Ask how much other folks in the area use. Compare their houses to yours both in size and insulation. Figure from there. Many people here in the mountains are able to get all the firewood they need from deadwood on public lands.

CLOTHING

All our needs are products of the land. Clothing too. Like cooking, sewing is a subject apart from producing fiber and hides. I hope readers won't mind if I avoid the subject altogether. I don't know enough about tailoring to do it justice. Being covered and being clothed aren't the same things.

I'll at least say that unless you raise sheep or devote some land for textile crops, you'll *have* to get clothing through commerce. One distinct advantage of not having to hold "a job" is that one needn't always wear "nice" clothes. As a matter of fact, it is often a fine feeling to wave off the salesmen, evangelists and other intruders from out your dungareed tatters, with the counsel that "The boss ain't here!" Works every time.

HAZARDS

Landslides, mudslides and avalanches are sometimes hazards in hill country. Many places underlain with limestone are also underlain with caves; sinkholes have been known to appear from "nowhere" swallowing up anything built over them.

West Virginia's Greenbrier Valley recently had two great floods within ten years of each other, each of which was "the flood of the century;" beware of flood zones. Tidal waves are a danger in some parts of the world. Earthquakes and volcanoes cause them.

Wind and fire hazards often go together. In dry regions accumulated brush, even with a wide clearing, can consume a small holding with astonishing rapidity, a matter of minutes! Being surrounded by beautiful forest has its risks. In mountain and valley regions, blustery weather is very localized. A farm 2 miles away from me constantly has high winds, even when it's calm here.

Governmental menaces have been increasing for decades. Zoning or planning ordinances are more often used as weapons than for beneficence. Find out who must be appeased and how much they demand. Or better yet, avoid zoned areas. Same for other intrusive "codes."

Electricity, telephone and gas lines are a plus, even if you don't use 'em. The first allows instant lighting, refrigeration, and labor savers like washing machines and power tools. You could phase these out if you like, as it becomes convenient to do so. Phones are better insurance than a paper policy, because an emergency call may save a life, a policy (after suing the insurance company to honor its commitment) can only give the promised money after a tragedy.

I hope readers don't need to be told that the worse the access to the place, the less your chances of success. I'd rather be right on a good road with all its annoyances than on a goat path a mile from it. Emergencies and freight hauling are without end. And if you change your mind about self reliance? How will you ever find someone as gullible as you were to buy the inaccessible place?

Buy your land outright if even remotely possible. Better have a smaller place than take on a bank note. I know we would never have survived here if we'd had a monthly payment to make--even if it were small. There have been times we went a whole year with no more annual income than three ounces of gold would buy. Never mortgage your homeplace. Never.

Summary

Consider all your human needs as you decide on a parcel of land. Those needs you cannot provide yourself may one day terminate your economic freedom. Atmosphere first, water second in importance. I know rural folks who live with the constant stink of a paper mill, and who buy all their drinking water; their house is heated by oil! Rural security my foot. Even if you can't live entirely from the land at first, see to it that it's not the land's fault.

6 BUYING LAND

It is *naught,* it is *naught, saith the buyer: but when he is gone his way, then he boasteth. Proverbs 20:14*

Most land probably changes hands through the work of real estate agents. My experience with these people is they know very, very little about land itself. Most don't even know their way around the development of unimproved land. Some know far more than they need to sell property, their primary goal. But all of them work for the seller. Precious few will even take a retainer to locate property on the buyer's terms.

Finding It Yourself

Every prospective husbandman should familiarize himself with courthouses (they're all nearly the same in operation, though not in size, staff or equipment). A good place to learn is in your present locality.

Start with land records offices. Find the indexes to grantors (land sellers) and grantees (land buyers). Look up the seller's (or buyer's) name of the place where you now live. The staff is paid to assist you, so ask questions. Find out who had it before now, and before them. Look up all those names in records of judgments, liens, divorces, wills, etc. Write them all down, in order, like a family tree for the land parcel. For extra practice, go all the way back until the county was formed! You have just conducted a "title search."

Then search other parcels, writing down the details in a notebook. The tax assessor's office will have "tax maps" which will enable you to find specific parcels of land by comparing them with ordinary road maps. Some tax offices have aerial photos, too.

Having learned to do such title searches (and in the process having found out possible clouds in the title) you may explore country roads, taking note of places which interest you. Go back to the courthouse, locate the parcel in a tax map--by comparing to roadmaps--and find the present owner. Then find what he paid for it, if possible, in the records (transfer taxes, and often property taxes are figured on sales value.) Find out what neighboring properties sold for recently. Seemingly abandoned land is often "for sale" though not listed by agents. Absentee owners are the most likely to sell, especially if they have no resident tenants on the place.

You may attempt to make offers

on land at this point. Let me offer the suggestion that in any offer you submit, you should include a "weasel clause." This is a "subject to" provision which allows you to withdraw your offer in the event some conditions are not satisfied. Some examples are "this offer is subject to buyer's ability to obtain satisfactory financing" or "Subject to buyer's approval of any liens, easements, rights of way, etc." or, subject to buyer's approval of present means of ingress and egress." I will bow out here, and leave that to your own business acumen. But I will mention certain terms you ought to be familiar with: Deed of Trust, Quitclaim, General Warranty, Special Warranty, Land Sales Contract, Covenant, Surface Rights, Earnest Money, Adverse Possession, Fee Simple, and Riparian & Appropriative Rights. Look them up at a library.

Money saving methods to investigate are: Partnerships in land buying; subdividing larger tracts, selling off what you don't need; estate auctions; trades for land; and tax sales. Look these up too.

Legal surveys are sometimes needed for property transfers and they are expensive. Get estimates before making binding offers or commitments.

Husbandman's Land Checklist

1. Climate/health of location, growing season, rainfall, temperature range.
2. Good water.
3. Tillable ground enough for gardens & fields, clear pasture land--at least 3 acres.
4. Orchards, groves and/or vineyards in good/reparable condition.
5. Mineral, riparian and timber rights
6. Good access; good buildings; good fence; proximity to established farms (to learn and buy from in times of failure);
7. Emergency services;
8. Extent of governmental menaces, tax rates.
9. Utilities and Mail Delivery
10. Nearness to market, or to paying work, etc.

7
HUSBANDING LABOR

There is *nothing better for a man,* than *that he should eat and drink, and* that *he should make his soul enjoy good in his labor. This also I saw, that it* was *from the hand of God. Ecclesiastes 2:24*

You can see conserving labor had become important if I was to make a living in 3 hours a day. Labor itself became a sort of study with me. I observed that planned convenience made onerous jobs like dung-work or wood-splitting more likely to get done. Laying out the farm with the heavy tasks in mind saves much needless drudgery.

Gardens and fields ought to be close to livestock pens, so dung isn't moved far.

All everyday heavy-lifting jobs ought to be scrutinized to see if loads could instead be slid, scraped or toppled. For example, old time husbandmen often kept their hay or straw on a floor above the winter pens in a barn. The insulation from the straw kept the stock warmer--so they needed to eat less to stay warm. Also, such an upper floor was often entered and filled with the summer's haying by way of a banked-up earthen ramp--hence the name "bank barn." This way hay might be dumped from a wagon; and again dumped into troughs below as needed. Finally, the dung was scraped, not lifted, through a chute or gutter into a dung-wagon waiting at a still lower level. If it wasn't a mechanical dung-spreader wagon, it might have a lift built in so that the dung could be scraped off the wagon onto the field. Thus the first lifting of the hay onto the cart was the only lifting, and the frequency of dung-spreading depended mainly on the daily accumulation and on the weather.

Many of us load the hay upon a truck and stack it in the barn. In winter, we lift the hay back onto some type of carrier, tote it to the fields where stock are kept. There it is fed, on the ground, to be trampled and dunged itself, wasting half of it, and the labor that went into it.

A thinker would try his best to rather extend the grazing season, so less hay is needed to begin with. Then the hay might be kept in much smaller buildings in each field. This way hay isn't hauled as far, either coming or going, and the animals are rotated in winter, dunging the fields more uniformly.

All that is moot for the husbandmen who can't or won't keep hay-eating animals. But compost making for the garden consumes much labor, too. I once build a tower with three compartments of convenient height. The highest compartment was at about eye level, and raw vegetable matter and dung were forked into it as they accumulated. The floor of this compartment consisted of a series of steel rods about 2 inches apart, forming a grid. These could be pulled out, one by one, when the top compartment was full. The heavy mass, which by this time had begun to really heat, dropped into the middle compartment which had the same type of floor. The "one-by-one" rod removal caused the mass to turn itself, and it became even hotter in the second compartment. This heat in turn helped heat the new deposits being added daily to the top compartment, so it'd break down faster. Finally, a dung cart was parked in the bottom compartment, and would be hauled afield only when the top chamber was full and needed to be dumped. Thus on my flat land I lifted my compost only once, and that in very small and convenient daily dollops. I never got an aching back from it. Better yet, the heating and aeration it got did away with flies and stink. The floors of the pig styes are concrete, and the dung which accumulated there, together with what absorbents I put in it, could be scraped, rather than lifted. Using gasoline-powered lifters for all this would have encouraged me to let the manure lie until it was positively filthy in the poor pigs' pens. And, if the machine broke, I'd have to pay for repairs. And if I couldn't get gasoline because of snow or other weather problems....

It pays to set up the homeplace so that much of the work does itself like that. Minimize labor, and make what's left more convenient.

Do this with everything. One finds such mental puzzles a real joy to solve--real satisfaction in this, you know.

Know Your Labor Force

One man can't lift as much as another. Men and women aren't equal either. It's foolish to put small women to work clearing stones and stumps from the field while a big man tends the wash.

Likewise it makes no sense at all to want to "equal" men, when women are clearly superior at many jobs. And if those jobs are done in the comfort of home and yard, how much better than in the dung heap with flies and sweat?

Children too, are useful where jobs are close to the ground. I remember (with a distinct lack of fondness) digging peanuts in South Carolina as a boy. The hot sun was dreadful--I'd have much rather been dragged from bed early and be digging peanuts in the cool early light. Men sometimes must labor in hot sun, children ought not, for they learn to hate work.

Horses and oxen can be calculated as having the same power as men in their prime on a weight basis. That is, a 1500 lb. draft horse has about the same power as ten 150 lb. men. Though animals must be trained to each new task, men can learn through explanation alone. On the other hand, horses have more of something called glycogen stored in their muscles which enables them to temporarily generate great power. In fact, a horse excels all types of motors and engines in its capacity for overload.

The best rule of thumb for calculating potential labor capacity is to divide the weight of the laborer by between 8 and 10. For a 150 lb. man, this means between 15 and 22½ lbs. This is the forward moving force. This is enough to pull a load of about 1000 lbs. (including the wagon or cart) on level pavement. On common earthen paths it takes from 1½ to 5 times this force. Of course it takes more power to start a load than to keep it going. But one man can momentarily exert one full horsepower; if he doesn't have to repeat it often, the above load isn't out of the question.

Lifting is another matter entirely. All forms of repetitive lifting are best just avoided unless they use only leg and/or arm muscles.

Human Strength

The force of one healthy countryman, unaided by power machinery, and working to his best advantage equals roughly the power required to raise 70 lbs. one foot per second for a ten-hour day.

The maximum power of a strong man, exerted for 2½ minutes, equals that required to raise 9 tons one foot in one minute. After that he's pooped for the day.

A man may travel, without a load, on level ground, for 8½ hours per day, at about 3.7 miles per hour, or potentially, over 30 miles per day. He can carry, in spurts, about 90 lbs. 11 miles in one ten-hour day.

I don't think I've met many city men who could do any of this. Those who can, exercise and aren't flabby.

Women are typically figured at roughly two-thirds the strength of a man, and children over 10 between one-fourth to two-thirds.

It takes a man about 12 to 16 hours to cultivate one acre with a wheelhoe of 8" width.

A man may plant an acre of corn with a jab-type planter or using a wheelhoe furrower and dropping corn in bunches (called "hills") of 3 or 4 seeds in as little as two hours, depending on how far apart the hills are spaced.

One man can cut and shock an acre of corn per day.

One man can pick an acre and a half of corn ears per day.

One man can dig with a fork roughly one-fifteenth of an acre per day.

It takes from 10 to 20 minutes to hand milk a cow, depending on how much milk she gives.

One man with a scythe can mow about one-fourth acre of hay daily, or about one-half acre of small grains.

Twelve to 14 bushels of potato seed can be cut by hand in one 8-hour day. Dropping potatoes into the furrow by hand, one can plant 1.2 to 1.5 acres in 8 hours.

A 3 foot deep trench 80 feet long may be dug in one eight hour day by hand, in good soil. If 4 feet deep, about 48 feet may be dug in one day.

A man can set out up to 1000 mangels or rutabagas per hour, or about a half-acre per day.

Strength of Horses

Horses travel about 400 yards, at a walk, in 4½ minutes; at a trot, in 2 minutes; galloping, 1 minute. They may carry up to one-fourth of their weight up to 25 miles in 8 hours.

On a big old-fashioned treadmill, the kind where the horse walks inside a sort of giant hamster-wheel 25 feet in diameter, a 1000 lb. horse exerts the power of about 4½ horses. 'Tis a horse's most efficient way of working.

A large draft horse may plow up to one acre of already loose ground with an 8-inch plow in one day.

A large pony may pull a 5-inch plow, and plow ½ to ¾ acres in one day in loose ground.

A horse pulls a ½-row walking cultivator, cultivating ¼ acre per hour. An ox will do the same, except only two-thirds as fast.

A two-horse team will pull a 1-row riding cultivator, at a rate of about two-thirds of an acre per hour.

A one-horse, 3-foot wide mower cuts up to ¼ acre per hour.

A two-horse 5-foot mower cuts up to ¾ acre per hour.

Some Rules For Labor

1. Non-strenuous methods, even if they take longer, are better. If you are rushed in your need to complete a task because of weather, etc., then you didn't plan the job properly, or took on more than you should have.

2. A daily task of 10 minutes' duration is more likely to be completed, and seems to accomplish more over all, than the time equivalent tackled once a month.

3. Labor spent preparing tools is generally less and easier than working with poor or dull tools.

4. Some jobs require waiting, and force us into greater labor all at once. For these, extra capital investment is often justified. The following come to mind: Root/potato crop harvest--special plows, lifters, etc., using draft animals or other power. Butchering--hoisting equipment, special work tables, meat saws, etc. Sheep shearing-special pens, clippers, etc. Manuring--lime spreaders, dung carts, etc. Fences--wire stretchers, post drivers, props or jigs for sharpening posts. Buildings dedicated to specific use--hay barns, corn cribs, ice houses, spring houses, etc.

5. Being independent means

your labor supply is not bought, but rather maintained upon the products of that labor from the land.

6. Comfort in labor increases output, endurance and morale.

Comfort in Labor

Mornings and evenings are very often the time when winds abate. When working with dusty or chaffy materials, this is the best time. It is also the best time to burn brush or other refuse.

Summer thunderstorms are least common between 5 to 10 a.m.

The coolest temperatures of a summer day are from the false dawn (earliest light) to about 9 a.m., beginning to cool off again in most places about 3 hours before dark.

The warmest times of a winter's day in sheds and outbuildings are usually from 4 hours before dark to one hour after.

The coldest winter hours in sheds and outbuildings are usually from dawn to early afternoon.

The most productive of all work is almost always done seated, under cover.

The "economies of scale" apply to rural people just like to urban factory owners, but the economies of proximity are often overlooked by both.

Convenience encourages labor; inconvenience sows sloth.

Convenience & Economy

Routine should be encouraged so mistakes are not made and progress is expedited. Wandering minds don't interfere with efficiency when tasks are routine.

All livestock animals perform incidental labor of some kind. Employing this is very thrifty. Neglecting it interferes with the economy of the holding:

Chickens main use is for eggs, then fresh meat, then dung, then incubation and brooding all sorts of eggs, and lastly as scavengers and garbage consumers.

Cattle's most efficient use is for dairy products, then labor as oxen, then manure production, and also for land conservation (they can be kept indoors happier than most other livestock.) They are least efficient as meat producers. And yet beef cattle utilize poor land most productively of all livestock, and make the best draft oxen.

Swine provide the most versatile meat products, especially cured hams, bacon, sausages, etc., are the best recyclers of garbage, the best dung producers, and best clearers of new land for cultivation.

Sheep are singularly the best textile producers of all livestock, and secondarily provide good meat on good pasture.

Horses' primary use is draft, then transport, "hot" dung to heat hotbeds, then lastly, recreation.

Geese can thrive on good pasture alone, unlike other poultry, and are sometimes used to weed certain types of cultivated crops.

Goats, because they are browsers rather than grazers, are better for nomadic self-supporters than for settled husbandmen; their main use is dairy, but serve secondarily as brush killers and for textiles.

Animal labor is cheaper than human labor.

Synthetic labor (motor power, etc.) fosters dependency on others, while natural labor encourages independence.

Commercial husbandry favors expansion, eventually necessitating hirelings, and then, monoculture, an unhealthy practice.

Labor Saving Devices

Labor savers often turn out to be labor modifiers. For example, in order to pay for a large tractor, we must ordinarily drive the tractor far more hours than we used to have to work when we didn't have equipment to pay off. So it doesn't save us labor; it adds to it. Though the labor may now be less strenuous, we can no longer keep a leisurely pace, because the bill collector must be paid.

If we save our pennies long enough, we may not have continual payments for the tractor. But we might instead use some interest from such savings to pay others to do disagreeable work. The capital itself is a better labor saver than the tractor, which wears out in time. But neither truly SAVES labor.

Buildings made to outlast our lifetime defer repair work until someone else owns it.

Living fences (hedges) require moderate labor at fairly frequent intervals. Stone walls require very great labor only once, if properly built, with rare repairs. Wire and wood fences require strenuous labor about every 15 or 20 years, with frequent minor repairs.

Water transport moves the greatest loads with the least cost.

Tight insulated buildings economize fuel gathering work.

Masonry is more durable than wood, plastic, and most metals. When permanence is important, it is often worth consideration.

Summary

1. Before undertaking strenuous work, always weigh health risks, hazards and less demanding options. 2. Use your legs more than your back; shoulders more than hands, and brains more than all. 3. Better to have many routine small operations than one huge job each day. 4. Time spent preparing tools is rarely time wasted. 5. Comfort and convenience are companions to diligence. 6. Animals often benefit by being beneficial to us. 7. Commerce is ultimately the only tool officials have to deprive us of the full fruits of our labors; we must therefore use it wisely and not become dependent on it.

8
WATER

Drink waters out of thine own cistern, and running waters out of thine own well...Let them be only thine own, and not strangers' with thee. Proverbs 5:15,17

The Indian Well

The easiest way to get drinking water is with an Indian Well. It is simply a hole dug some distance from a creek, pond, or other body of relatively clean water (no sewage treatment ponds, please). You'll strike water not too far from the surface in these cases, unless the stream runs on a bed of solid slate or other bedrock. The sand and gravel bed of most streams is an ideal filtration unit, but most of the water you'll take from it isn't coming from the stream, it's coming from the same underground water tables which feed the stream.

Dig down until a couple of feet of water is in the hole. The pit should be big enough to stand and work in. Bail out the water which accumulates, and try to keep ahead of it. A powered pump may be in order here. When a point is reached where water can't be taken out fast enough to dig any more, the hole is lined with flat rocks or stones laid in a circle like an old-fashioned well. A better way to line it is with a concrete culvert pipe, with an 8 foot diameter concrete pad around it at ground level. It may be capped off with something to keep vermin out. If this apparatus is installed uphill from your house, gravity can be used to convey the water home. That way there's no dependence on electric companies or mechanical gadgets.

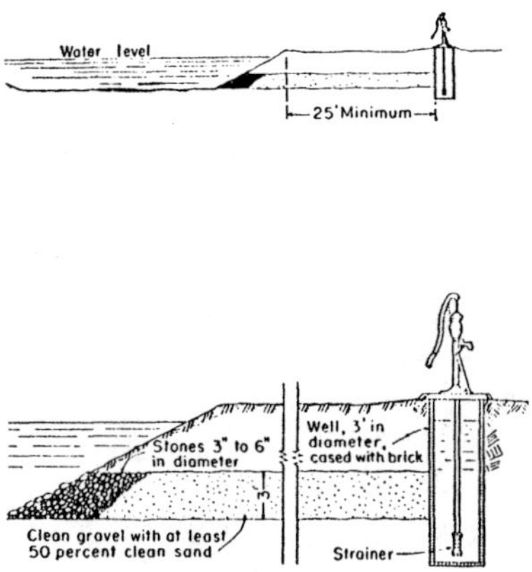

Details of Filter Installation

Seeps and Springs

While an Indian well is a homemade spring, subsurface soil layers of gravel and sand usually carry water even if no stream is near. When sloping ground abruptly levels out, gravelly strata are very near the surface. The water traveling in them merely follows the path of least resistance, and the ground is wet there. Very commonly water runs freely over the surface, even in the driest part of summer. Then it soaks back into the ground, or finds its way to a creek, whichever resists it the least.

These wet spots are springs. They often supply hundreds or even thousands of gallons per day. All you do is dig down into the wet ground to where most of the water flows. The deeper you go, the more water you'll find. Usually there is clay under the gravel, sometimes impervious slate or bedrock. Stop digging when you hit it. Here you can either install a large bottomless tank, which could be a concrete culvert pipe again, or whatever. I've heard of folks burying corrugated plastic water tanks, with holes drilled in the bottom. If the water layer is shallow a large diameter reservoir can be cast by hand. Lots of folks just line it with stones or bricks. An outflow pipe takes the water downhill to the use site. As long as surface water is kept from getting in, everything's safe. And really, the water in such a device is constantly moving in and back out of the tank. Your pipe just supplies the path of least resistance.

Wells

Wells all need their water hauled out either by hand or with a pump, so they aren't as desirable

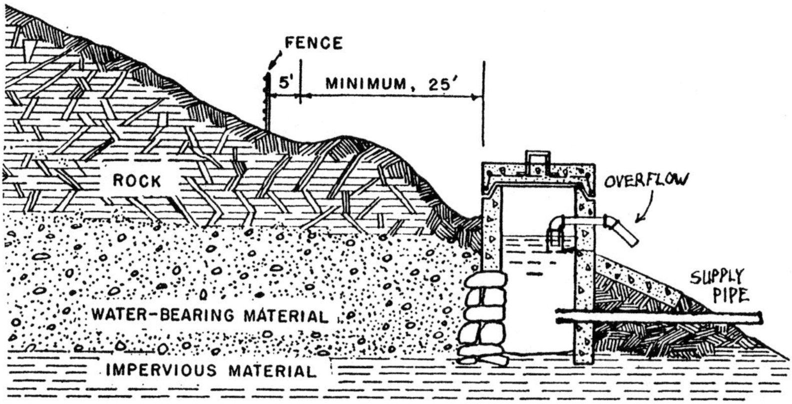

for water sources as other methods. If the electricity goes off for some reason, you cannot wash, drink, flush wastes away, irrigate or fight fires. If the fire is electrical in nature, it may very well eliminate your water pumping ability and burn the place down.

Drilled wells cost lots and lots of money. You pay folks who own very expensive equipment to install them for you. Drilled wells usually have hard water with an iron taste. In limestone districts at least, folks drinking such well water keep doctors busy with their kidney stones.

Driven Wells

Where water is abundant, and boulders and stones rare, you might drive a well. A driven well is what it sounds like--pipe with a point at the bottom, driven downward into water bearing gravel and sand. The top is protected by a stout *drive cap*. The best way to drive it in is with a weighted pipe which is dropped onto the drive cap like driving fence posts (q.v.). Other folks just use sledge hammers. Either way, your arms get dreadful tired doing this, and your fingers seem to take a set; please take your time to avoid hurting yourself. *Drive points* are perforated and are therefore not very strong. You can't expect them to go through rock. If the screen becomes corroded or plugged the well will have to be abandoned or if you can, you can pull out the point and re-drive it. When a 2 ft. joint is driven down, a coupler is added, and another 2 ft. joint. All the couplers should be coated with red lead or graphite and oil and screwed tight with pipe wrenches. If you don't, threads will strip or couplings will split. Use a weighted string to tell when you've hit water. Drive the point until there's 5 to 7 feet of water in the pipe: It is possible (and foolish) to drive it past the aquifer!

Digging a Well

Dug wells are usually shallow. Simply dig a hole from 4 to 20 feet in diameter. Huge diameter wells

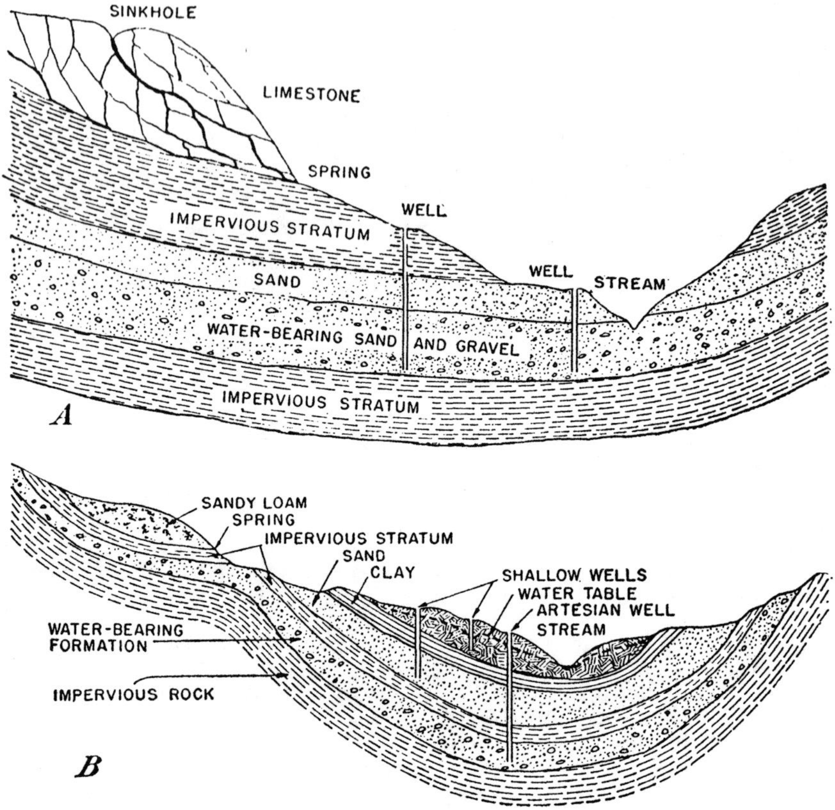

—Earth strata, showing water-bearing formation. Horizontal distances are foreshortened and slopes exaggerated. *A*, Sinkhole shown as a likely source of contamination of water moving through rock seams; *B*, earth strata, showing possible sources of good water.

like that are made with a view to large-capacity water storage. A 4-foot diameter well with 6 feet of water in it stores 564 gallons. It may take a long time to fill to this level, though. That means one can use over 500 gallons at a time, but sometimes the water can't flow back in as fast as it is removed. A 10 foot diameter well of the same depth, though it will also refill slowly, holds over 3500 gallons.

Wells are most easily dug with a hired crane using an "orange peel" bucket. This way comparatively large wells can be dug quickly and surprisingly cheaply.

Most dug wells were done by hand. Pick and shovel are used and after 4 to 6 feet, curbing is installed to prevent dangerous or fatal cave-ins. Except in areas of quicksand, the top 4 to 6 feet of the hole are usually all that need to be shored up. The weight upon

the earth beneath that is enough to hold soil fairly tight. Still, one ought not be in the hole while hoisting up heavy buckets of spoil (excavated dirt) or large stones.

In the event blasting is needed, one might check at local quarries for experts to help with this. In the old days black gunpowder was used, but it is expensive, and fools are not to be trusted with it. It's nearly useless when water is reached, except with special fuses and wax coatings. High explosives are much better and often cheaper.

Still, explosives ought to be avoided if possible. Just learning about them makes one suspect in the minds of law enforcement officials. Even displaying curiosity about them may bring the wrath of frightened urbanites, the most common target of dreadful crimes of terror. Besides, with all the time it takes to round up experienced help and the necessary materials, you could probably just refill the hole and start somewhere else. Also, stones can be drilled with a mason's star drill (check with a hardware merchant on how to use this) and sometimes split apart with steel wedges. I even know of folks who have made these themselves.

After the hole is down to water, start pumping or bailing it out. Keep digging. If using a powered pump, keep the exhaust far from the hole; dangerous gases sink down. Sooner or later you won't be able to keep ahead of the water. Stop there, and start lining the well. The upper 10 feet should be watertight. Below that, porous lining is o.k. Usually stone or brick is used down there, laying it up with all joints between stones or brick overlaid by another stone--this is called "breaking the joints."

Lots of well diggers tamp clay or concrete between the soil and lining in the upper 10 ft. A flat concrete apron cast as shown helps keep out surface water, too

Rainwater

It is a fact that a large storage tank (cistern) set in the ground can be filled from the runoff of a roof. Even desert dwellers get enough water to go for a whole year from one or two rains, with a large enough roof area for collection. Rain water may pick up some impurities from the air while falling or the drops may even collect around dust particles. But it's usually relatively free from contamination until it hits the roof. The first portion of roof water is diverted away. Once the roof is rinsed, the remainder goes to the cistern. An inch of rain furnishes 623 gallons of water per 1000 square feet of collection area. Ordinary gutters and pipes are used to convey the water to the storage site.

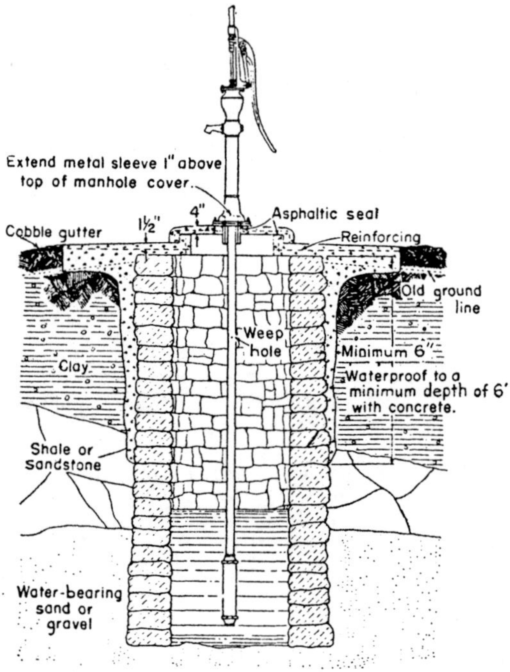

Filtration

A common sand filter is shown. Sand is still the most reliable means of filtering water, but it's slow--only 3 gal. per hour per square foot of filter area.

Rusty water is remedied by either cascading it over and over or running it through aerated sprinklers. This requires considerable fall, or two pumps: one for the cascade or sprinkler and one to pump the accumulated water into the house. These problems are mostly the result of deep drilled wells. Rusty water is much less common in dug (shallow) wells and in seepage springs.

Why Bother?

It's important to remember most of our water need not be of drinking quality. Washing clothes, bathing, flushing toilets, watering the garden, etc., can all be done with moderately clean to questionable water. The amount needed for drinking, cooking and washing dishes is only a few gallons a day. A steady drip every second pro-

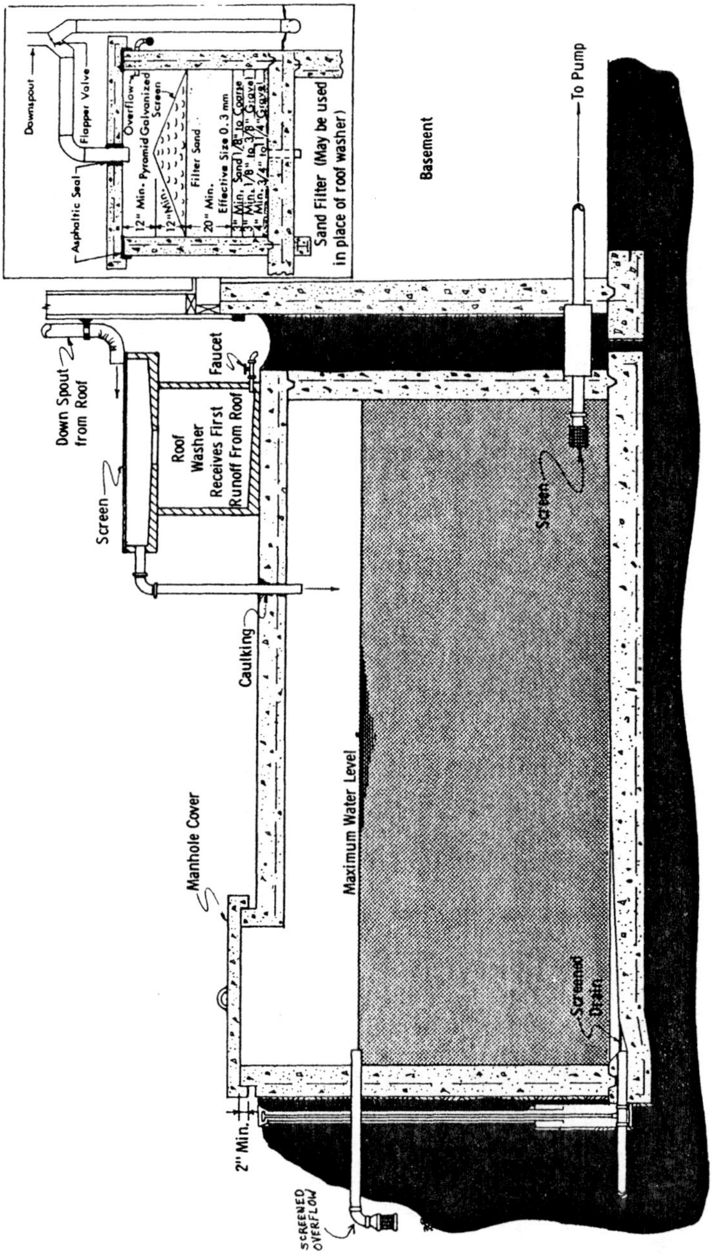
Downspout
Flapper Valve
Asphaltic Seal
Overflow
12" Min.
Pyramid Galvanized Screen
12" Min.
Filter Sand
20" Min.
Effective Size 0.3 mm
3" Min. Sand 1/8" to Coarse
3" Min. 1/8" to 3/8" Gravel
4" Min. 3/4" to 1 1/4" Gravel
Sand Filter (May be used in place of roof washer)
Basement
To Pump
Down Spout from Roof
Screen
Roof Washer Receives First Runoff From Roof
Faucet
Caulking
Manhole Cover
Maximum Water Level
Screen
Screened Drain
2" Min.
SCREENED OVERFLOW

duces 45 quarts a day. If this little amount were collected in a tank and led downhill to a tap, a family of four would have drinking/cooking water. The rest of the water might easily be supplied by creeks, etc.

Pipes

The above methods of developing wells, Indian wells, and springs are as simple as the sound. But they are not often soft labor. The difficult part of establishing a water supply is conveying it to the point of consumption without substantial pressure loss.

The first 15 ft. of actual fall is best kept on a steady decline, so that no high spots of any kind are permitted. These high lumps develop air locks. That is, an air bubble's flotation power exceeds the weight of water uphill behind it. The only remedy for such a high spot is an air bleed valve or stand pipe so the air cannot obstruct it. An inch lump in each of several places will shut the whole system down permanently. If the whole pipe is filled with water, the suction/siphon action will pull minor air locks out, keeping the pressure up. Still, it's better to install it right to begin with.

Rigid pipe is used for the first 15 ft. of actual fall (called "head"). The trench is dug to below frost level. If it runs through heavy woods a few inches is sufficient, especially if the pipe runs at the base of a hill. There it will get deeper with time, and leaves will cover it just before winter every year. Plastic pipe has the least friction loss. Iron pipe has advantages, mostly strength. It is expensive, heavy, and rusts through eventually. But even when it'll no longer convey water a flexible plastic pipe could be run inside it, benefiting from its yet-considerable strength. In general every foot of fall equals about 0.45 psi. (lbs./square inch pressure). Usually inclines are deceptive without measuring devices, but on the whole, naturally flowing water is another reason for houses to be in or near rolling land. Very flat land is often very breezy; a windmill might fill some sort of reservoir dependably from a well.

It may be wise to review these methods before even purchasing land so that the water supply could be considered first. In an emergency, a safe, dependable water supply may mean the difference between life and death. No other single attribute of a parcel of land is so important.

Hydraulic Rams

The much-touted hydraulic ram has its place on some holdings. It is a water pumping device which uses only the momentum of the water itself to pump a portion uphill. To demonstrate the principle for yourself, try the following: 1. Start some water running through a garden hose. 2. Suddenly restrict the opening of the hose with your thumb, causing a concentrated blast of water. 3. Leave your thumb in place, and notice how the blast shortens its reach slightly once the pressure in the hose equalizes. This phenomenon is caused because the weight of water moving in the hose embodies a force beyond the simple pressure of the water. All that force must go somewhere when suddenly constricted, so it pushes the water out harder to get rid of all that momentum. Even if your source only gives a gallon a minute, and only falls 5 feet, you can fill a 100 gallon tank daily, even 50 feet uphill.

There are few places to buy these pumps, and you'll probably have to get 'em by mail. The best I've seen are those made by Rife Hydraulic Engine Mfg. Co., P.O. Box 70, Wilkes-Barre, PA 18703 and though they seem expensive, they'll last a lifetime without much maintenance.

I recommend bolting the body of the ram to a heavy block of concrete, to absorb the considerable and repeated shock. I further recommend you take into account winter freezing and the exit of waste water from the overflow valve. Somehow place the whole works into a building or underground site where the water can run off without freezing in winter.

9
PREPARING LAND

Prepare thy work without, and make it fit for thyself in the field; and afterwards build thine house.
Proverbs 24:27

Though woodland ought to form part of the holding, especially hardwoods (I speak of hardwoods as those trees, hard or soft, which have broad leaves; they're more versatile than softwood, or coniferous trees) sometimes trees must be removed to make way for other operations. Always think on this, since trees are quick to remove, but slow to replace.

Sometimes standing timber will pay for the cost of its removal, and subsequent stump removal, and still have enough left over to pay for a house. Assuming the trees are gone, the tops make as good firewood as the boles, and the twigs can be used for charcoal making, or to burn for their ashes--a fine source of calcium and potassium--to feed to pigs as a mineral supplement, or to the soil for the same reason. But then you have the stumps, of course.

The quickest way to get stumps out is to hire mechanical excavation crews. It is expensive, but often you can do more stump removal by sweeping floors for a pittance to save up and pay the contractors. I have seen skilled operators clear a half-acre of stumps in an hour or two, for as much silver as one might get in twenty or thirty hours' menial labor. I can just about guarantee a man with hand tools will in no way clear a half-acre of stumps with hand tools in thirty hours!

But suppose menial jobs aren't available (or skilled contractors). Then hand remove we must, and depending on the species of trees it can be done in several ways. Trees may be cut down close to the ground, and the ground between them used until they rot enough to remove with a pry bar, chains and maybe a draft animal or two. Or they could be blasted out with explosives, if you can get them. When blasting, punch or drill a hole beneath the main body of the stump, place the charge under it, and either allow yourself plenty of time on the lit fuse to WALK to cover, or fire it by remote control. There is a knack to knowing how much powder or dynamite to use to simply "lift" the stump instead of blowing it to bits. Best get an experienced blaster to help.

In former days special machines were made for lifting stumps and large rocks from the ground. They took various forms, most based on a tripod with a large screw-jack built into it, or a tripod with a large lever equiped with chains and hooks. The latter can be made by handy husbandmen by first constructing the tripod of very stout poles, hanging a chain from their apex. From this chain a lever is hung, balanced in its middle. This makes two handles for the workers, when one pulls down, the other end lets up. Try to balance the lever as well as possible. Two inches from the center, in each direction, a chain with a hook is hung from the lever, so that when one handle is down and one is up, one hook is down and one is up. If the lever is long enough, say 16 or 20 feet or more, each man's power is multiplied tremendously. When one man lowers his handle, one of the two hooks dips lower; this is fastened to a chain attached to the stump or rock. When the other handle is lowered, the fastened hook hoists up the load, and the other hook is now hanging low. It is fastened lower on the load-chain, and the process is repeated until the stump or rock is out.

It is also possible to use a high-lifting jack to hoist up the main side-roots one at a time, then the tap root.

Some folks burn stumps out. Sodium chlorate, which used to be a fairly common weed killer, was poured on the cut stump, into a depression or holes scooped out. It was covered, and after standing a few weeks, the thing would burn like dry wood, right down to the rootlets. Many stump-removing products are on the market nowadays, but they all seem to be nothing but saltpeter. This will make them burn, but not as well as the chlorate, and if they're not burnt, they'll rot more quickly than if they were untreated. Use about a cup or two of the saltpeter. By the way, nitrate of soda is Chilean saltpeter, and works fine. You can find it in any good garden store.

Stones are often harder to remove than stumps but there are a few ways they can be gotten up. The small ones aren't much of a problem: after either plowing, or spring tine harrowing a few times, just load 'em by hand onto a sled or cart and tote them where you want them. Big ones may be hoisted as with stumps, or levered out by lifting one side with a large lever while a helper places rocks under it. Then lever up the other side, placing rocks under it. Repeat until the boulder is at ground level, and roll or otherwise remove it. If it's really too big for this, I have heard of folks excavating a hole beside it in the field, but a couple feet deeper, and levering the big rock over into the hole. Then it's covered back up with the dirt, and is no longer an obstruction. Then there's the method of building a large fire beside it and when the rock is well heated, pouring water on it to crack it. I have tried this, and I don't think it works--that is I think you'd need a whole lot of wood for it, and it may only chip a bit off the rock, if at all. Lastly, one could blast the rock. This is different than for stumps in that the charge goes atop the rock instead of under it. If high explosives are available, try a "fast" one. This means the impact is a shattering one, not a "lifting" one. The charge is covered with a few inches of mud pack. Water doesn't compress, and the force is almost entirely directed through the rock. The chunks can then be removed manually. Again, get experienced help.

Clearing Brush

The easiest clearing jobs involve cutting briars and small trees, and burning or composting them. Seven years later there will be practically no roots left in the ground.

Many small trees and brush can be pulled out with levers and chains or with draft animals. Goats kill trees by stripping their bark. I have seen them kill two-foot thick locust trees within a few weeks; smaller trees are killed within a few minutes, but you still have to get the remains out.

Hogs penned on scrubland will commonly root out most of the brush, and derive vitamins from it. As soon as they have done about as much as they can, they should be confined on the next plot. This way the husbandman simply cleans up after the swine. This is one of the reasons why swine were kept by races of men who didn't eat them, such as the ancient Hebrews.

Hogs will root up even large stumps sometimes, and many, many stones. The best sized hogs for this work are adults or near-adults. My experience shows that swine from about 150 lbs. on up are the best size; the bigger, the better. A stone boat or sled can be then filled up and pulled away. The stones are commonly dumped where permanent fences will be (property bounds) or near building sites.

This ground work done by the hogs will leave the plot in tillable condition, unless you let them get it really muddy or hard packed. In those cases, wood ashes or burnt or slaked lime liberally scattered on the surface will help loosen it. These substances, in addition to their alkalinity (which helps neutralize overly acid soils) also have a chemical action which pushes each particle of earth away from its neighboring particles. This is helpful to drainage and tillage. You need enough tillable ground for a large garden and perhaps for field plantings of grain, roots, etc.

Fences

The enclosure of land is what made self-sufficient husbandry possible. Prior to that, only the nomadic herdsmen and hunters were autonomous; but even they were an easy prey to two-legged marauders on the wide open range. Fences define "mine" from "yours." They also confine stock on new parasite-free ground, and protect field crops. And all this while you sleep.

Other uses of fences are: Blocking out ugly views, discouraging interlopers, breaking high winds, storing heat, and making a home for wildlife (or keeping it out!).

Perhaps the most common type of fence in our time is the wire fence. This is because it takes little time to put up, and can be placed nearly anywhere. Wire fences cost quite a bit of money, and usually last about 15 to 20 years, longer in arid climates.

Wire is sometimes a hazard to livestock. Barbed wire tears udders and other skin. Square meshed (hog wire) fences killed a ram of mine years ago, when neighbors were supposedly counting the sheep daily in my absence!

While they can be made to turn all sorts of livestock, from poultry on up to large horses, my wire fences never worked for goats. Others claim to have no trouble.

One way to keep cattle and horses from mashing down the top of the wire fence is having the fence tall enough to prevent their getting their heads over. This means about 5 feet or higher. It sounds high, but unless the fence is topped with dangerous barbed wire, that is the only way. In fact, it works better than any barbed wire, and is cheaper, too.

I shall not discuss wire fences further since there is so much other information available on building them--even from the wire merchants themselves.

Driving Posts

Steel posts are set rather easily by either driving them with a large hammer (mine is 16 lbs.) or--better--using a pipe-driver. This consists of a heavy 2-foot length of iron pipe with thick flat iron welded over one end. Handle rods are welded onto the sides sometimes. They usually weigh 20 to 30 lbs. The pipe is slipped over the steel post to drive it in.

Wooden posts are similar. They are mostly driven, with corner and gate posts "set." Driven posts last about as long as ones whose holes are dug and tamped. This is because the only rot occurs within about 6 inches of the surface of the soil, where all the bacteria are. These posts are tapered at least 18 inches on four sides, to a wedge point. A hole is started with a "spud bar." This is a bar of inch-thick steel, with a point or chisel taper at one end. The bar is driven in, strongly worked from side to side, then removed to leave a tapered hole as deep as the post is to be driven. The post is either driven with a very large mallet called a *beetle* or with a pipe-driver of large diameter and weight--usually about 35-40 lbs., and about 6-7 inches inside diameter. In some soils, a hole need not be started. If a mallet is used, the workman must stand in a wagon or on some other elevated platform. Steel hammers tend to considerably shorten the life of the post, by splitting them, or by mashing the wood at the top so that water flows in easier.

Setting Corner and Gate Posts

Gate posts must resist leaning. Setting them is always difficult: Dig a deep hole to set the post. In boulder-filled soils, 2½ feet is deep enough. In other soils, up to 5 feet is not too deep, depending on the size of the gate.

To dig the hole, use the spud bar to cut the earth loose, then use a "clam-shell" digger to remove the spoil. Sometimes, a hole may be "jetted." This involves turning a garden-hose vertical against the ground, allowing it to rapidly erode a hole. Keep feeding the hose down the hole as it deepens.

To set the post, place it in the hole (some taper the bottom to prevent frost heave--hardly necessary if set below frost penetration depth) then fill in a small amount of earth, tamping well with the point (not the flat) of the spud bar. Add a little more, tamp it also. In tamping, we strive for a very hard pack before adding more earth. Very often there will not have been enough dirt removed from the hole to completely tamp it full. Find some elsewhere, but avoid using topsoil--it rots the wood.

Wood for Posts

Independent husbandmen avoid those factory-made, poison soaked, softwood posts so commonly seen in gardening shops. Not only are they expensive, but they are not as strong as naturally resistant wood species. I doubt they last as long as forest-grown locust *(Robinia pseudoacacia)* or osage orange *(Maclura pomifera)*. Other good woods are: Catalpa *(Catalpa speciosa)*, mulberry *(Morus rubra)*, Eastern Redcedar *(Juniperus virginiana)*, white cedar *(Thuja occidentalis)*, cypress *(Taxodium distichum)*, post oak *(Quercus stellata)*, white oak *(Quercus alba)*, and chestnut oak *(Quercus prinus)*.

Post and Board

Posts are set about 8 feet apart, and planks are nailed onto them. The gaps between the boards are closer at the bottom than the top. Planks can be gotten from a sawmill as low-grade or "out" lumber for a fraction of the "good" lumber cost in timber cutting regions. This kind of fence is often expensive anyway, but will hold swine and horses much better than wire fence. It looks better too.

Post and Rail

There are two ways to do post and rail. One is to nail split rails, or poles with their ends tapered thinner, to the posts. The other way is to have holes or notches in the posts at the required distances from the ground. Boards or split rails or poles are fitted into these holes. There is no better post for this latter method than a concrete one. It will not rot, and helps keep the rails from rotting. In fact, with concrete posts, this is perhaps the longest lasting type of fence.

Snake Rail Fence

Rails may be stacked criss-crossed, without using posts, zigzag fashion. Stack about 7 or 8 rails high. Keep the wood off the ground, each corner on a rock. Often they are nailed to prevent sliding. These are then braced with *stakes and riders,* which are merely short rails leaned into the *obtuse* angles at the corners, crossed at the top, and two or more rails (called riders) added on top. This fence needs lots of straight, durable wood, but sometimes lasts a lifetime with moderate upkeep.

Hedges

Hedges are truly permanent fences, since they can last many centuries. Replacing a fence thirty years from now may be more work than you're willing to undertake at that age.

Hedges can be grown just about anywhere, and can be made so tight as to even keep hens in. The best plants for them are the cockspur hawthorn *(Crataegus crus-galli)*, the osage orange *(Maclura pomifera)*, and honey locust *(Gleditsia triacanthos)*.

Stretch a string where the fence is to stand, and place the hedge plants nine inches apart in two staggered rows. Pack the ground firmly around the plants, and mulch the ground with straw or tree bark (don't dig it in, though) to conserve moisture. Keep stock away from it for two years, which may be done by spraying the plants about once in ten days (or after each rain) with a diluted (10:1) dung or hog urine solution, or temporary fence.

Osage orange is good for the Middle and Southern States, but when neglected, it gets beyond all control. In Northern and Eastern States it is likely to be killed by frost. It turns all stock and all pests, especially two-legged ones.

Honey locust grows more rapidly than the others, being sown from seed along the string marker line in some districts. It is also easily started setting plants 6 inches apart. It requires 2 annual prunings to keep under control, in June and September. It has no superior except where Osage Orange grows. Doesn't spread by runners like Black (Yellow) Locust *(Robinia sp.)*. Honey locust pods are good stock fodder.

The hawthorn will probably harbor more wildlife and insectivorous birds than the others. It has beautiful flowers and in winter, red or yellow berries, or *haws,* which feed the birds, and give you something to plant elsewhere.

The best species is one which grows wild in your area. Select for the biggest and wickedest thorns. Dig 'em up, trying not to damage the rootlets too much. Prune the top down so that it won't draw more nutrient than the now-disturbed roots can supply. It isn't ordinarily harmful to prune the plant right off just above the ground level. New plants may be gotten by *layering,* which is the practice of covering with soil a branch. Once rooted, it is severed from the parent plant, and placed as desired.

Then after 5 years, *lay* the hedge, which is done by cutting about ¾ of the way through the stems with a small hatchet or bill-hook, and leaning them into each other. This laying of the hedge causes the plant to send out more thorns and close the gaps. Then an occasional "buck-heading," which means cutting off the tops to encourage them to spread, is all the maintenance needed. It is hard work, but satisfying. Again, they last for centuries, and there is no rusting and dangerous scrap--as with other types of fences--left to choke cattle grazing nearby.

A hybrid sort of hedge works in some mountainous districts. The thorns are laid up, and then the rocks pulled from the field are thrown around their base. If some thorn plants die out, it is difficult to re-plant, and ultimately these things have a tell-tale barbed wire on top doing its nasty business to milking ewes or goats which happen that way.

Hedges have the added benefit of blocking the view of outsiders, and making the outsiders feel very conspicuous when they do trespass. They form wonderful windbreaks. It also blocks ugly neighbors from your sight.

Stone Fences

Stone fences can be made in two ways. The first method uses smaller, flat stones carefully laid as a wall. The foundation thickness of the wall must be at least half the height, if the stones are laid up dry (without mortar between the stones). The weight of the stones above pinches the stones below, so that they are not dislodged. The weight needed to pinch them requires at least about 4½ feet of height. Otherwise swine, rubbing against the sides, will dislodge some stones, and the whole thing will come a-tumbling down. Higher is better; stones must be laid carefully and closely with tie-stones, called *thorough-bands* at frequent intervals. These stones go all the way through the wall. All joints must be staggered. The foundation should be dug about a foot into the ground, and the dry stones laid up from there. The drainage this affords helps prevent water standing between stones (which may freeze) heaving and toppling the wall.

The other method of stone walling requires laying up large stones, of about 500 lbs. and up. These will be 2 to 3 layers high, whatever is needed. Such walls rarely topple, even if laid on bare undug ground. Getting the stones in place is the problem. Industrious folks always find a way. Stone walls generally improve the climate immediately beside them. They reflect heat to tomatoes or small fruit trees on their south side all day long, and at night release heat which helps protect from untimely frosts.

10
HENS

All clean fowls ye may eat.
Deuteronomy 14:20

Some of the red laying hens we brought with us from Maryland were *sex-links*. I had no idea at the time what that meant, but I reckoned the name alone would be enough to get anyone to buy their eggs. Alas, it means their sex is linked with their coloration. At birth, white chicks are cockerels (males), pullets (females) are red. They are hybrids. For example, a cross between a white Leghorn cock and a Rhode Island Red hen will become sex-linked.

They were superb layers, but the crossbreeding will not breed true on subsequent generations: When we eventually had to order day-old replacement pullets from a hatchery, the price had nearly doubled! Eureka! A farm product keeping ahead of inflation! The bad news is the only market for it is farmers.

We kept hens for several years with no roosters, so replenishment from far away suppliers was a must. That's the price you have to pay for forcing them into your own private convent. Then one year, when it came time to induct more novices into the Sisters of Eternal Chastity we ordered day-old chicks from a different place, and all the hens began to lay very poorly. In fact, from then on, though we ordered from our good source, we still had poor layers. It seems we had bought something besides pullets: Disease.

We were keeping the hens in an ample house with an attached run, and gave them plenty to eat and drink. We supplied them with the necessary oyster shells or limestone sand and grit to replace the calcium lost to the eggshells, and to aid digestion. We tried giving them worm medicine, thinking parasites were the cause of the drop in laying. All to no avail. When we put "clean" pullets--those raised in quarantine--which were laying well, in the main hen house, their laying dropped off. We tried medicines, feed additives, other breeds. Nothing worked.

Hens which scratch about, chase bugs, eat grass, bathe in the dust, flap their wings, and have clean roosts rarely slack off in laying. Declining monastic enrollment might take a lesson from this. But if hens run loose all the time they can make a nuisance of themselves. The little monsters eventually get into the garden, destroying small plants, eating tomato seeds just as the tomatoes begin to ripen. Ask me, I know. They get all over the porch (a hen's favorite haunt) and you know what that means! Let's just say your work "keeps piling up."

Their peaceful cooing is so very gentle sounding, but it's pure subterfuge. They hide their eggs in poison ivy patches, thorns, or in places you can't reach, but where skunks and 'possums can eat them. They roost very high in barns or trees, so that even at night it is a risky business to try to catch them. Ask me, I know. If moved to a new dwelling place, even if you entice them by giving them a rooster, some will attempt to return to the roost of their youth. And they will do this for months, sometimes years. Even after breaking all their vows they still head for the cloisters.

Then again, I too have difficulty leaving behind the irresponsibility of my youth. Experience keeps a hard school, but a fool will learn in no other. And do you think I've gotten wiser with years? Please don't ask.

Coops

The simple answer to all this is to raise hens in little flocks of no larger than twelve. Their houses may be light enough to be easily moved. There should be a penned-in run attached, so that the hens don't roam until you let them out late in the day. This way they won't have time to get into mischief. One might keep multiple flocks this way.

Such a coop (for a coop is a *portable* enclosure) ought to contain about 3 to 4 square feet of total floor space per bird. It shouldn't be wider than 4 feet, nor higher than about 3 feet, total height. The top should tilt open so you can get at the feeders, waterers, eggs, etc., from outside. About one half its length ought to have a roosting compartment whose floor is a foot off the ground. Said floor should

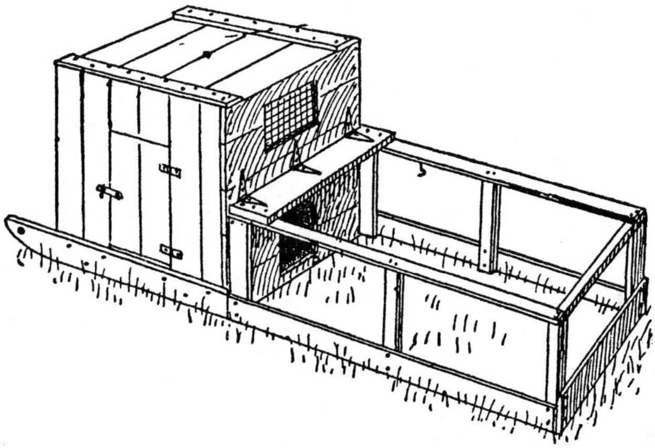

PORTABLE COOP AND RUN

Packing case coop and wire fence covered run.

be wire mesh with cross slats for the hens to walk and roost on. There must be a door communicating with the rest of the pen, whose dimensions are 6 inches by 8 inches, and can be latched open or shut. The door needs a perch in front so the birds can get in at night. The sides of the coop may be of poultry wire, except the roosting house, which needs to be lightweight, but proof against wildlife--I favor sheet metal like old roofing tin. The roof should also be of tin and should be designed to shed water from both house & run. The runners, for the coop must slide easily, need to be rot proof wood. Two-by-fours are plenty big enough. All other structural members of the frame should emphasize lightness, relying on the support from the roosting house for overall sturdiness. Inside the roosting house there must be a box with a solid floor and some straw or hay for laying nests. Feed and/or water troughs may line one or both sides (attached to the runner, and fillable from outside through the wire.) A grit/shell feeder should also be placed like that. I like a flap door 1 foot high at each end, and a tow rope.

Portable coops almost never have unhealthy fowl in them. But they must be looked after! Move them to new ground every day. Make sure the hens have plenty of water, except when there's snow available--snow is safer for hens than water on very cold days. Let the hens out to run free when the day starts cooling off. They'll be back in at night. Close 'em in to

READY FOR MOVING

thwart predators, and open and feed 'em first thing in the morning when you move the coop. On snowy cold days we put straw on the floor of the house portion to cut the draft. It is sometimes necessary to remove snow to move the coop. Hens shouldn't have to stand in snow. Try to admit as much sun as possible.

In their evening romp, they'll scratch cow plops flat so they decompose fast, and eat lots of maggots and other revolting things from your farmyard. There is an astonishing difference in the quality of the grass or other crops where the coop has been so place the coops where they're needed.

There is alas, a grave drawback to coops: the moult. This is the time, ordinarily in the fall of the year when hens slack off laying, lose feathers, and then grow new ones. Confined in a coop much after sun-up the hens pick at each others' newly emerging feathers, ruining their chances of a new and warm coat in the approaching winter. You must either be very vigilant in letting the hens roam early each morning, risk predation by just leaving their coop open all the time, or move them to a large stationary hen house. Extra protein, especially of animal origin, helps too.

Hen Houses

Most chicken keepers use stationary hen houses, sometimes mistakenly called coops. A hen house can be quite small provided it has outdoor runs. If the hens are to be kept indoors for any considerable length of time, they need at least 4 square feet of space per bird, and 6-8 inches of something like sawdust, shavings, ground corn cobs, bark, or peanut shells for floor litter. They need 9 linear inches of roost space per bird; we use poles or other scrap wood. The best way is to have the roosts over a dung pit, screened to keep the hens out. The dung can accumulate until it is required for fertilizer. I like a hen house whose floor is elevated enough to park a dung cart under the "dung pit" for ease in cleaning and manuring. This is the easiest way to obtain and load dung for your gardens and fields. About half of a hen's dung is in this pit. The rest is mostly in the outside yards (you need to have a minimum of two yards). Grow vegetables or other crops in one yard, while the hens roam in the other. A better way is three yards, so a longer rest is possible. This way parasites can be kept down. If the feed hoppers aren't kept indoors, wild birds will feed occasionally from them, depositing worm laden dung where the hens might inadvertently consume it. Such wild birds also bring mites, lice and diseases. Best keep 'em away by not tempting 'em with feed.

Make your hen house at least tall enough for you to walk around in, and have a window on the upper half of the south-facing side for winter warmth. The larger breeds are fairly winter hardy if the hen house is tight on at least three sides. I find it helpful to plug all cracks, especially in the upper part of the house, and hang a "curtain" of feed bags just in front of the roosts to hold in the warmth of the hens. The difference made by the curtain feels to me like a good 20°F warmer in winter. And the winter lay rate improves dramatically.

Eggs from the best layers are saved for hatching. This means you'll need to keep some roosters. You need at least one rooster for 20 hens; one per dozen hens is better.

Select roosters with the broadest backs, biggest breasts, soundest health and least aggressive behavior towards people. Usually the best looking roosters are the best natured.

Perhaps the most objectionable feature of a stationary hen house is the fact that sooner or later it may become infested with mites. Poultry mites must be destroyed by dusting the whole house with proprietary dust from a farm supply store or they'll annoy the birds all night while they're roosting, and prompt them to seek outdoor laying nests rather than their provided

ROBUST AND INFERIOR TYPES OF FOWLS

boxes. Worse, when you go in at night to get the eggs, you'll get 'em crawling all over you! They don't bite, but they drive you nuts.

The Way of Chickens

Some eggs are pointy, some are more blunt tipped. The pointy ones are supposedly more likely to be males; blunt eggs, female. Fertile hens' eggs incubate at 102°-103°F. They develop into chicks in 21 days. Their mother turns the eggs several times per day. After three days, if the eggs are *candled* they show signs of development. Candling means holding the eggs up to a bright beam of light in a dark room. A spidery web of veins appears as dark lines on the yolk in the translucent egg. If no such lines appear, either the egg is infertile, or there is something wrong with the hatching temperature. In China eggs are sold as fresh if they candle "clean" by the third day. If the temperature is too warm, the chicks will hatch too soon, and stick to the shell, or just die. If it is too cool, they'll hatch late, or die.

At later stages of incubation, hens are seen rising off the eggs more often, because the eggs are giving off considerable heat of their own. Like an unborn baby, they are already alive and well. They are fed by the substance of the egg white. Indeed, just before hatching they swallow the last of the egg white, which sustains them up to 3 days before they need to eat or drink. This fact allows hatcheries to mail chicks by air almost anywhere.

Small chicks, not fully fledged, need heat even more than they need food. This is called the "brooding" age. A normal hen will go broody, which means she wants to sit on eggs, and hover over her hatched chicks. When day-old chicks are bought, they must be given 95°F shelter for a week. The second week may be five degrees cooler. And each subsequent week, five degrees cooler yet, until the peeps have a complete covering of feathers.

It is easier to just let hens go broody. They do this most often in summer, but if you should be lucky enough to have one go broody in spring, say March through April, her young pullets will begin laying in the fall and lay right through winter. When a hen goes broody, she'll spread out on her nest, and make purring broody noises. If you remove her she'll return to the nest.

The broodiness can be "broken up" by placing her in a wobbly cage suspended above the floor so she can see lots of activity, but can't really "set."

But to use her as an incubator, place her in her own *broody coop*. Remove whatever eggs she has and place 9 clean, average-sized eggs with normal looking shells under her. This way they'll all hatch together and she won't leave any behind when she leaves the nest. Feed her a handful of grain a day, and give her access to water. The broody coop not only protects her, but when the chicks are small she can be confined until the cold dew is off the grass; otherwise the chicks may die of chill. Some old-fashioned breeds are much more broody, but then, they lay fewer eggs.

All fowl need drinking water at all times. Perhaps I should clarify that by saying they need to *eat* water when there's snow available. I have found winter chickens much more susceptible to the cold if they must drink cold water instead of eating snow. When the hens have unlimited access to snow, there is

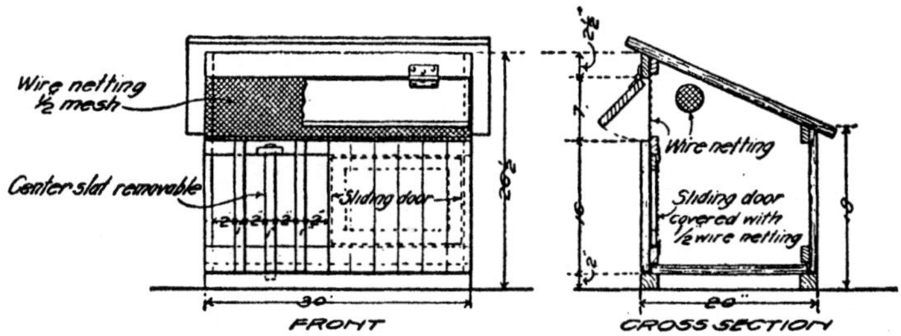

—The working plan of a box-shaped brood coop that can be easily constructed.

no need to water them, and in fact they are better off without.

Different aged birds have different food needs. Small little chicks really ought to have 22-24% protein feed. The protein has to be of high quality, with some animal-origin stuff like meat scraps, skim milk, whey, buttermilk or fish meal. A British poultryman of great renown in the 19th century, George Kennedy Geyelin, advised keeping what he called an "artificial vermin nursery." It was really nothing more than a compost pit near the chicken house, and contained all manner of horrible things from rotting garden refuse to guts and dung. But then, his fowl were not given outdoor runs. Adult chickens need 15% protein.

If a coop is placed in a paddock where a cow or horse is kept, when opened in the evening the hens will run from dung plop to dung plop, scratching them completely away in the search for new-hatched bugs and undigested bits of grain. This serves the triple duty of providing animal protein for the birds, relieving part of the fly problem for the cow or horse, and preventing the all-too-common sight of mature grass going ungrazed because of the dung being too close by.

But grass and bugs alone do not an egg-basket fill. Hens also need vegetable protein, like soybean oil meal or sunflower seeds.

The mode of feeding hens is either: 1. An "all-mash" (complete) diet satisfying all feed requirements. 2. A grain and "mash" diet, in which whole grains are supplied *ad lib* with the higher-protein "mash" mixture. 3. Mixed grains supplemented with kitchen garbage, pasture and foraging.

"Mash" is the poultryman's word for a ground mix of grains like corn, wheat, barley and a smidgen of oats, together with either soybean meal, linseed meal, sunflower seed oil meal, or other high-protein source, and perhaps some dried meat scraps, fish meal or milk. These mashes usually include a small bit of salt and steamed bone meal. Costly mashes really call for special feeders to minimize waste.

Some breeds of hens don't respect fences. Leghorns, Minorcas, Andalusians, Hamburgs, and other Mediterranean breeds are very flighty. They are impossible to catch unless well confined. They can fly over an 8-foot fence. Many people clip their wings to prevent this. The big feathers at the end of one wing are cut with scissors. This way they can't fly straight. Some still fly out. Just try getting them back in though!

The bigger the breed, the less flighty. That's another reason for the American breeds like Plymouth Rocks, Rhode Island Reds, Wyandottes, etc. They don't roam as much.

I do protect the garden with 4-foot "garden fence." This is a standard woven-wire fence whose bottom horizontal wires are only an inch apart. They gradually widen to about 4 inches at the top. It wouldn't protect the garden from flighty breeds.

At 22 to 26 weeks the hens are old enough to begin laying and then the cycle from egg to hen can start all over again. The meat of *cockerals* (young roosters) begins toughening at this age, too.

Predation

Neighbors' dogs killed some of my hens. They would lie in wait until the hens were close enough to catch. Neighbors rarely accept having their dogs shot and killed, even when the dog is trespassing and has a hen in its mouth. Though they don't care enough about their pets to keep them safe, they often make trouble if something bad befalls them. Still, the best solution is to ruthlessly liquidate such dogs every time they begin killing chickens. They have all day to kill your hens. Better to risk the wrath of irresponsible neighbors than part with your livelihood. Also, make friends with your official dog catchers; it's surprising how much assistance they can give when they really like you. Rural dwellers are more likely to stomp their own snakes.

Other pests include the fox, mink, weasel, rat, skunk, possum, and hawk. I have caught raccoons killing my hens, but it's rare.

Foxes snatch hens anytime, so at night they need to be shut in. I have had gray foxes attack pullets roosting in trees right next to my front porch! I was sitting just inside the door playing music at about 10:30 p.m. in early July. I heard only a sudden cackle of a hen. Running outside I saw a fox run by with the chicken in its mouth. I went for a shotgun, but in the weak flashlight beam, I missed. I snared the fox later. Then one day its mate started stealing hens from a stationary henhouse while I rested on a rock not fifty feet away! I snared that one that night in an opening I made for the varmint to enter the poultry pen. (See the chapter on vermin.)

Coyotes don't always come back day after day like foxes.

Rats invade chicken houses when there are places they can live without fear of capture by cats or other predators. Rats need cover more than they need food. This is another reason for having the floor of the hen house a couple of feet above the ground. Portable hen houses don't harbor rats.

Weasels can enter any hole a mouse could enter. The way to tell you've got weasel problems is the way the hens have been attacked. If you find a dead hen with her throat cut and/or bloody, but no further apparent injury, it's probably the work of a weasel. This is especially likely if several dead hens are found like this. The work of a mink is similar to a weasel, except that several hens have been killed in most cases, and their heads are missing.

Skunks almost universally eat only the eggs.

'Possums are hens' and eggs' most common threat. They hardly ever come in daylight and are easily thwarted by shutting in your hens at night. (See the chapter on vermin.)

In many areas it is illegal to kill hawks. In those which aren't, the best way to remove them is to set up a pole about 6 feet tall with a steel trap set on the top. Staple the chain to the side of the pole. The hawk will land first on the trap, thinking to ready itself for the kill. This works well even in places where it is illegal, so beware.

Fresh Eggs

Store bought mixes sometimes have drugs and other chemical growth stimulants added. Since the commercial market never wants deep orange yolks, they avoid yellow corn or alfalfa meal. These contain xanthophyl, a substance found in certain plants which colors the yolks a deep orange. City people don't think they like rich-looking eggs, never having eaten them before. They certainly don't like fresh eggs, because they think they stick to the shells when boiled.

Well, fresh eggs do stick to the shells when boiled, but not if cooked properly. Bring the water to a raging boil (a little salt added to it will raise the boiling temperature, making this trick even more effective). Place an egg on a spoon and quickly lower it into the boiling water, so there's no splash to burn you. Then wait a couple of seconds until the water is raging again and add the next egg the same way. Continue on until all your eggs have been added; not one will stick to the shell.

How Many Eggs?

A Rhode Island Red hen, which lays brown eggs, or a Leghorn, which lays white eggs, will normally do all her labor of laying in six days, resting on the seventh. Other breeds may lay every other day or even less. The Reds and Leghorns should get all the feed they can eat. Other breeds can be fed less. Also, the former do better on a complete mash diet. The latter are cheaper if they get at least half of their feed in whole grain.

It was determined long ago that an average hen will lay 600 eggs in her lifetime. In the wild, or in less-developed breeds, the second, third, and fourth years will be the biggest laying years. In more highly bred birds the first two years will total about 600 eggs, and then you get very little, or the hen simply dies from wear. But some will go on and on year after year. It's best to select replacement pullets from these, if you can.

Culling

"Replacement" has a definite and unlucky connotation for some hens. And it brings in the subject of culling. Knowing which hens to cull is a knack every hen keeper ought to learn.

Raggedy looks don't necessarily mean anything more than hyperactive roosters. Or too many roosters. A hen which isn't laying, and isn't going to start back up again, often has a shriveled comb and/or shriveled wattles. Layers have healthy pinkish-red combs and wattles which appear to have a "waxy" cast to them. Hens cooing their gentle songs, are usually layers; non-layers tend towards silence. The coloring of the legs and around the vent fades, getting more pale as a hen lays eggs; most hens quit laying during the molting season, and their color comes back:

If they've got a rich yellow color on legs and vent, haven't molted for some time, have shriveled combs & wattles, and are taciturn, start eyeing them for soup. And they make an excellent soup! The best way I know to determine their laying status is to check how wide their pelvic bones are. If the bones are far enough apart for an egg to pass through (3, sometimes 4 of my fingers' width) the hens can still lay. Check 'em when you know they're laying, then compare when you suspect they've stopped. When those bones narrow down and lose their flexibility, a hen has called it quits, and it's time to "fire" her.

Mixing Feed

For the first few years I had to get a paper and pencil and recall all that multi-variable algebra from my school days. What a chore. Then, I came across an old farmer's trick to determine the correct proportions for feed ingredients.

Say you wanted a 15% protein mix, and your hybrid corn was only 8% protein, while your soybean meal is 43% protein.

First, find the difference in the target protein content and the actual protein content of the corn. The difference is 7.

Next, find the difference between the target protein content and the protein content of the bean meal. The difference is 28.

43% bean meal ↘ ↗ 7 parts bean meal needed

15% (target protein)

8% protein grain ↗ ↘ 28 parts grain needed

Now, those two numbers are applied to the opposite ingredient. That is, we'll use 28 parts corn, and 7 parts bean meal. Or, a ration of 4 parts of the corn and 1 part bean meal.

Mathematically skillful husbandman can quickly use this method to determine ratios of more than two ingredients. I'll leave this as an easy puzzle for you to toy with. Who says husbandry is the province of the dull and ignorant?

Parasites

Worms and skin mites are mainly controlled by proprietary mixtures. The worms are mostly cured by mixing wormer with drinking water, and denying access to other drinks. The mites or other bugs are gotten rid of with dusts or powders. It is important to note that mites live in the litter and in the cracks in the walls & floors of the hen house. They "hibernate" during the day, and crawl up on the roosts at night to infest the hens. A good precaution to break the growth cycle of mites is to have the roosts set on legs whose bottom ends are set in a can of oil. The mites then cannot reach the hens.

Battery Raising

I have no moral objections to raising poultry in confinement. My main objections to battery confinement methods are that it is costly, and keeps drawing one further into commerce. What with so many birds ready all at once it is quite a job to kill and clean them. Then they must be either frozen or consumed right away. If the former, you must trust an electric company and your refrigeration equipment more than I like trusting 'em. Daily cleaning of the pens is the rule. But about the only feeds which really work well with such a system are commercial mixes, because the birds can't get the bugs, grass and dirt they crave from early on. This all means dependence on commerce, which is what autonomous husbandry is supposed to obviate. For handymen and those who think they want to battery raise their chickens--an admittedly faster and tenderer product than more natural methods--here's a plan for a battery I found in an old poultry book. I think each deck of the battery will hold 40 to 50 fully fledged young chicks.

Geese

Geese are the hardiest poultry. They have eggs several times the weight & size of a hen's egg, and are commensurately bigger upon hatching. They don't begin growing feathers right away, so are spared the drain on their reserves. Their mothers are the most protective of all domestic birds. They live on grass and clover, and really don't even need grain. But they'll grow faster and fatten better with all the grain they can eat, plus access to clover and grass. They are pretty respectful of fences; at least our Embdens are. They don't mind coming inside a high-ceilinged shed

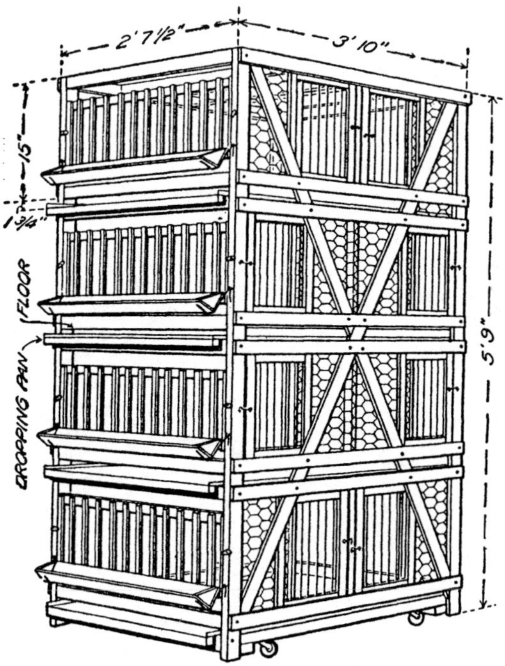

—Working plans of a portable fattening battery.

if it's open on one side, but they don't like coming indoors. They prefer to sleep near the house, or at least near animals they somehow sense as protective, like the ram. They are the smartest of all farm animals, quickly "deducing" new dangers, obstructions, or inconveniences, and avoiding them, seemingly without trial and error.

Goose eggs are laid starting in February normally, and sometimes will continue until late May if the hatch is disturbed or if the eggs are removed. Don't let a goose see you carrying a goose egg, for she'll soon move her nest to a place you may not easily find. As for nests, they like locations where they can hide somewhat, but where they can see through or over in all directions. The two biggest enemies of geese are canines and possums. Eggs hatch in 31 days, and a goose can sit on 7 eggs.

At about 16 weeks, if the geese have been well fed, they are ready to butcher. If not well fed, one can feed them all the grain they'll eat for 3 weeks. But it's no use trying to confine them in a small pen, and certainly not in a coop, unless they are to be force-fed with a funnel. That's so much work I won't bother explaining it. If your young geese can be separated from the older ones, feed them with a feeder--all they can eat. They need at least say, 4 to 9 square rods, which is to say a pen 30-50 feet square, depending on how many you're fattening. They make pretty good holiday gifts, but only if you do the (considerable) work of killing & cleaning them.

Ducks

These are superb eating, some are great egg layers (surpassing even the best chicken breeds) they eat snails and other nasties and are good fun to see on a pond or lake. But the disadvantages are many. Ducks are abysmal mothers. They lead their newly hatched ducklings all over through the cold wet grass, and wear them out or lose them. If that doesn't kill them, they lead them to swift streams where the little ones are swept away downstream. They're not particular where they lay either. I have seen them walking along, drop out an egg and hardly break stride. There the egg sits until someone, or something, finds it. They are a pain to pluck unless they are precisely 10 weeks old (full of bristles after that) and they make a pasty-mud mess of duck butter (dung) in all places you'd like to walk. They are difficult to fence and their young are nowhere near as hardy as goslings. I no longer keep them and am quite glad about it.

Turkeys

Don't bother. These are among the hardiest birds in the wild, but are such wanderers that no fence can really hold them unless their wings are severely clipped. The young are almost unbelievably stupid, refusing to eat just about anything unless it's white. They can't figure out a water supply without help, and even then have been known to drown in an inch of water. They pile up on one another suffocating whoever is on the bottom. And they immediately begin growing feathers, taxing an already frail physique almost beyond its ability. Most commercial operations are very heavy on artificial heat, appliances, and drugs. Neither I nor any of my neighbors has ever made a success of turkey keeping, so I won't trouble you further with my description of them.

Exotics

I have yet to find any exotic which is worth growing for its innate benefits. The very best bird to grow is the common laying hen, and that's that.

11
GOATS

...Who feedeth a flock, and eateth not of the milk of the flock? 1 Corinthians 9:7

Before I left Maryland I got rid of my goats. Yes, I too had been wooed by the magical stories of how much more efficient they are than cows, how their milk is not only as good, but better for health, how they transform poison ivy, brush and old tin cans into high quality food, how they give twins and triplets, how they need such small pens, etc., *ad infinitum.*

Learning the rest of the truth in the "school of hard knocks" was an experience I shall not forget.

Goats are more efficient than cows if the former are kept by "soiling" (q.v.; chapter on cows) and the latter by pasturing. Cows are more efficient when the *vice* is *versa.* But cows are much more contented than goats in confinement. Goats are much livelier than cows, and they love to climb. Mine loved climbing on the cars, destroying windshield wipers, convertible tops and paint jobs.

Commercially, goats' milk is often healthier and more easily digested only because many states exempt it from their unreasonable pasteurization regulations. Pasteurized milk is not only less nutritious than untreated milk, it is less safe. It is a dead substance which rots with age. Untreated milk sours with age, protecting itself from dangerous germs by its own enzymes and lactobacilli. Tuberculosis, you say? No case of tuberculosis being passed from cow to man through milk has ever been authenticated. It doesn't happen. The use of milking machines, which prevented tubercular dairymen from coughing into the milk bucket, became widespread at the same time the pasteurization edicts were enacted. Investigate it for yourself, and you'll discover the truth.

But whereas cows' milk takes on the pleasant flavors of yogurt or cheese when it sours, goats' milk usually begins to smell a bit like the way a buck goat smells during breeding season. Visit a goat farm, and take a whiff. Sometimes goats' milk doesn't need to age to smell like that.

Goats' milk requires an expensive, hard to clean centrifuge to separate the cream. Is this a simpler life?

Often goats' milk is recommended for babies, especially those allergic to cows' milk. It cannot be disputed that this is a plus for the goats. But there are many people allergic to cows' milk, nearly always pasteurized, who are also allergic to goats' milk when it has been pasteurized. Likewise many folks who are "allergic" to cows' milk are not allergic to untreated cows' milk. Something to think about.

As for feeding, it takes the same feed to get milk from goats as from cows. Some goats are very impressive in daily yields, but so are some cows. Each has its slackers. No goat or cow will spin straw into gold.

Now for the things goat enthusiasts dislike discussing.

Goats cannot be easily fenced. Whereas a single strand of electric wire will hold most cows, I have yet to see the goat which can be contained by hot wire. In fact, my goats had 3 strands of hot wire surrounded by a 4 foot woven wire fence, surrounded by a piled up visual barrier of logs and brush wood. The goats got out sometimes, and headed straight for the orchard. When they killed all the fruit trees (a goat's favorite food is fruit tree bark) they went for the ornamentals (their second favorite). Then up onto the cars for a tap dance. And then there are the irate calls (lawsuits) when your goats do this after entering the neighbors' fenced land.

One solution was to tether the goats. Being about as intelligent as sheep (but much more active) they soon tangle hopelessly in the poison ivy and briers. Then you've got to untangle them. Hourly. Yes, they can graze grass this way, but they really are browsers, not grazers. They'll be less happy, and you'll get less milk. Then of course, they're sitting ducks to predators such as coyotes or the neighbors' dogs. I once came back after only 15 minutes' absence to find one of my goats quite lacerated by a neighbor's German Shepherd dog. The dog was off as soon as he noticed me, which was before I noticed him. The goats were terrified, and hurt. I sewed them up the best I could, and put them into their "prison" with some hay. I followed the retreating canine to his owner, who hastily made it clear that "his dog is too gentle to have done that." Every dog owner says this, and will make real trouble if you abate the problem in a final way.

Goats don't commonly breed out of season. You'll have difficulty getting them bred from April through August. There are modern injections to help with this, and a very rare one will breed out of season without it, but is this really a simpler life?

Buying a very good milch goat is quite expensive. Indeed a *very good* Jersey cow, freshening (beginning her milk cycle) will give about 6 to 7 gallons of rich milk a day. A *very good* goat will give about 1¼-1½ gallons. But a goat which is giving that much presently costs around $200-300 locally. Our heavy milking Jersey cow cost us $600 with registration papers. So let's see, that's about 5 times the milk for 2 to 3 times the price.

Not all goat news is bad news. They are superior to cows in some other ways. For non-settled people (nomads) they excel because they won't wander far from the goat herd's wagon. They are somewhat cleaner than cows, which sometimes lie down in their own dung without thinking about it (tho' neither goats nor cows will eat near their own excrement).

Goats are better in another very important respect. If you have enough goats to equal a cow's production, say 5 of 'em, you can justify keeping a buck. It'd take at least that many cows to justify a bull, but then you'd be awash in milk! Which leaves artificial insemination (A.I). This is certainly neither self-supporting nor a simpler life! But no law says a dairy cow must be bred to a dairy bull. Beef cattle abound, and a beef bull will do the job, and usually fetch a better price for the calf, if you want to sell it. A local dairyman friend comes and performs the A.I. on my cow. Most are quite willing to do this for their neighbors. Beef bulls are ordinarily very calm and gentle. Dairy bulls are meaner than snake spit.

If you keep a buck goat, you need to keep him well separated from your does. Otherwise that bucky smell in your milk will be there all the time. Bucks often turn mean as they age. I once stood next to a stall door when a buck rammed it, smashing it apart! I caught him before he escaped, but then what? My hands were full of goat. Which brings up the next advantage of goats.

Goats can be grabbed by hand. I'd hate to see anyone try to grab a bull like that. If a goat "goes down," which is to say it's unable to rise due to some ailment, one or two people can lift it into a wagon or sled to bring indoors. Cattle require special machinery. Most full time cattlemen have such tools lying around, and will lend them if you don't bust things up when you borrow. But goats can be brought home or to market if need be, in a car's back seat. They can be thwarted in their waywardness, whereas a rogue cow is a job for true cowboys.

Goats' dung is drier, and in some ways easier to handle. But then, unless you're really blessed with straw it tends to fall through the tines of a dung fork, which is very aggravating at times.

But enough of goats. You can see they've got some advantages and many disadvantages, and the debate among small holders on this subject is endless. If dairy goats had been better for American conditions, they'd have long ago displaced cows here, like in other parts of the world. It is not mere prejudice governing this, since profitability has always ruled the marketplace.

Feeding, milking, and health problems are almost identical to cows, though the animals themselves are worlds apart.

12
COWS

And he took butter, and milk, and the calf which he had dressed, and set it *before them.... Genesis 18:8*

My first cow was quite an education. I got her because I could no longer afford my gallon-a-day milk drinking habit. A musician's income is more suited to bread and water. But I had a bit of grass, and was working to clear more land. Milk is mainly a grass product, and most of my land is too rocky to even think of growing stuff like corn, wheat, or beans.

Dairy animals have the unique advantage of providing a harvest every day from the time they're brought home. In 1636 Thomas Hooker's company of followers survived the overland trip from Massachusetts Bay Colony to what is now Connecticut by relying on their dairy cows for survival. Again, the local herbage was already in place, and the cows quickly turned it into food for man.

But today's musician often must live "on the road" in order to get enough jobs to pay living expenses. Frequent travel is incompatible with cow keeping. Sure, an occasional trip could be covered by getting experienced folks to do the milking for us but these people seem to be uncommon any more, even in the country.

On the other hand, when our "living" doesn't depend on money, we musicians no longer need so much mobility. A wise man once said "When we find our lives are worthless we may either commit suicide, or travel." I don't recall ever seeing a contented person with the wanderlust.

When I brought my first cow home, I was dismayed that she didn't give as much milk as when I milked her before at the seller's. I loader her up after three days, and took her back. The sweet old man I bought her from assured me she was just nervous. He said "Try to milk her here." I did, and she gave plenty. I hauled her back home. I was lucky the poor dear didn't get "shipping fever," a common ailment associated with uncovered trailers and trucks. The next day and thereafter she gave a full bucket every time.

The Way of Cows

When a calf is born, it struggles to its feet within an hour and fumbles around until it gets a teat into its mouth. It is encouraged and envigorated by the cow's licking the birth fluids off. This initial licking helps animal mothers identify their own by smell. Once dry, most new-born farm animals are surprisingly hardy.

Cows like to "hide" the calf in some sheltered location until they return hours later from grazing. The calf drinks, and then hops about until it's tired again.

Within 2 weeks the calf should be nibbling at grain and grass. I like to see them eating well as early as possible, so from the beginning, if it's winter and there's no grass, I provide really green, soft leafy hay like alfalfa taken from the third or fourth cutting of the season.

Calves sometimes get diarrhea, or "scours." They get dehydrated in a hurry so it's good to get 'em back to normal as fast as possible. Spend a few minutes at each feeding to inspect the calf. If it's standing pitifully with a humped up back, check its rear end. If it has "the squirts" and its rear end is a mess get help from an experienced hand with cattle forthwith. But it is unlikely your calf would ever be jeopardized because you'd keep the barn mucked out, and let the little sport have all the sun and air it wants. And for at least the first full day ONLY the calf gets milk. To avoid her getting milk fever, don't milk the cow at all for 24 hours after calving; milk only about half the milk the second day; milk her dry the third day.

Incidentally, another healthful aspect of untreated milk is that it may be made to combat colds and other infections. The germs in a calf's saliva work their way up the cow's nipple to the udder. There they stimulate considerable antibody production, so that the next time the calf drinks, he's treated. This is why calves drinking from

the mother seem to never get sick, even when her udder is caked with the dung and urine she was lying in all day (which should never happen because you keep her stall clean!). Hand milkers may, at the conclusion of milking, lick a finger, and touch it to the nipple. Cow owners almost never get ill.

Dairy cows have been bred to give much more milk than nature intended. Young calves can get too much, causing a disease called *enterotoxemia*. This is also known as "pulpy kidney." The abundance of milk makes the animal grow amazingly fast, but causes an overload on its organs. A poison suddenly gets released without warning as early as two weeks of age and within minutes of the appearance of symptoms, the calf is dead. It should get enough milk to bulge its belly satisfactorily, but not enough to satisfy its appetite. Observe and govern your calf. After a week *never* allow the calf full access to the cow.

The calf grows right along on grass or hay and grain. Give it whole milk up to 3-4 weeks. Skim milk fed until 6 months will make the fastest gains. An excellent grain mix is equal parts of corn, ground oats, wheat bran, and soybean meal. Feed the calf all it can eat up to 4 lbs. of grain a day, but don't give it more.

If the calf is a bull calf it will probably need to be castrated as soon as the testicles can be felt to have descended into the scrotal sack. The easiest method is to use the "elastrator" or rubber band method. The strong, tight rubber band is placed above the testicles. Circulation is thus cut off, and the testicles dry up and fall off in a few weeks.

Calves of both sexes are disbudded, that is, their horns are removed before they begin to grow much. There are several ways of doing this; get an experienced cattleman or vet to show you how.

My experience with dairy calves is they are the least trouble if taken from the mother after a day or two. The longer you wait, the stronger the maternal bond becomes. When you finally do get the calf away the cow bellows all day and all night for some time. If the calf can get to the cow's fence, it'll find a way to suckle, you just wait and see! Never wait more than a week to take the calf away from its mother. Determine how much milk to feed it by observing its belly, and its appetite; it's just common sense. Start with say, 4 feedings, reducing this to twice a day after about 2 weeks. Keep the calf away from milk for up to 12 hours, to get it hungry. Then, when you get it sucking on your fingers (it'll do this naturally) lower you hand into the milk bucket. At first, it will fail. Persevere. It doesn't take long. Now it's bucket trained.

The growing calf, or any age cow for that matter, will need access to water at all times, and a salt/mineral mix available free choice. Plain sifted hardwood ashes (2 parts) and salt (1 part) are a good mineral mix. Or ground limestone and salt in the same proportions. Or equal parts limestone, steamed fine-ground bone meal and salt.

As young as 3 or 4 months old the calf, if a heifer, may come into heat. Keep her away from bulls! Jersey cows can be bred as young as 15 months, but larger animals are better left until about 18-24 months depending on the size of the adult cow.

During the gestation period, and even before, a heifer can be fed nothing but good pasture or good legume hay. For that matter, a mix of legume and grass hay is good. Two and a half lbs. of hay a day per hundred pounds weight is what they need. Lots of farm stores carry weight measuring tapes.

Along with the hay, the heifer needs 2 to 4 lbs. of daily grain. Split into two feedings of course, and increase this by a pound 3 to 4 months before calving.

Gestation in cows takes about 280 to 285 days. Be ready ahead of time, and make sure to mark both breeding date and due date on your calendar. If she comes in heat about 21 days after breeding, it means you missed. She needs to be bred again. Sometimes cows have "silent heat." This is mostly due to heavy milk flow, causing a suppression of hormones.

Signs of heat include swelling of the vulva, lots of mooing, a crazed look when you're near, attempts to jump on you (beware!) jumping on other cows as if mating (called "bulling") walking the fences, drop in milk production for a day. Mark the date of heat on the calendar even if she's not bred. In 21 days, she'll show signs again, and you can be ready for her then.

Signs of imminent calving are tight swelling of the udder, enlargement of the vulva, a springy sort of walk, looking back at her rear end, getting down and up a lot, looking for a sheltered place, and when the moment is at hand, a mucus discharge from the vagina.

A "water bag" breaks first, and a minor gusher issues forth from her vulva. The front hooves then appear, and the nose, mouth with tongue usually hanging out. If the mother can pass the calf's head, she's home free. Sometimes a different presentation occurs. Sometimes one front leg is back, or worse, the head is turned back. Someone will have to ease a clean hand inside to correct this. It gets pretty tight when her contractions come. When she eases, proceed. Be careful, and it's best to get an experienced hand, at least for your own peace of mind. After the calf is out, an awful looking fleshy blob called afterbirth comes out. The cow will often eat this. Just let her, it's good for her. If not, clear it away to keep the flies down. It's just the "bag" the calf grew in when it was still in the "calf bed" or uterus. If the afterbirth is retained, it will need to be taken out. It's best to get help the first time you do this, so call in an experienced cattleman to help you.

First and second calf heifers rarely get *milk fever,* but full grown cows, especially if they are heavy milkers, often do. Milk fever is a rapidly occuring condition which oddly includes no fever. It is caused by a calcium deficiency right when the demands for calcium suddenly increase--just after calving. The most common cause of it is feeding high calcium feeds such as alfalfa, clover, limestone, bonemeal, wood ashes, etc., during the last two months of gestation. You see, when you stop milking your cow (as you should 2 months before calving) her calcium utilizing ability goes dormant. She must be forced to draw it from her body reserves until she starts producing milk.

The second most common cause of milk fever is milking the cow too soon after calving. Only the calf should get milk, and even then, not much. Some dairymen don't even let the calf suck for a day, but that's a bit excessive, I think. In 2 or 3 days you can start milking. The bulging udder seems to help prevent the condition, and indeed, one old time remedy which really worked was inflating the udder of afflicted cows with air. This is not as good as intravenous injection of calcium gels available at farm stores or veterinarians, because it may cause *mastitis,* the second most dreadful ailment of cattle.

Symptoms of milk fever are, first, a bit of discomfort and a "distant look" visible in the cow's face. Then, incoördination, and when she lies down, and inability to rise. The nose is no longer moist, and the ears and other extremities get cold. The neck assumes a strange z-shaped configuration, and the cow's eyes are glassy and oblivious to almost everything around her. Soon she becomes comatose, and shortly after she dies. This all happens in just a few hours' time. If you see any of these symptoms within 3 days after calving, get immediate help--local dairymen are much more reliable than veterinarians generally, since the money is mostly in cats and dogs these days. They'll be amazingly kind and patient with you if you don't get smart with 'em. Their lives are hectic enough as it is. Offer them coffee and a snack, buy 'em some extra calcium from where ever they get it. Treat 'em good, because they are the best friends you could have. They're always at home (after all, they've got dairy cows!) and they've seen it all. Besides, a vet will cost you plenty just for showing up--even if he doesn't save your cow.

The cow will give colostrum for about 5 to 7 days after you start milking. This is a high-protein, antibody-laden milk of a darker color than normal. It can be sugared, and baked to make an excellent custard called "beestings." But sometimes the cow is so flush with milk that a few small blood veins burst and then there are varying amounts of blood in this early milk. But the pigs love it. All ill animals can be treated with colostrum frozen into ice cubes. New borns, especially if their mother lacks milk, or if she dies, do well on colostrum. We always freeze as much as we can for future use. It keeps for a couple of years if frozen. You can also buy powdered colostrum for outrageous prices at farm stores. The best thing about them is they don't affect us like antibiotics--there's no withdrawal time if you want to sell or slaughter a treated animal.

The cow will come back in heat in a couple of months. Record the date of the first heat; be ready the next time. Plan breeding so the next calving date is convenient. If the cow doesn't come in heat, check with your dairyman friends.

Dairy cows don't like being out in the cold and wet. Warm rains don't bother them. Snow, sleet and cold winds do. Give them a shed closed on at least three sides to exclude drafts--one which is well drained, because a cow dispenses many gallons of urine. I like a stall at least 10 feet square, though many folks get by with as small as 8 by 7 feet. In inclement weather the cow will spend considerable time there, and the dung accumulations between milkings are more apt to get all over the cow in a small pen. Keep plenty of straw on the floor, so the wet bottom stuff is packed about 6 inches beneath a comfortable and clean-smelling dry layer. Whatever plops are deposited are easily picked up on a dung fork with a bit of the straw they rest on. Then straight to the heap or dung cart with it.

A cow made to lie down in urine-soaked straw for want of more comfortable bedding is a real headache to clean for milking. One made to lie in snow is sooner or later going to lose a teat or two to frostbite. This is common sense but it is remarkable how many folks refer to such lost teats on their own cow as inevitable "bad luck." Please be kind to your cow.

Feed and water

A cow in milk needs a very large amount of water. It is nothing strange for a cow to drink 10 gallons at one draught! But if the water vessel is only half full, or if you're relying on 5-gallon buckets, your cow may not get enough. Once she breaks the concentration of drinking to change buckets, the typical cow will lose interest. Cows drink a lot, but not many times a day. Milk production will suffer if water is in the least held back. In winter, warm water is very appealing to a cow. A 750 lb. Jersey cow drinking 10 gallons of water is the same as a 150 lb. man drinking 2 gallons. Imagine how cold it'd make you feel if it were ice water in winter!

The lactating cow can have all the hay or grass she wants. If grazing, 2 to 3 acres of rich grass/legumes will supply all her fodder for the year, including hay. When there's no more grass, she'll eat about 2½ lbs. of hay daily per hundred pounds of her own weight. Plan for this when accumulating hay for the winter. The best hay is good enough for your dairy cow. Lots of legumes (clover family, like alfalfa, trefoil, lespedeza, and of course, clover). Feed her in a clean feeder. She'll waste a lot if you feed her hay on the ground, the way most folks do. She can also get silage, or mangels, carrots, or potatoes. Turnips, rutabagas, cabbages, onions, kale, and sweet vernal grass all cause bad flavors in the milk. Oddly, kohlrabi, a member of the turnip and cabbage family, doesn't seem to cause milk taint. Of these succulent feeds, a Jersey cow may get up to 40 lbs. a day. It can cut your hay consumption by as much as 90% and keep the cow looking great. No matter what you feed your cow, do not abruptly change her diet. Gradual change, over say, a week or two, will avoid any severe digestive troubles.

Soiling

In the 19th century it was discovered that dairy cattle fed fresh fodder, cut from the fields daily, were more efficient. To quote an eminent farm adviser, George E. Waring, Jr.:

"Concerning the saving of land, the fact that four times as many cattle, or even more, may be kept by soiling as by pasturing, is sufficiently conclusive.

"The saving of fencing is a matter of great consequence; for the large amount of money, or what is equal to money--time and labor, expended in building and repairing the interior fences of the farm, is often appalling.

"[Of] the economizing of food, [Josiah] Quincy says, 'There are 6 ways by which beasts destroy the article destined for their food,--1st, by eating; 2d, by walking; 3d, by dunging; 4th, by staling; 5th, by lying down; 6th, By breathing on it. Of these six, the first only is useful. All the rest are wastful.'

"Of course, it is not to be inferred from this that animals avail themselves of only one-fifth of the food produced by the field on which they are pastured. But any farmer will at once admit that the amount destroyed in the various ways referred to is very great; and it is probably even greater in rich pastures than in poor ones, for the reason that after an animal has once filled itself it seems to devote a large part of the remaining hours of the day to the destruction of luxuriant food for which it has no immediate use.

"With regard to the better condition and greater comfort of the cattle, there is a sentimental idea that the advantage lies on the side of pasturing. [But] the amount of exercise required to maintain the health of any domestic animal is not great, and if we observe the conduct of cows at pasture we shall see that under the most favorable cicumstances they take only so much exercise as is necessary to enable them to fill their stomachs with the choicest grasses within their reach, and that, being filled, they invariably remain quiet until they need food again. On poor pastures, where a half-starved cow is obliged to walk nearly the whole day to pick up a scanty subsistence, there is no doubt that one of the principal causes of her poor condition is...the fact that she has been obliged to take more than the proper amount of exercise.

"Ample experience the world over has clearly demonstrated the fact that with proper facilities for exercising in the yard, cattle fed regularly with nutritious food only in their stalls, are in better condition, and live longer in good health than do those who are exposed to the vicissitudes of the weather, and to the more precarious subsistence that natural herbage usually affords.

"The product of milk, as is very ably stated by Mr. Quincy, is greater throughout the whole season than under the system of pasturing.

"The attainment of manure, which is so greatly facilitated by the soiling system, may be regarded as second in importance only to the saving of land. In the field the dung is dropped principally on those parts which are chiefly selected by the animals for their resting-places. More will fall upon the rich land than upon the poor." Book of the Farm, 1877

In my comments on goats I pointed out that because of the difficulty of fencing them, they're almost always kept by soiling, and thus are more efficient (theoretically) than cows. I might here add that cows are much more contented in confinement than any other animal, and especially goats. If goats were happy being kept like that, they'd wouldn't constantly be looking to escape, and we'd all have goats instead of cows.

But the reality is soiling is really a great deal more human labor. Unless you really must keep a cow on only a fourth as much land, it's no good to work so hard just feeding an animal.

Breeds of Dairy Cows

The Jersey tops the list. It gives the least amount of milk of all the dairy breeds, except maybe Devons and Shorthorns, but the richest milk. It not only has far more cream, but far more solids. If you took a day's milking of Jersey milk, added the same amount of distilled water, and mixed it well, you'd have the equivalent of a day's milking of Holstein milk; half as thin, but twice as much. The only difference is Holsteins eat twice as much as Jerseys. Guernsey cows give milk almost as rich as Jerseys. Both make good cows for family use.

Many breed associations tout "dual purpose" cattle. By this, they mean dairy and beef. But they aren't beefy enough to bring the prices of Angus and Herefords, and usually don't milk enough to compete with the common dairy breeds. Holstein and Brown Swiss are perhaps exceptions, but they are huge, certainly more meat than a self-supporter would want at one time.

Still, it isn't unthinkable to split up the beef between yourself and one or more other buyers. I am amazed at how fast we have sold portions of cattle we couldn't fit into our freezer. We have done this even though we didn't raise the beef. We bought beef we knew was clean of drugs and hormones and sold the half we didn't keep to several other buying partners. We all got good beef cheaper than supermarket prices.

Unless you have an icehouse, or really trust rural electric lines to run a deep freeze at all times, beef is out as a small farm food. The only way around this is to split beef among lots of folks on a regular basis, and that requires more organization than I can muster.

Jerseys and Guernseys have good enough beef, but it isn't in

A FINE JERSEY COW

demand commercially because it won't "marble." Which is to say no fat is deposited into the meat itself--it builds up on the outside where it must be trimmed off to be salable. Because of this it often lacks tenderness and flavor. It cannot produce "prime" or even "choice" grades. But it's still good to eat, there's just more hamburger, that's all.

Stick with a Jersey or a Guernsey, and go for milk. A good one will give you anywhere between 900 and 1500 *gallons* of milk a year, which translates into as many *pounds* of rich cheese. This is far more food than any beef breed will give you in a year. And you'll still get the calf! Dairymen measure milk in pounds, not gallons, and the above figures make about 7500 to 12,500 lbs. The more they give, the more careful treatment they require, which is why most dairy men know so much more about cattle than most beef men.

Milking

When you get your first cow, you may be lucky enough to get the present owner to show you how to milk. But it is surprising how few house-cow owners are really good at it themselves. Commercial dairies all use machines, and are usually too busy to show you hand milking. Yes they must hand milk sometimes if they have put their trust in rural electric companies.

Begin by giving the cow some feed and tying her to a "leash" so she won't leave when she's finished eating. I like to thoroughly wet her feed, because when it's like pudding she won't waste as much. While she's eating wash her entire udder with warm water, even her whole lower rear end is a good idea, so loose bits of hair and dirt won't fall in the milk bucket. Many folks never wash their cow unless she's really filthy, and don't seem to have trouble.

Dry off the cow's udder, and squirt a few strokes from each teat unto a separate vessel for the cats. This first milk has many germs in it. The squirts are made by clamping the teat in the crotch of your thumb right up next to her udder. Slide your clamped hand on down the teat, forcing out the milk contained within it. Now do the same with an adjacent teat with your other hand. Alternate both hands this way. This is called "stripping." You need not fear much hurting the cow, if you pinch her tight enough to hurt her, your hand will be clamped so hard as to prevent sliding down the teat at all.

The other method, which ought to be employed most, is as follows: Clamp the teat as before. This separates the milk in the teat from the milk in the udder. While continuing to thus clamp the teat, press the index fingertip firmly on the teat just below where it is already pinched, right to the palm or heel of the hand. While holding it there, add the middle finger in like manner. Now the ring finger, and last, if her teat is long enough, the little finger. Done right, this method is at least as fast, makes less friction and therefore is potentially easier for the cow. When your hand is tired, switch back to the stripping method until you can milk the other way some more. It takes far longer to describe than to do. I once counted up over a thousand squirts on the cow's 4 teats before she was finished! At about 3 strokes per second (between two hands) a nice leisurely pace, it'll take you about 10 minutes to milk your cow.

If you compare the gallons of

milk thus obtained, and the quarts of cream, with prices at the market, you'll see how much you saved by getting it fresh, instead of pasteruized. The only reason there are pasteurizing laws at all is to make safe the blood, dung, and germs which "mega-dairies" must carefully cloak through homogenization to get folks to buy it. Sadly, there's no other reason for homogenization, and no other reason for pasteurizing. Tuberculosis is not transmitted through the milk unless someone with TB coughs into the bucket. The distinguished organic-farming champion Lady Eve Balfour always won first-place in agricultural fairs for low-bacteria count in her milk, but was always disqualified when the agri-biz judges discovered she milked by hand in a straw-covered stall.

Your cow will soon come to regard you as "her calf." She may even lick you while you milk to encourage you. She'll run to the milk stall when she sees your bucket, and she'll come when called. She'll always be fond of you, as you will of her.

Always milk your cow completely out. At first, she may not "let down" the milk. Cows are said to "let down" the milk, because of an involuntary reflex action. The udder's milk cells contract, squeezing out the milk. The faster you can get the milk out, the more completely she'll be milked, and you'll not only prompt more production on her part, but avoid mastitis problems, too.

Milk at least twice a day at first; towards the last month of lactation, once a day is probably enough. Twelve hours apart is best, of course, but not absolute. No need to rise at 4 a.m. like the commercial dairymen, because you're not meeting the "milk truck." Early and late enough to beat the flies, that's all.

Mastitis

After milk fever, mastitis is the second worst disease of cattle. It is the result of a bruise, cut or other irritation in one or more "quarters" of the udder, causing an internal infection. The milk from infected quarters can be immediately detected in that it smells, and tastes, like it came through an old rubber hose. That rubbery smell appears before the characteristic stringy milk and yukky stuff begins coming from the udder. The sooner you tackle the problem, the better the chances that quarter won't be ruined.

The irritation will almost certainly be tender, and the cow may want to kick when you're milking it, but keep at it. She'll calm down in time. Just keep milking it. Get it all out. In fact, I have heard of folks milking all night at that quarter, and the next day the mastitis was cured. In fact, Eva once massaged with "bag balm" and milked 6 hours, and the next day 4 hours, and the next day the mastitis was gone. And you wondered what folks with so much leisure time do for fun.

Antibiotic injections are the next step. They will show up in the milk for several days, so you can't use even the good milk from the other quarters, unless you don't mind all those drugs. Get your dairy friends or a veterinarian over to have a look at it and advise. But treat it immediately, or it'll get worse--a lot worse. You'll almost certainly lose that quarter for milk production if you delay at all. You could lose your dairy cow if you delay too long, then she'll only be a low grade beef cow.

Now you see why so many house cow owners are reluctant to let others milk their cow. We always want to see the helper milk the cow completely before trusting her health and life to them. Mostly, cow owners arrange to have their vacations during the two months or so when their cows are dried off for calving.

Also, thoughtful cow keepers avoid briers and barbed wire, whose nasty scratches can be the cause of mastitis. Treat all udder scratches and cuts with disinfectants like iodine as soon as noticed.

Bloat

Cattle sometimes get bloat, a fatal condition. The first stomach, or rumen, gets frothy with the foam from dew-wet clover or other legumes, or lush grass. The cow can't belch, and the gases build up. After a while the cow becomes unable to rise. Her left side, and sometimes her right side too, will bloat up like a balloon from all the frothy gas. The first remedy for this is only good if she hasn't gone too far with it. Get a heavy cola bottle and put some vegetable oil in it. Stick this into the side of her mouth and *slowly* get her to drink it. The oil will subside the foam. If she's farther along than that, she may need more drastic measures. I have heard of smearing a rope with dung and working it in her mouth, so she'll have to chew to try to get the horrible stuff out of her mouth. The chewing helps cause her to belch. I don't think this'd work if the oil method doesn't work. If the cow is fading into a coma, you have no choice but to stab the rumen with a "trocar" and hold the hole open with a "canula." I've heard of folks using a simple pocket knife, and then twisting it to hold the incision open while the smelly goop runs all over their hands. Here's where experienced dairymens' help is real good to have. Vets often can't make it to you in time if you live too far out from the 'burbs. Some vets are afraid of cows.

Bloat always occurs under the same conditions, viz., 1. Lush pastures, especially when the cow has just been moved from the stall or sparse pasture. If the cow has been on the grass for some time, then the following; 2. Dew on the grass. 3. A barely perceptible cool breeze. 4. Clover, alfalfa or other rich herbage. To prevent bloat, be on the alert. Feed the cow some hay first before letting her into these conditions. If you wake up to these weather conditions and you know your cow is in the pasture, go have a look, watching for that first inflating of her left side up in front of the pelvis. Give her a bit of corn meal mixed with a half cup of oil as a preventive. Many folks wisely keep a "bloat block" in the field when they go on trips. This looks like a common 50 lb. salt-and-molasses block for your salt feeder, but it contains a bloat inhibitor, usually *poloxolene.*

Hardware Disease

One last ailment deserves mentioning, because it is so bad, and is so preventable. Hardware disease is caused by cows eating small nails, fence staples and the like inadvertantly in the grass around fences. The staple sits in the rumen until it finally works its way toward the gut. It either lodges there, because its sharp points catch on the tender linings, or it tears a hole. Either way you may lose the cow. Magnets are made in the form of large pills, which you induce the cow to swallow, keeping all such trash from getting beyond the rumen. Or, you can just do the right thing and not leave your old nails and trash lying around.

13
HOGS

...And he sent him into his fields to feed swine.
Luke 15:15

When I lived in Maryland, I had such a smashing success with hens that I decided to try sheep. I kept my eyes on the newspaper classified ads until sheep appeared there for sale. I drove about two hours away to look at the man's flock, which were of very mixed breeding. The shepherd was from England and had settled in a rural part of Northern Virginia with what is now called a "small" farm (about 100 acres or so.) He had poultry, pigs, and sheep. While there a penful of piggies, about 25-30 lbs. each, caught my attention. I inquired as to price and negotiated in the pitiful way only a novice can do, and then the man informed me he took all major credit cards, of which I had none. He was disappointed when I told him my personal check was all I could offer, and after some hemming and hawing he reluctantly gave in.

I transported my piggy and young sheep home in the back of my little car, which thenceforth smelled of pig and sheep. More on the sheep anon, but the pig was great fun. It was of the Yorkshire breed which is to say it was white all over, with a long body and pricked-up ears. It was a *barrow,* which is to say a castrated male.

As with practically all pigs, it was very skittish while young. I turned it loose in the fenced fields of our rented little farm, and it did all right. I don't seem to recall buying much feed except shelled corn occasionally.

My pig knowledge consisted only of hazy recollections of hog work on my favorite uncle's farm in South Carolina. But that had been at least 15 years before. As a mere stripling, I rather spent more of my time playing in the hay, shooting the BB gun, fishing, riding ponies, etc., than observing pigs.

So corn was all this pig got from me. Still, along with such paltry fare, it was able to rummage for bugs, eat the grass which the landlady's horses wouldn't touch, and horse dung too, as well as all our kitchen waste.

The fences, as it were, lacked substance; my pig went a-roving. Or I should say, my hog went a-roving, because by the time it'd gotten the wanderlust it certainly was a good 300 lbs. It reached this size in a mere 8 or 9 months. Not willing to shut it in a tiny pen, nor to expend considerable sums unrequited on the owner's fences, I began bribing the hog to stay at home with more corn.

One morning, as I was leaving for work, I spied my hog, two neighbors distant from home. I stopped, got its attention, and it dutifully followed me the several hundred yards back to (yes!) ever more corn. Unfamiliar with judging weight of animals, I figured my hog was "at least 200 lbs.," and I decided it was time to make pork of it.

I invited several friends for the big doings; all were excited to be included. The big hog was closed in a stall in the stable, and everything was ready, or so I thought. At the appointed hour, with no one yet arrived, I decided to go ahead and slaughter it, to avoid all the gushing sentimentalism usually met with when townsfolk are present. Then they could simply help with the hard part: cleaning the carcass.

Hah!

In fact, none of 'em ever showed up, and I was left to do the job myself, with only trusty (and longsuffering) Eva to assist.

What a job! After quickly killing the hog with one well placed .22 cal. shot to the head, and just as quickly sticking it properly to bleed, we were left the staggering task of moving the hog to our makeshift hoist. It was then we realized the hog was no mere 200 lbs. weight!

We were at it until 2 a.m. just cleaning and gutting it. What a fiasco. Now I know how utterly ill-equipped we were, and what difference a few small conveniences would have made.

Despite having ruined some of the meat by spilling a very minute amount (!) of the contents of the guts on it, most of the meat turned out not only very well, but superbly. I cured the bacons and one ham. I did the cutting up during the next two days; it being February the meat could be taken from the barn as required, so I didn't need a huge kitchen. This is a very good thing, because I didn't have one. I had the dining table and one 2 foot square counter.

My next attempt at butchering came the following year and it was not only much less work, but much less hog; I did the job on time this time, and had prepared better.

The Way of Swine

Pigs are *farrowed* (born) in litters of from one to about 15. The world's record litter is 36. Average is about 8. Good mothers have at least 12 good teats, although 14 or more are much better, just in case. The piglets are commonly from 2 to 4 lbs. at farrowing. Well cared for, they double in size the first week, and double that the second week. Then they should double again by the fourth week. By this time, they'll be eating solid food in addition to milk, because the mother's milk begins to slack off. But they start eating dirt from the day they're born, which is a good thing, because without clean dirt they get anemia and die by the end of the second week. Factory "pharmers" give them iron shots to prevent anemia. Healthy forest humus (or from under your fence-rows) is far better because it seems to prevent *mycoplasmal pneumonia* (also known as enzoötic or virus pneumonia) and to aid the piglet's digestion and appetite. Indeed when pigs appear unhappy, a bit of humus often brings 'em around lickety split.

Most commercial hog keepers wean their piglets at 21 days, a bad time and much too early. The piglets' immune systems haven't developed until 5 weeks, so they must be heavily drugged to survive. But it gains a couple of weeks' not feeding the *sow* (mother) and she can be put back with the boar to be bred for the next go round. Some wean at 10 days, or even less! In this way they get 3 litters in one year per sow. Then she's used up, and goes to slaughter while still not finished growing herself.

This is a waste, because a well treated sow lasts through at least 3 years of bearing piglets; many used to last up to 4 or even 5. The present ways were made necessary when commerce became the sole purpose of pig farming. To their credit, city people often pity the animals, but to their shame, their only solution is to either harass the already tightly squeezed farmers, or pass silly laws, conceived in arrogant ignorance. The inevitable result is always to enlarge the size of the hog factories and discourage the "little guy."

For my part, I think it's wicked to choose coercion over persuasion.

I rest easy knowing my hogs do what hogs like doing: enjoy a varied diet, dig & root, wallow in mud, procreate. When their time comes, it's quick and relatively painless. I won't be led by crusading townsfolk into abusing farmers who must cut costs to compete, after all, for the town shopper's business.

After pigs are weaned, they grow at a rapid rate, and are fed all they can eat until they are butchered. If there's a chance they'll be used for breeding, then they're fed until they are able to eat 4 to 5 lbs. per day. That's how much they get from then on, with very few exceptions. This way they don't get too fat, nor do they reach sexual maturity too fast to be durable and prolific breeders.

Females come into heat about every 21 days. The signs begin with a swelling of the vulva. *Gilts* (females less than 1½ years old) may swell up as early as 5 days before actual heat. They remain in *standing heat* for a day or two. This means they will stand stock still while the boar mounts them. Or while you sit on their back, trying to determine if they're in heat or not. If a boar isn't available, you at least know you'll need one in 21 days, and can plan accordingly. Also, you know when she'll come back in heat if the boar didn't *nick* her (get her pregnant).

Gestation takes about 114 days. I have seen it take only 106 days, and sometimes as long as 125 days. A week or two before farrowing the udders begin really swelling up. Sometime within the 3 days before farrowing milk can be squeezed from the sow's nipples. When the sow begins making her nest she will farrow within 24 hours.

Fencing

Hogs can be fenced best with post and boards or rails. The best recommendation I ever got was from the great self-supporter John Seymour, of Great Britain. He said get pigs with drooping ears instead of the pricked up ear breeds. That means Chester White (which I keep) Duroc, Hereford, Landrace, Spots, Poland China, etc., in America. With these breeds adult hogs can be confined with a visual barrier such as a single board nailed between each post, and a single strand of electric fence. I have kept many dozens of hogs with pricked up ears, and two strands of electric wire with a woven wire (hog wire) fence behind it, but it wouldn't reliably hold them in! Small piglets are held with a board fence or woven wire stretched tight and with no gaps underneath for them to shimmy under. I prefer to keep pigs up to at least 250 lbs. in a sty.

Housing

All the books and magazines depict A-frame houses for swine. I have used these, and modified forms of them. I don't like 'em. Without going into all their many shortcomings, I will simply state I like rectangular pig houses. For full grown swine (500 lbs. & up) they need 30 sq. feet of floor space for the first hog, and half that for each additional hog. This assumes at least one side is left entirely open. In places where the temperatures get to -25°F several times during the winter, it's not a bad idea to partly close in this last side, too. Planks laid on the ground stay warm, but must be fastened so the big hogs won't root them up while grubbing. If they do that in their house, it will eventually fill with water and be useless for keeping warm in winter.

Worse than being cold is being hot. Hogs have no sweat glands. I once had two big sows die of the heat while I was away. They need a wallow, or much better, shade. This is best supplied by tall trees, but some folks make shade arbors of poles and brush. Adult swine will form their own wallows if left on a piece of ground during the time of year when the soil is soft. They nose out a depression to sleep in in cool fall or spring nights, sometimes even in winter, and sleep outdoors with their bellies to the warm earth. These excavations will fill with water during warm, rainy days, and the swine wallow in it, mashing the dirt together like clay, so the water stays a long time before it evaporates.

If you want to clear up an old wallow on ground the hogs have cleared for you, just heavily manure it with wood ashes or hydrated lime. Toss a quart of it evenly right in the standing water. The reaction of these substances with soil causes *floculation*, which means the soil particles push apart from one another. This method is sometimes used as a short term substitute for drainage in clay soil areas.

I have had large swine pass through dreadful cold weather unharmed without shelter of any kind except the windbreak of trees and brush. I would certainly not recommend this way, because they require about half-again or double their usual feed to stay warm.

Pigs under about 250 lbs. need shelter, and with all swine the optimum temperature will save feed, avoid disease, and is just kinder.

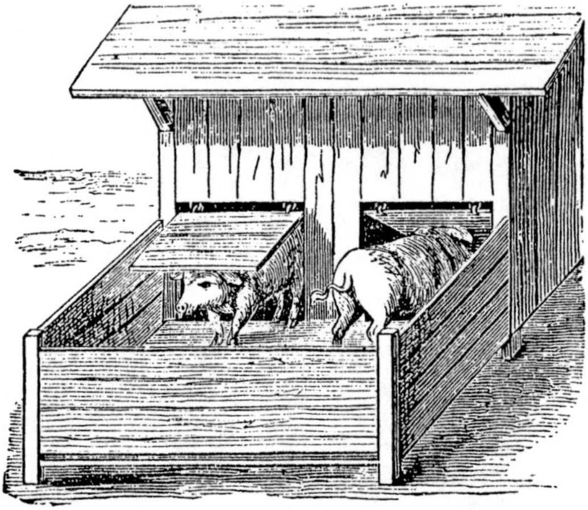

—SELF-CLOSING PEN DOORS.

Quite by accident I discovered that swine keep very warm and dry in the coldest weather when given tight houses. And I mean tight. Plywood or matched boards, with a tight floor of same, and a doorway which is hinged from the top to swing in and out, has kept pigs of all sizes very cozy indeed at sub-zero temperatures. A common fault is to make the ceiling too high. For really small pigs (up to about 75 lbs.) a foot and a half is plenty high enough, and for shoats (75-175 lbs) 2 feet is good. For hogs (175-300 lbs) 3 feet is plenty. Hogs above 300 lbs., up to 700 lbs. never needed higher than 3½ feet of ceiling height on my place. In fact, 4 feet was too high for them to stay warm and too low for me to get around in comfortably. So in my larger hog buildings I built a roof high enough to walk under without anyone having to crouch, but I made false roofs about 3 ft. or so high for the hogs to stay warm under. These low ceilings form the pig pens into the type of kennels so popular with swine growing "factories" in England.

The floors are uncomfortable if concrete, so planks are laid on my concrete pigsty floors. This makes for happier and healthier animals whether piglets or brood hogs.

Outdoor Pigs

The autonomous husbandman need not raise pigs in a sty at all. Many folks in the past raised pork every year in a pen of about 8 poles (rods) per pig. A pole is a land measure equal to 16½ ft. square. So for one hog, an outdoor pen ought to be say, 33 ft. by 66 ft. The feeder pigs were bought at about 25 to 40 lbs., fed until they were a desirable size, and then slaughtered. Many folks like to get pigs in mid summer, when the garden is starting to really produce, feeding the hogs on garden waste and other slops until November or December (or any winter month, really) and then slaughter the hogs while the weather is cool enough to chill the carcass without refrigeration.

Other old timers used to buy their piglets in the late fall, when piglets are cheapest, and feed them corn in the winter, when corn is cheapest. Then they'd be bigger in the summer, and bigger at slaughter time, around Thanksgiving. These always have better hams and bacons, because they are not so dry as the younger pigs, but are still young enough to never have stringy meat.

In the South winter days are often not cool enough to insure a complete chill, so slightly salty ice water in a barrel helps the process. More on this later. Let it here suffice to say one may butcher a hog by the Southern methods at any time in the year as long as the work is done in the morning before the day heats up.

Farrowing

Probably the vast majority of small time pig raisers don't keep breeding hogs. It takes far more knowledge of hogs, and little piglets are fragile things. Consider this: A baby is usually about 5 or 6% of its mother's weight. A calf is perhaps the same ratio with the cow. So is a lamb or kid to a ewe or doe. Even a hatched chick is about 3% the hen's weight. But a piglet is commonly less than .5% its mother's weight. It is the easiest animal to die from the clumsiness of its mother.

Also, a piglet is practically naked. Even grown hogs rely on the fat layer beneath the hide instead of fur or hair to stay warm. Piglets have no such fat.

Then again, sows normally need to lie down to nurse the piglets. Lying down at least once in 2 hours puts the piglets at risk of crushing 12 times a day.

When a cow is delivered of a calf, or a ewe her twin lambs, there is cause to rejoice. If a sow has 2 piglets, it's a disaster, because the two cannot possibly pay for the feed she consumed while carrying them to term.

Small pigs, under say, 40 lbs., are subject to a variety of ailments, many of them fatal, and most very damaging. Over 40 lbs., they are subject to very few fatal diseases, and damages are far less detrimental.

But just in the last 10 years, governmental regulation has made obtaining feeder pigs very difficult in many areas. This is one reason I keep sows. Besides, under natural conditions, and with common sense, feeder pigs are the most profitable animals a small holder can raise, after the milk cow and hens. They assure you and a good many eager customers a supply of feeder pigs, they provide year-round clearing and ground breaking services, and they still make superb sausages and bacon when they've gotten too old to breed well.

Farrowing Outdoors

It is probably safe to say the sows whose pigs are born from about the latter half of May until October need no supplemental heat, nor really much more shelter than a roof and whatever brushwood, vines, weeds and straw they drag together to make their nests. Watching a sow make her nest is one of the most pleasant and satisfying rural events. If she has no brush, briers or straw, throw some where she'll have to transport it at least a few yards to her nest site. She rakes with a front foot, arranges the wastes "just so," often destroys it all and starts over until it's "just so." When she farrows there, in moderate weather, she's just as contented as she can be.

By contrast, a sow who is seen to be acting fretful, who has a drop or two of blood at her swollen vulva, who is hastily ushered into foreign surroundings in a "proper" farrowing unit often jumps up at the birth of every pig, does a tap dance from fright, loses her temper at the squealing of the little sweetheart she just stomped to death and can sometimes even be afraid to nurse the scary things. This is especially true in the cold months.

Indoor farrowing needn't be this horrible, but it very often is, so for a leisurely pork grower, outdoor farrowing in good weather is best.

Best plan to have the first outdoor litter a little before the time of the last killing frosts of spring. That is, count 114 days backward from that date, then maybe 2 or 3 weeks further back, and let the boar breed her then. And if she's not in heat? Put her with the boar anyway, and sometime in the next 3 weeks she'll be bred. RECORD THIS DATE!

Before you do anything more with her, count up to the day when she's due, and have her in her own pen away from other livestock two weeks ahead of time. That way she's at home on the happy day. Don't fuss over her on farrowing day, except to feed her about a double portion that day. For young gilts it's good to feed 'em about 25-50% extra during the last 2 to 4 weeks' gestation. If possible, keep things quiet at the farrowing site, so the sow won't be getting up and down much. Older sows go into a kind of trance, and often don't rise for as much as a day or two after the pigs come. Don't feed her for 24 hours after farrowing; the less she's disturbed, the better.

The piglets "hit the ground running" and often eat dirt before they suckle. The sow croaks to them gently to reassure them and encourage them. They often sleep on top of her when it's chilly.

Young sows never seem to have milking problems, but old sows sometimes do. This is almost universally caused by constipation (costiveness). This can be somewhat prevented by making at least a third of her slop wheat bran or ground oats starting 3 days or more before farrowing, and continuing several days after. Give her plenty of water at all times.

Older sows have more pigs and bigger ones, but they are clumsier than younger sows, so they may mash more piglets.

Newborn piglets often eat dirt before they suckle. This is perfectly natural, since it is practically the only way they receive iron and avoid anemia. Sow's milk is remarkably deficient in iron. At about 3 weeks, they will begin rooting around for something to eat besides dirt. Cracked corn can be placed where they, but not the sow can reach it. At first, feed only a handful a day, and when you see it disappearing faster, supply all they'll clean up. In time you can give them a protein supplement like soybean meal, sunflower seeds (dehulled), peas, or peanuts, etc. Kept this way, they will have practically no setback when weaned off. This practice is called *creep feeding*.

Indoor Farrowing

Indoor farrowing offers some flexibility in bad weather. It also makes it easy to wean and give emergency care to sows and pigs, if needed. After many years of successes and failures with both indoor and outdoor farrowing, I have settled on the following method:

My experience absolutely convinces me the best size of a farrowing area, whether a pen in a house, or a portable hut, is 5 feet by 7 to 8 feet, by about 3 feet high. In one end is a door for the sow to come and go, and which can be closed when needed either for weather or to work with the piglets apart from the sow. In the opposite end from the door is the piglet creep, formed by walling off the last 1½ to 2 feet of the inside of the house, except for 8 inches at the bottom where the piglets may slide under, but not the sow. A ceiling of corrugated roofing tops the creep at a height of about 15 inches, to keep all the heat down around the pigs. This ceiling may be insulated with something fireproof, like fiberglass, rock wool, etc. I like to use a kerosene lantern for at least the first 3 days after farrowing, to heat the creep. The best way is to make a little rigid fence around the lantern so the pigs won't knock it over. A lantern at each end isn't too much in cold weather. Be sure to give the lantern the "sniff test" to make sure it's burning so clean you can't smell anything but heat. Keep the wick trimmed periodically to avoid fumes. A number 2 lantern burner is the right size, and needs filling only once a day. It will take about 4 gallons of kerosene to cover the whole 3 weeks. The warmth makes them grow faster, and keeps 'em away from the sow while they're sleeping so they don't get mashed the first 3 days.

In addition to the creep, I consider farrowing rails a necessity. These are just rails, sound oak 2 x 4's, or 1¼" galvanized pipe affixed no more than 8 inches from the floor and 8 inches from the wall in the sow's sleeping area. They don't prevent her from turning around, but sows like to lie down by first leaning on a wall, and then sliding down it slowly. Piglets love to rest next to the wall, even if they have the option of a heated creep. If they're there when she slides down, they're done for. She won't even hear their cries of terror as they die.

The best floor is of planks laid on concrete and held together so she can't flip them around. Bare concrete is too cold. Simple wood encourages drafts from beneath if attached to joists or runners.

If the farrowing area is in a building make a false roof only 3 feet above the stall's floor. If it's a portable hut, make its roof the same, only weather proof. An indoor stall can be made non-drafty by placing old feed bags (or what have you) to cover all ceiling and wall cracks. The door must be closed in real cold weather, say below about 20°F at night. It's best to avoid farrowing at all during parts of the year when subzero temperatures are possible.

A sow seems to nearly always

prefer to sleep with her nose out the door. This is very good, because I usually have a heater in the creep area (behind her tail end). This way the piglets are lured to safety early, and only venture to the sow to drink milk. For the first 3 days the piglets are rather uncoördinated, and thus are then most likely to be mashed to death. In fact, it is rare to have a mashed piglet after the third day. This is especially true with the use of farrowing rails. I use these, and they save piglets. Some folks use a shelf of some sort, to keep the heat in, but I'm not sure it's really good because of cleaning problems.

Sows should be placed in the farrowing sty at least a week before the farrowing date. That way they can get used to it. Before they are put in, put feed (but not water) in the indoor part, because otherwise they'll defile it with dung, and worse, it'll be an unbreakable habit until the next time they farrow. After three days, the sow may be fed in a corner of the outdoor half of the sty. By the way, all sties should be sloped about an inch for every 8 feet of pen length, for drainage. Feed on the uphill side.

The planks on the floor help insulate the sow, preventing illness at farrowing. They raise her a little, keeping things dry and sanitary. If piglets fall asleep at the teat, they'll lose valuable body heat on bare concrete. The first day they need heat more than feed; the next 3 days they're equal in importance. The next 2 weeks planks will still greatly improve survival and growth.

Clean the dung from the sow's pen twice a day.

At 2 weeks of age the sow may be moved with her pigs to more spacious quarters. The farrowing sty may be cleaned and rested for a few days (preferably with full sunlight on it to kill germs) and another sow moved in. Even if moving isn't needed, the animals will stay cleaner in a bigger pen, though I've kept 'em in the farrowing sty up to 6 weeks of age.

It is important to supply rich humus in the piglet creep from the day they're farrowed until they're well started on solid feed. This will keep them from getting anemia and mycoplasmal (viral) pneumonia. Indoor piglets get anemia at 2 to 3 weeks of age, because sow's milk is (like most milk) deficient in iron and copper. The piglets look too pudgy, sort of "bunched up." Their skin has a faint yellowish color if they're white pigs, of not, the whites of their eyes don't show healthy veins when the eyelids are pulled gently aside. They chill easily, and generally die without other symptoms. It may sometimes be cured by a shot of injectable iron, sometimes not. Much better to just feed a quart of humus or clean dirt in their creep from your fencerows daily.

Weaning

The piglets are shut up in a nice clean sty when they're at least 5 or 6 weeks old. It's best to remove the sow, leaving the pigs in familiar environs. It is exceedingly important not to change ingredients in the pigs' feed for a week after weaning. So much the better if you can feed leftover milk, buttermilk, whey, etc. from your cow. Sour milk is better than fresh, as long as it's not pasteurized or otherwise cooked. This way they'll hardly notice the weaning, and won't get scours (diarrhea). Scours at weaning is the most common reason for setbacks in growth of feeder pigs.

But at least it's not a terribly serious ailment. You may need to drug them if it gets bad though. A bit of antibiotic in their (diluted) milk or water is the easiest way to get at the bug.

Upon weaning it's better to under-feed than over-feed. Try to give 'em what they'll clean up well before the next feeding.

Four days after weaning the sow will come into heat, and if she's not worn too thin from all that milking, she can be put back with the boar. But if you're going to do this, make sure she eats as much as she wants in the 4 days before breeding. When a sow is gaining weight rapidly, she ovulates more freely, and you get more pigs. As soon as she's bred, her feed should be cut right down to 4 lbs. a day plus grass or hay.

If the sow is looking too thin (some bumps of her backbone visible, maybe some ribs and shoulder blade too) you may do better to keep from breeding for 3 more weeks until the second heat. Her udder will have dried up by then too, so she'll be safer from teat damage.

If all goes right, the next farrowing can take place up to about a month after the first fall frost outdoors without closed housing. If your seasons are spaced properly, you can plan the second litter to come 6 months after the last, and have spring and fall pigs to raise and sell. The remnant of the litters usually bring enough money or trade goods to pay for all the cost of keeping the sow, the keep of the pork hogs, and even her share of a boar, but not if you have fewer than say 5 or 6 pigs out of the litter to sell. Sows and gilts eat about a half ton of feed from the time they are bred to the time they are weaned off the litter of pigs. In other words, they eat about a ton of feed per year.

Castration

Here is where indoor farrowing really shines. I castrate between the third and fourth week, so the pigs are big enough to be evaluated for future breeding stock. Piglets are much easier to catch with the sow locked outside her house with a bit of feed. The piglets are taken out of sight and out of earshot if possible. They are best lifted up by a hind leg to minimize squealing. Castration cuts heal much faster if the piglets stay with the sow for at least a week after.

There are two common ways to hold a piglet for castration. The first is to simply suspend him by his hind legs. There are many special devices on the market for this; I use one with a treadle which activates two leg straps to hold the hind legs. Others use a helper for this, but it's more fiddly. The helper holds the hind legs, the pig's belly facing away from him, and uses his own legs to steady the pig's abdomen.

The other way to restrain a pig is to place it on its back in a V-shaped trough. I have made these troughs of wood, with a cross bar or rope to fold the hind legs forward on the belly. This way isn't quite as good, in my opinion, because the testicles (seeds) "float" toward the anus, and the cuts have a harder time draining to heal. But the method works.

With the vertically suspended pig, the only danger is to avoid cutting too near the abdominal wall, which will form a rupture, and a near worthless pig. The best place for the cuts is halfway between the anus and where the abdominal wall meets the hind legs. The cut should be directly in the middle of the hind legs.

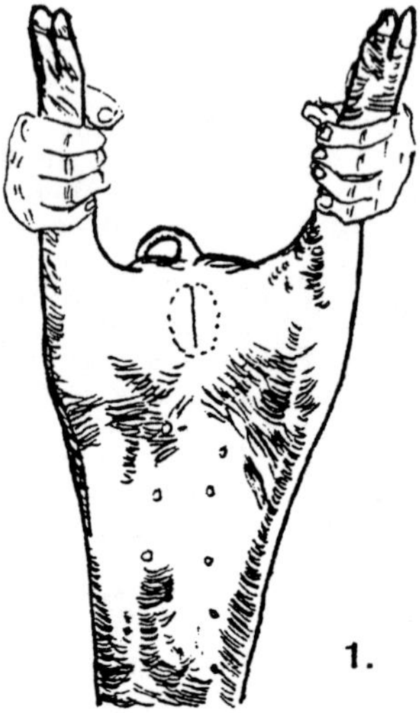

1.

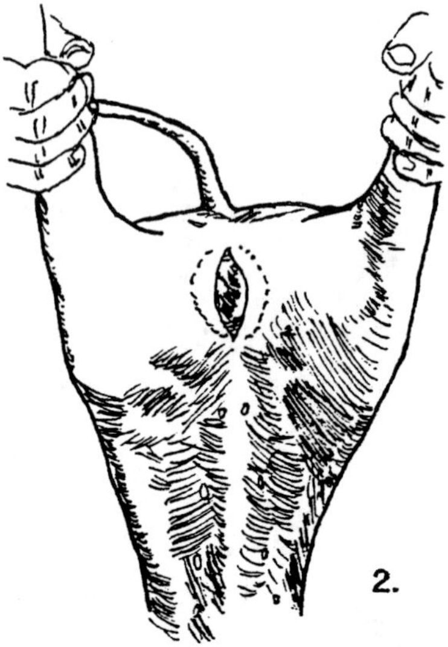

2.

The seeds drift towards the belly when the pig is held vertically. On its back, they still may drift that way sometimes. Just firmly use the fingertips to slide the seeds to where the cut will be. A very sharp knife, razor or scalpel makes the slit, about an inch or two long, and about ½ inch deep. Sometimes it bleeds, sometimes it doesn't. Even what seems like profuse bleeding is sometimes encountered. Continue on. Squeeze one testicle out through the incision. Cut right across it and the testicle will pop right out, remaining attached to two cords. Place the index finger between these, and grasping the seed itself make a smooth pull until the cords snap loose. This way the rough ends will recoil up into the body, helping to prevent blood loss or infection. Find the other seed in the same way with the finger tips, and take it out similarly through the same incision. You may have to make a second incision inside the first incision to expose the second testicle. Don't leave any portion of the testicle inside, because it will result in a "slip" with boary smelling pork.

No need to worry about sewing up, or flies, but it is nice to dust the wound with hydrated lime mixed with about 2% iodine. There are proprietary mixes available at veterinary and farm supply stores. A castrated pig is called a barrow.

To castrate a large boar is more difficult, but mostly for the animal. I will relate the way I do it, from an 1897 text I learned from:

"When the time arrives for him to be superseded as the head of the herd, and it is desired to make him a barrow, it can be done by one active man operating as follows: After drawing up one hind leg, and fastening it securely to a post, or stake, fasten another rope around the upper jaw, back (of) the tusks, draw it tightly, and fasten it to another stake; in this position

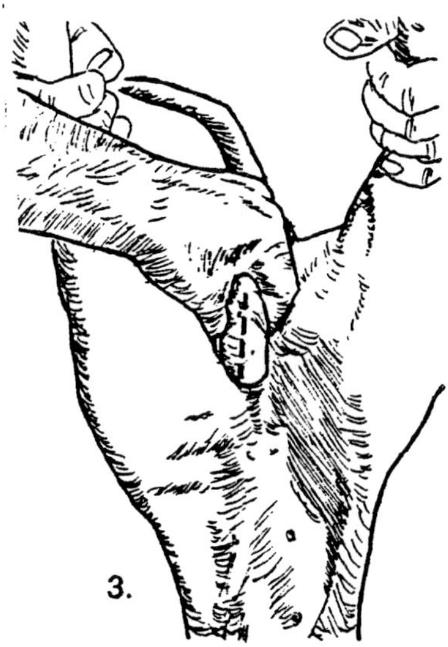

3.

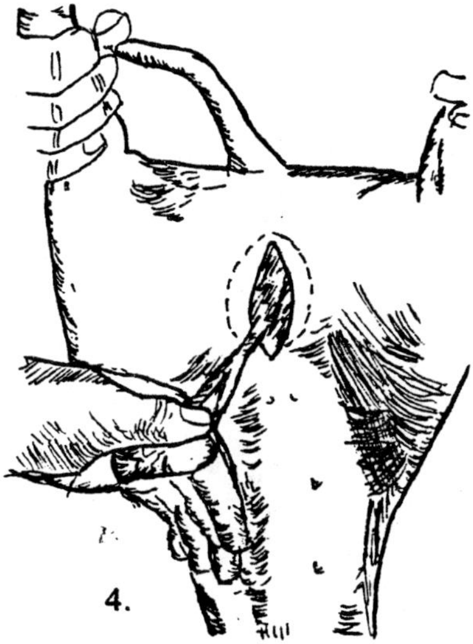

4.

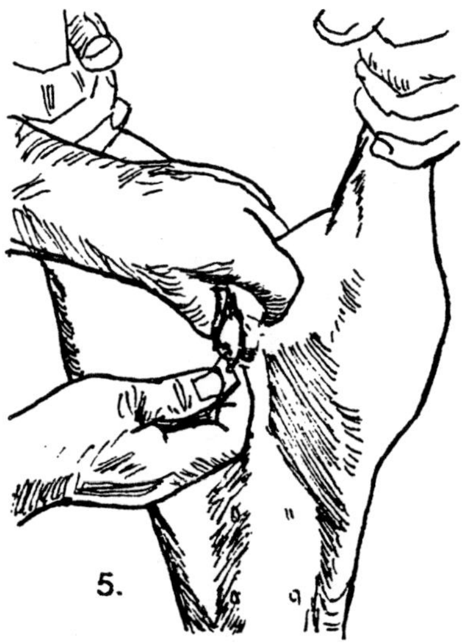

5.

1. Holding for castration.

2. Make incision.

3. Squeeze testicle out through incision. Cut across membrane as shown.

4. Pull testicle free of cords.

5. Press remaining testicle towards incision; cut to release. Continue as before.

the animal can offer no serious resistance. The cut should be low down, and as small as possible; the low cut will afford a ready means of escape for all extraneous matter, and allow the wound to keep itself clean, there being no sack, or pocket, left to hold the pus formed during the healing process. It is not best to perform this operation when the boar is very fat, or the weather too warm, as the risk is much greater." Swine Husbandry, by F. D. Coburn

Large barrows will have lost the musky odor and flavor by the time the wounds heal. It takes roughly 3 weeks. I can vouch for the fact that pork begins to have objectionable flavor as soon as boars hit puberty at 4 to 6 months. I also know sometimes they have no odor. It seems to be a factor of their surroundings, and whether they're excited by nearby females. Boar's breath smells just like the smell of cooking boar's meat.

Boars

Like the sow, the boar can be bred at 7 months' age. Eight months is better. He will eat about 1300 lbs of feed per year. He needs a big enough lot to get exercise, protection from cold, and he needs wood ashes (or a mix of ¾ ground limestone and ¼ steamed bone meal) mixed half and half with his salt for strong bones. He'll be too small to avoid punishment by big sows if he's only 7 or 8 months old, and they'll be too small if he's over 3 years old and they're only 8 months old. So after a good breeding it's a good idea to separate disparate sized brood hogs.

A boar takes up to half an hour to complete a mating. It takes me between 4 to 5 pigs to pay for the sow's feed for half a year. Two more pigs will pay for a boar's feed for 6 months, and of course the more sows a boar can be made to service, the more pigs you'll have left over to sell or eat.

I am convinced that during the 46 hours a sow is in standing heat--the time she'll allow a boar to mate her--early breedings produce more females, while late breedings yield more males. If you get all females, your boar's semen count may be getting low, if the next litter does the same or nearly so, replace him soon.

Raising & Fattening

It's easy to feed swine. They eat practically anything organic, and some things inorganic. If they get poisoned, they'll usually vomit it out before serious damage is done. Someone once poisoned two of my hogs, and though they looked nearly dead, they came back. One was just bred, and she produced a few deformed piglets amid her next litter.

Besides that, pigs need water. Give 'em all they want; they'll drink about a gallon a day each, maybe more in summer. The more you can get 'em to drink the healthier they'll be, which is why slops are cooked up with water as a swill by old time hogmen. I always mix any leftover dairy products with an equal part or more of water. Even 5 to 10 times as much water is o.k. if there's not much milk to go around. That way they'll drink. If they know they're going to get milk, they won't drink any plain water at all!

Growing pigs do best with protection from cold. I like about 3 square feet of sleeping space per head in the house for pigs between 30 and 75 lbs. From 75 to 150 lbs., 4 square feet are enough, and from 150 to 220 lbs., about 5 square feet. It's small, but it's only the indoor part.

I fatten my swine by feeding them all they can eat up to butchering weight. If you must have 'em really lean (but tougher) feed 'em all they'll eat up to 135 lbs., then feed them only that amount, no more, from then on. That's the best way to feed breeding stock for the next generation.

They should get 16% protein until 150 lbs. weight, then plain shelled corn is o.k. with some grass, or spare slops. The usual commercial way is to feed 14% protein from weaning to slaughter.

One important feed pigs should always get if they don't have plenty of greens is alfalfa hay or meal. Pigs need 5% of it, by weight, in their diet; sows and boars need 15% or more. If piglets won't eat it, mix alfalfa hay chaff (leaflets) with their slops or ground feed.

When pigs are fed watery feeds, like tomatoes, or grass, or table scraps devoid of baked goods or animal products, they hardly grow at all. Their stomachs get distended, so they can digest the poor quality stuff, and you must feed them much more. Butchers pay much less for them because they are buying a lot of belly there.

Pigs like their potatoes cooked--they cannot digest the starches well if raw. Soybeans, and other beans too, must be cooked, or they can't utilize the protein well. Soybean meal from the feed merchant has been cooked. Peas, sunflower seeds, dried meat meal, fish meal, etc. are good, but expensive in the U.S.

Jerusalem Artichokes

I hear Jerusalem artichokes are a fine diet for brood swine before farrowing. They winter over in the ground, and the pigs can dig them out anytime the ground's not frozen. I know they like 'em, but I haven't gotten around to this method of keeping them. My artichoke patch does well every year, and crowds out all weeds. Never are they touched by disease of any kind, and they are ready to feed by late September. They can be fed until June, when the tubers "go hollow" supporting the next season's growth. Unbelievably, sows given about a sixth-acre each of 'chokes will eat all winter and spring, and the plants still come on as strong as the year before if you move the sows to other ground. I plan to keep swine in artichoke patches in winter. Even if it doesn't supply all their feed, it'll cut out at least a third of the cost. Feed costs are the most substantial part of swine raising.

Mast

A once-common cloud in the title to a piece of land was "pannage rights." This was the right to graze hogs on the fallen nuts, berries and other products of the forest. These tree crops are called *mast*. With hogs, nuts come immediately to mind. The best among these is the chestnut *(castanea spp.)*, which is nearly the equivalent of corn for fattening pigs. Turned in to a wood where this tree is dominant, a 100 lb. shoat will become a 200 lb. hog in about a month! Smaller pigs grow significantly slower, presumably because of the low protein content in chestnuts. In with the chestnuts is the chinkapin *(castanopsis spp.)*, which is probably more palatable to swine, but they are smaller, and only contain one nut per burr casing.

The second best nut is acorns.

Among these the Yellow or Chinkapin Oak *(quercus muehlenbergii)* stands supreme. After this comes the Swamp Chestnut Oak *(quercus michauxii).* Then come the rest of the white oaks. Red oaks don't bear every year, and mostly their nuts are bitter. Hogs will make poor gains on them, because they'll only eat 'em if they're starving.

Hard shelled nuts like hickory, walnut and butternut are useless for pigs.

Many folks overlook the common mulberry. It bears an abundant crop--especially the everbearing varieties--from as early as late April in the Deep South to as late as August. In our region, common mulberries bear during July and August. They are so sweet they keep the birds off your bush fruits--gooseberries, currants, raspberries, etc. It was once very common in the South to keep mulberry orchards for swine, to cut the corn consumption by two-thirds!

When the mulberries are spent, there should be some field crop coming ripe. A few wise husbandmen dispense with the labor of harvesting and storing these, instead turning the swine into the field to get it for themselves. Corn can be treated this way, but it won't have made its full growth by the time the mulberries are through. Field peas were once highly esteemed in the far north, but summers are too hot for it if temp's reach much above 80°F.

In late fall, after a couple of frosts, persimmons *(diospyros virginiana)* and honey locusts *(gleditsia triacanthos)*--not to be confused with the near worthless fruit of the black or yellow locust *(robinia pseudoacacia)*--drop their sweet fruits, and hogs avidly graze and gain on both. Rotten apples or other orchard fruits, yellow-jackets and all, are wolfed down with great gusto by hogs, but they won't put on much weight with 'em--too much water. Once the cold weather sets in, only the Jerusalem artichoke is reliable, since it remains in good condition right through heavy freezes. As soon as a thaw comes, the hogs can have at them again until June.

I haven't raised my hogs on mast because my place harbored none of these trees, and to establish them for the pigs alone, well, it hasn't been feasible for me yet. But I had relatives whose pigs did pretty well in the swamps of South Carolina, as long as they stayed out of the way of the 'gators.

Troubleshooting Swine

Most problems with pigs come before 8 weeks of age and after 2 years old.

Coliform Infection

The first hazard of piglets appears within a week of birth. They become paralyzed, lie on one side, and paddle until they die. This is a coliform infection, frequently not appearing until a sow's second or third litter. It may be prevented by farrowing on clean pasture, or feeding or injecting the sow with an antibiotic (oxytetracycline works best) a day or two before farrowing. Some sows never have the problem. Save breeding stock from them. If any pigs show symptoms, all should be dosed right away with spectinomycin--you'll save the rest of the litter, but you'll also know which sow to cull. Rest your pens!

Hypoglycemia

Very often piglets get chilled if

born in cold weather, even when indoors. They can be "revived" if completely submerged, (except nose and mouth) in warm water until they come around. They can be fed a teaspoon of honey mixed with an equal amount of water.

Nervous Sows

First litter gilts are sometimes afraid of the piglets and get very excited each time one is born or makes noise. Be on hand at the first farrowing if possible. If the sow leaps up in fright, keep her away from the piglet with a rod or poker, moving the piglet to safety with a dull hoe, a walking cane, or shepherd's crook. She'll calm down some towards the end of farrowing, and if the pigs are kept warm and dry in the mean time, they can be let to nurse. If she's well treated, she'll be soothed by your voice, and caresses. I have heard of cases requiring protecting the litter for 3 days--cull these sows.

Sow Constipation

Sows sometimes get constipation after the second litter. Their feed should be at least a third wheat bran, ground oats, or some other bulky stuff for 2 or 3 days before farrowing. If you don't see normal soft dung (it's hard and unusually dark) give her some mineral oil and ½ cup Epsom salts, or ¼ cup psillium seed hulls in her feed. Give wet feed. Don't feed the day after farrowing. She won't have much of an appetite anyway.

Agalactia

The sow has no milk, even though she may have had some at the time of farrowing. Usually caused by constipation, and only seen after the 6th litter in most sows. Also occurs when the sows are too fat. Check By milking teats with 2 fingers (see "Milking" under Cows) twice a day after farrowing, for 3 days. Inject oxytocin--1 cc. under the skin behind the sow's ear. And again hourly until milk flow starts normally. Keep the pigs warm.

Sow & Boar Foot Problems

Weak joins, difficulty walking or standing, hoof separates from foot. Don't keep brood animals on hard floors for periods longer than a month or two--especially in winter. Urine contact causes hoof separation on adult hogs.

Mashed Pigs

Pigs are either stepped on, or lain on. Sometimes they survive, but they're usually dead or dying when you find them. Indoor pigs need very little litter. A large bird-like nest is the cause of many such deaths. Use a half-bushel of sawdust, ground corn cobs, or (best) dry dirt. Supplement it as needed, but don't allow a large build-up. Have a heated creep with a low ceiling (1 to1½ feet high) for the piglets. The creep may contain deeper bedding.

Salmonellosis

Young pigs get drowsy, ears, tails, and/or bellies turn purple. Pig dies within 24 hours. Caused by rodent bites and/or ingested bird dung. Keep cats, but don't let 'em in the farrowing pen. I have had only two of these deaths in 14 years and hundreds of pigs farrowed.

White Scours

White or yellow diarrhea in

piglets 2 to 3 weeks old. Caused by piglets eating germ-laden feed about the time they're starting to eat solid feed. Sanitation, and feeding humus early helps avoid it. They'll get over it in a few days anyway.

Bloody Scours

Caused by unsanitary sties, rodents, and bringing in infected pigs. Mostly the latter. Quarantine all new animals for a month. Don't get new animals unless necessary.

Congenital Abnormalities

Pigs which get ruptured bellies, prolapsed (everted, ruptured) rectums, ruptured scrotal sacks, inverted nipples, etc., are usually caused by a defective boar. Occasionally the sow's fault. Replace these animals. Ruptured bellies usually get better if the rupture isn't too bad--if the pigs are kept safe from further rupture by other pigs. Prolapsed rectums may be replaced, and a purse-stitch sewn in the anus. Special devices are available, success rates vary. Best slaughter such animals and have a barbeque. Don't choose pigs with bad teats as breeders.

Swollen Ear

Usually comes in cold weather, when an ear is stepped on by another pig. It's an infection, and will eventually pop, then the ear will kind of shrivel up. It doesn't affect the pork quality. I haven't had this happen in years.

Poor Growth

Wiry hair, humped back, daily coughing, sometimes scours (diarrhea). This is most likely worms. Worms live for years on pasture, so it's best to rotate hogs like any other crop. Or cultivate the ground during the months you don't have a pig. Pigs may be wormed with any of the proprietary mixes, or pure soap savings mixed with ashes in their feed. I've used both, and prefer the store-bought mixes. Wash sows before farrowing, move to clean pens (sloshed with boiling water on floors and walls, don't walk 'em through pig droppings). Break the cycle of worms by all means. Adult hogs show very little signs of worm infestation, but it can kill small pigs!

Skin Parasites

Hogs rub a lot. Lice are visible on the pigs--especially around the ears. Mange mites aren't visible, but hogs without 'em hardly ever rub themselves. You can kill lice with buttermilk (unpasteurized) poured along pigs' backs. Mites & lice are best totally eradicated by injecting all your swine with ivermectin, and then again 3 weeks later.

Trichinosis

This is caused by tiny worms (trichinae) which create and inhabit small cysts in the muscle cells. It produces a dreadful, sometimes fatal disease in man and beast. We hear much about this from nanny-state regulators, who with all their laws can't seem to do much to prevent food borne illnesses. I believe there were fewer diseases when folks knew to beware, instead of receiving assurances.

Well, to beware, that's just it. Animals can only get trichinosis from eating live-trichina-infested meat. That means don't feed your pigs uncooked meat, nor let varmints abound where hogs can eat them! If you want safety, don't eat your meat uncooked either. Trichinae are also killed if frozen at -25°F for a day, -6°F for a week, or are cured and aged 60 days. Isn't that easy?

Mycoplasmal Pneumonia

Coughing. That's the only symptom, and if you've wormed 'em and they still continue to cough, you should have fed humus when they were young. It'll cost you 10% more feed, but rarely much else. Still, the pigs aren't as healthy, and are likely to lay on more fat than usual.

Rear Paralysis

Can't get up or move hind legs. Usually caused by a sprained or broken back, mainly in long meaty hogs with lots of muscling. This is caused by deficient calcium. It cannot be cured. If the hog is not on its feet in a month, better slaughter it. I haven't had them go down for more than a month unless a sow got arthritis right at farrowing time. This won't happen if you don't keep 'em on hard floors. Make sure your hogs get plenty of limestone of wood ashes while growing up.

Streptococcal Meningitis

Pigs between weaning and about 150 lbs., after mixing with filth-raised pigs, get ill, paralyzed, comatose, and die. There's nothing you can do for a pig with this paralysis. It will die. The germs carried by the filth-raised pigs (factory farmed ones especially) infect your clean-raised pigs and kill them. The carrier pigs got immunity from the sow's antibodies until they developed their own, or they were drugged up until they developed their own. If your clean pigs are to be mixed with someone else's filthy or doped up ones, play it safe and inject yours with a long-acting antibiotic before they're mixed in a pen. If they survive 14 days they'll live.

Summary

Pigs are rarely affected with any disease if they're kept clean, comfortable, and properly fed. And if they're kept away from germs from other pigs. If someone comes with pig dung caked on his boots, don't let him in your pig pens. I wouldn't let him on my place until he rinses himself. With disinfectant.

Otherwise, pigs are as hardy as wild boars, give you very little trouble, are very efficient livestock and ought to be on every farm where dietary custom allows. They're the only meat which cures without becoming too dried out to eat, a real convenience food.

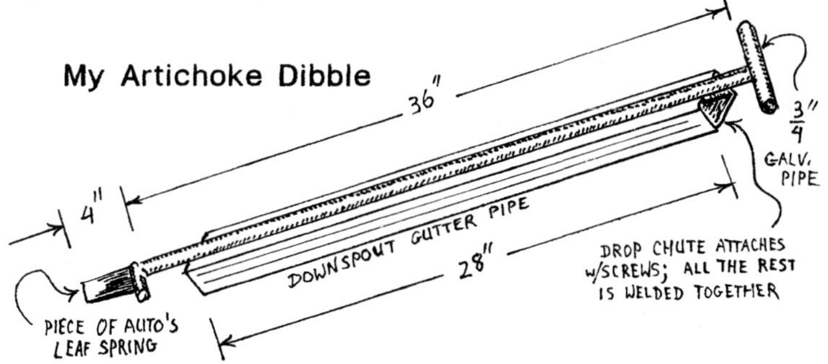

14
SHEEP

Be thou diligent to know the state of thy flocks, and *look well to thy herds. Proverbs 27:23*

My first sheep was nervous as a, well, a sheep. Not knowing better, I got one, figuring it'd do well with the goats, which were about the same size and shape. Wrong. The poor sheep cried out in loneliness day and night, I'm sure attracting the attention of all free-roaming dogs for miles around. Next day I went to a nearby sheep farmer, and she sold us a sheep to go with the one we had. The first was predominantly Suffolk in breeding, the other predominantly Hampshire. Both these breeds have black faces, and are probably the two most popular breeds in America. Both were as skittish as a wild animal. I have since repeatedly heard "shepherds" (I use the term loosely) declare "I like 'em wild; they ain't nothin' worse'n a dern'd ol' 'pet'!" After years of shepherding, I must here dub such comments as the height of stupidity in any but the wildest unfenced rangeland. Shepherds must always enjoy the confidence of their flock, unless they just have an inordinate desire to waste as much time and labor as they can. If you love to chase down and manhandle your stock every time you need to shear, medicate, cull or even just inspect them, that's your business. Such is nothing more than recreation, a form of it I can't enjoy. I can understand it though, because I was subjected to it as long as I owned those sheep and their offspring.

I later picked up some white faced sheep: Finns and Rambouillets. Finnsheep, known for multiple births had little else to commend themselves. They grew slowly, and never put much meat on their bones, and had quite inferior wool, too. The Rambouillets had a beautifully fine wool, meat that tasted like lanolin, and also were fairly prolific, giving twins and out-of-season lambs regularly. We tried several other breeds over time and were fascinated by the way blackfaced sheep segregated themselves from whitefaced. The racists.

Overall, the blackfaced had better meat, while whitefaced had better wool. But I suppose such generalizations are unwise. The worst wool was the the whitefaced Finn, and the best meat and wool was the whitefaced Cotswold. Speaking of Cotswold sheep, we loved these animals so much, we finally got rid of all our other sheep. They produce huge amounts of wool only equaled in quality by Lincolns. They produce meat of a clean, tasty character which is so good one has little craving for beef. Probably the two best features of this calm-natured breed are their hardiness and ease of feeding. The nearest thing to them, the Lincolns, had to be fed better than ourselves! They'd trample the best hay in their incessant search for "just another" grain. And withal they remained bony and thin, though they did grow wool! The Cotswolds are the most respectful of fences of any sheep we kept.

Aside from my partisanship on this subject, let me point out that sheep have a place only on farms with plenty of grass. The grass can be amid stony or steep ground, sheep don't care. They graze such land better than cows, whose hooves "poach" the soil, making a muddy and erosion prone mess of it in wet weather. And better than goats, who really aren't content to graze, preferring to browse. No ordinary fence is going to hold a goat where it can't browse.

The first mistake made by new shepherds is to assume sheep will eat brush. They don't. Instead, their wool gets tangled in it, and thorns and burrs become knotted tightly against their bodies. If you sell the wool, you'll get low prices. Without chemical carburizing, it is impossible to convert into usable clothing.

The second mistake is to underestimate the land needed. Sheep's worst enemy is other sheep. They are the least parasite-resistant animal on the farm. Stomach worms, keds, lice, nose bots, and blow flies all work their unseen evils. The sheep don't complain or act wimpy. Then one day they die. The difference between a sheep keeper and a shepherd is the shepherd sees what's wrong with his sheep before it becomes critical. A shepherd senses every nuance of aberrant behavior in the sheep, just like we can tell when those we love don't feel well before they tell us. A stranger can't tell any better than your average sheep keeper. This makes us all the more confident in the Good Shepherd, who sees the needs of His flock before they're visible to those who don't care.

The Way of Sheep

Ewes have lambs each year, unless there was some problem. Most sheep owners select for "twinners," ewes which regularly bear two lambs. A twin ram mated with a twin ewe stands a superb chance of bearing twinning progeny. It is almost certain if all the grandparents were twins, too. As long as they come from a twinning line, they're likely be twinners. The only real exceptions to this are where the ewes were bred when they were in poor condition, or when clover hay or clover pasture was fed during breeding season. Clovers have a very pronounced ill effect on hormonal balance.

In 145-155 days the lambs are "lambed" and soon get up and wobble around looking for one of the ewe's two teats. Meanwhile, the mother vigorously licks off all the birth fluids. If she doesn't do this, she's likely to reject her lambs, so don't dry 'em off yourself. Triplets are not all that uncommon, but the ewe's attention is sometimes so distracted in cleaning off the first two, or last two, that she neglects, and then disowns the other. So triplets aren't as desirable, because what you gain in numbers is lost in labor--and triplets are never as big as twins. Again, a ewe has only two teats.

The lambs very soon get as bouncy as popcorn; it's a delight to watch them "pop" in the air and then go scampering around the field with the others. Within a week or two they're nibbling at grass, and can be "creep" fed on cracked corn. Fancy mixtures with all sorts of "improved" chemicals have never proved their worth in my experience. I know personally at least two shepherdesses who have lost sizable numbers of their flocks due to poisoning from "improved" mixes. If you want a balanced mineral ration, mix steamed bonemeal and salt 50-50. Even pure, clean unleached hardwood ashes from your fireplace are fine and safe. It all depends on whether you're a shepherd or sheep keeper. The best grains for sheep are oats, corn, or clean rye, fed whole in every case. Little lambs need grain cracked or rolled at first, until enough teeth come through to chew well. Probably the best single feed is peas, but they don't grow well except in the northernmost areas where temperatures rarely exceed 80°F. Soybeans fed in the sheaf (with the straw and all) are very good. If you feed just plain green alfalfa hay and nothing else but water and salt, you will do fine. Alfalfa is Arabic for "father of fodders." It has been known as the single best sheep feed probably since the time of Abel. Fed in pasture, it's likely to froth the sheeps gut, causing a fatal, though treatable condition called "bloat." (See "bloat" under cows.)

The lambs grow amazingly fast. But they are even more subject to parasites than adults. They should be moved to clean pastures at least once every 3 weeks. Don't return them to the first pasture until 2 months or more have passed. Failing that, you may have to dose them with expensive wormers every month. Get an old hand at sheep to show you how.

Sheep will fatten on very good pasture alone, or on lesser pasture if they have enough grass that it never gets shorter than about 6 inches. They like the shortest the best, and will let some go to seed while they eat the rest right down to the dust. To prevent this, move them among 5 or 6 paddocks which are small enough for them to eat right down in a week or two.

Ewes will naturally wean their lambs at 8 months, at which time the lambs are so vigorous that they hurt the mother. Also, by this time the ewes are so run down in condition they can hardly produce any milk. They are unfit to rebreed in that shape. Good shepherds wean their lambs around 4 months. Some even earlier, but this is hardly warranted. Absolutely no change in diet, other than the cessation of milk, will be tolerated by them. They can get all sorts of heinous gut disorders, leading to pneumonia and certain death, if you change their diet at weaning. Even drugs don't always help.

Single lambs among a flock of twinner ewes very often get "pulpy kidney" which is also called "overeating disease." They are awash in mother's milk, and certain *clostridium spp.* bacteria take over their kidneys, ultimately spontaneously releasing a poison very like botulism into the bloodstream which paralyzes and kills the lamb within minutes of the appearance of symptoms. There are three ways of preventing it: 1. Vaccination of the whole flock before lambing--an expensive pain. 2. Supplying antibiotics in the water or creep feed (oxytetracyline is the drug of choice). 3 Segregate the twinner ewes from the singlers each year and feed the singlers less for the first 2 to 3 months. I like this method the best, because it is most self-reliant. The ewes getting less feed won't glut their single lambs with milk, and everything will be fine. I don't believe in using all those vaccines and drugs if it can be avoided, because we can't see which sheep to retain for replacement stock.

Little ram lambs ought to be castrated by 2 weeks old. "The sooner the quicker" as mom used to say. The easiest way is with the "elastrator," a strong rubber band placed around the scrotum between the testicles and the abdomen. If the testicles aren't in the sack, or are only partially descended, you'll have a "cryptorchid" a non-breedable, but still staggy looking sheep. All wool-producing sheep (yes, there are some "hairy" breeds) should have their tails docked (cut off) by the same method, if you like, just long enough to cover the anus. The wild sheep has no wool on its tail, and so isn't subject to all the dangers of undocked domestic sheep. Still it is heart-rending to watch the little dears flop around and bleat in fear the first 10 minutes after attaching the rubber band. Dock and castrate before 2 weeks old, and in any case, before hot weather, or flies will attack.

By 5 months, many young lambs are likely to attempt to breed, so they'll need to be separated by sex unless the males have been made "wethers" by castration.

The ewe lambs, called *tegs,* are sometimes bred their first year, along with the ewes, on early-maturing breeds. Most pure breed shepherds don't let their lambs breed until over a year old.

The ram's desire to breed is triggered by cooling weather. This is why most lambs are born in springtime. The safest thing is to keep the ram apart until the time you want the ewes bred. Then turn him in, and keep watch for when he's hanging around a ewe in heat. Write that one down on your calendar. In a tiny home flock I like to see at least 3 or 4 ewes with the ram, though one ram will happily service up to about 50 ewes. The ram is half the flock, so choose and treat him well. The best rams are twins, large for their age, have not hair among the wool, no black wool, nice long bodies (the rib chops are the best cuts, so you want more of 'em) small heads, relatively large testicles with both testicles in the sack (check by feel) and never having been played with or babied--no bottle-lamb rams! Rams become dangerous when they've been been handled too much while less than a year old--they're calmer and more tempting to pet than tegs, but you'll pay when they're older! Keep children away from rams.

Gestation takes about 5 months, but get the ewes to a protected lambing place at least a week ahead if you can. The best place for them to lamb is in a clean, well-strawed pen only 5 or 6 feet square, protected from wind and precipitation. In a barn is best. Many shepherds clip off the wool around the vagina, udder and belly, to keep clumsy lambs from suckling on the wrong thing. This is called "crutching." Others contrive to have the lambs born after shearing. Both methods work well, but pregnant ewes are the more easily shorn, being a little more sluggish and a lot tighter-skinned.

Shearing

Sheep need to be shorn before the weather gets too warm for their comfort. Otherwise they'll start rubbing on trees and posts. Then their fleeces will get matted into felt and be nearly worthless. Not only that, if the weather is wet flies may lay their eggs in the dung tags. The resulting maggots soon exhaust the dung, and begin burrowing into the flesh. The whole nasty business makes the sheep very uncomfortable and they walk funny and lose weight. You can't see the fleece worms until you really get close enough to smell their stench, and even then you often don't see them at all until you part the wool exactly where they are. They can be treated with proprietary wool maggot poisons, but it's best to prevent the problem if possible. Keep especially vigilant with your lambs.

If you get a professional shearer you'll save yourself a lot of backache and bother. The pro will use power shearing equipment and will finish a lamb every few minutes. I have hand shorn my sheep every year but one, after which I resolved to do it myself. I should say that Eva did the shearing some years with the sheep standing up, tied to a post. That is the way smaller folks wind up doing it, because the sheep are too hard for the inexperienced to control. Such shearers often never can learn to control the sheep without undue strain, because experience is about the only school. You can't do it without practice; you don't get practice unless you do it!

Hand shears look like really massive scissors joined by springy bows at their hinge point, so they open on their own. In order to do good work, they must be positively razor sharp.

The tips, or points, of the hand shears must be ground off so they're at least one-eighth inch "square-tipped" and then rounded off smooth as glass, to avoid scratching the sheep's tender skin. The blades come correctly set on such excellent makes as Burgon & Ball, the only shears which gave me really good results.

Notice they have two ground surfaces on each blade. One of these ground surfaces is rather flat, called the "hollow grind," extending into the blade about a half inch. The other is abrupt, say about 30°-50°, and is about a sixteenth-inch wide. This latter one is the sharpening angle. A good clean water stone is best. Oil clogs stones, and lubricates that which is supposed to cut. I always take a bucket of water with me to the shearing place.

In addition, shears really ought to have a *buck strap*. This is a small belt of leather or even twine, tied from one shank of the shears to the spring. The hand fits into it when you grasp the shanks of the shear. It helps you push the shears up into the wool without undue finger & wrist strain. It also makes sure you don't let go of the shears unless you want to.

I always use a piece of plywood to shear on when I can get it. It is smooth enough to keep the sheep from finding its feet and getting out of control. It is also easily swept clean of dung.

The shearer stands facing the side of the sheep, one arm around the sheep's neck somewhat loosely under its chin. Bending over, reach under the sheep's chest, and grab its front leg nearest your own feet. (If you're facing its left side, you'll grab its left front leg.) Using your knees, swing the sheep upward and its hips outward so it lands on its rump. It will jockey for position at this point, just don't fight it much. Absolutely save your strength! Striking the sheep won't help you, and you probably aren't much stronger than it, so just keep its hind feet clear of the floor and its front feet clear of striking you while it struggles. It means you no harm, it just wants to get loose. If you keep it from freeing itself through a sort of relaxed shifting to keep it from getting its feet, you've learned the secret of shearing control. It alone alleviates 50% of the work in shearing sheep. At times during shearing the sheep will probably attempt to change positions again, or try to get up. It may wildly kick with its front legs and scrape the floor with its hind legs, trying to get to its feet. This action commonly precedes urination or defecation. As soon as that's over with, the sheep will usually calm down.

It is a very good idea to practice flipping a sheep onto its butt by the above method, and hold it there for at least a few minutes, long before you'll need to be shearing. Lambs are easier to manhandle, but they wiggle much more, giving you much needed practice at non-exertion. With the sheep on its rump, stand with your right foot on the floor between its hind legs, your right leg against its breast. Yourother leg clamps against its vertically-poised backbone. Do this with all your lambs periodically, and feed them corn from your pocket, to keep them from learning to hate being thus handled. Use you fingers to learn the contours of the sheep's body beneath the wool. Observe the places which are free of wool or nearly so--the armpits, beside the udder, beside the anus, and around the nose & eyes. These places will be where you insert the shear's points along the sheep's skin under the more entwined wool fibers.

With the sheep on its rump, the belly is shorn first. Some shear it last, removing it from the rest of the fleece. But keeping the fleece in one piece gives you the option of leaving it on if you decide to just sell the wool to the (ultra-low-priced) commercial wool buyers.

After the belly, comes the sheep's left hind leg and rump. Next we shear the neck & head, then the sheep's left front leg & shoulder. Then comes the whole left side, including the backbone area; this is called the "long blow." Afterward the sheep is turned and the right side is shorn, starting at the head and going right down the neck & side until the sheep is finished.

At some points the sheep appears to be actually lying down on its side, but the shearer's feet, hidden under sheep's shoulder, keep it from trying to rise. The shearer's free hand NEVER pulls on the loose-lying shorn wool--especially not while the shears are moving or cutting. The sheep's skin is loose, and will be pulled up into the razor-sharp shears, causing a nick at the very least. Rather, the free hand presses and stretches the shorn SKIN, to keep it tight; its looseness invariably causes rolling wrinkles which are sure to be snipped into sooner or later, especially along the neck.

The Shearing Procedure

Prepare for shearing by placing your equipment out of the sheep's reach but easily within yours. You need at least one pair of razor sharp shears, a bucket of lukewarm water, a sharpening stone (just in case you find the shears aren't cutting perfectly) some pine tar to dab onto nicks. A cotton glove makes holding the shear easier. It protects from sharp nasties like splinters, prickles and burr-pins lurking in the fleece.

Feeding the sheep by hand from the corn in your pocket, capture the one you want to shear first, leading it to your shearing floor. Set the sheep on its rump; let it finish wiggling.

Once it has settled down, take up your rigged, razor sharp shears in your right hand (I will for purposes of clarity presume the shearer is right handed). Keep the sheep's head tucked by your left side. Insert the shears in the wool on the breast bone, sliding the back of the blade farthest from your thumb (like a sled runner) on the sheep's skin in that quarter-inch or so of straight, untangled wool right next to the skin. The blades may either be kept flat against the sheep's skin, or slightly raised to leave a bit more wool on the animal. You start in the wool-less armpit of the sheep, and clip right across to the other wool-less armpit, taking perhaps an inch wide row. Each pass is called a *blow*.

Soon you'll see why razor sharpness is so needful. Did you ever clip a piece of paper with sharp scissors? You sometimes kept cutting without opening or closing the scissors, didn't you? You could complete the job a lot faster that way. Wool is similar, only it's more prone to flop over one cutting edge if both aren't absolutely sharp. Electric shears were invented to allow less skillful workmen to clip sheep. With razor-sharp hand clippers, and a bit of practice, opening and closing the shears can be minimized as the blades glide easily along the contours of the sheep. It is quick, but unhurried. The better one gets, the less one's back and legs become fatigued--and believe me, you'll feel like you need replacements after shearing your first sheep! If you can persevere, clipping sheep is really quite fun, and if you can work yourself down to less than 10 minutes per animal, it begins to become quite profitable. There are many sheep owners whose sheep are far from the power lines. Besides, electric clippers spoil the tranquillity with their buzz and the growl of the generator and the stench of exhaust fumes. They'll pay extra for hand-clipped sheep, sometimes twice or more. You might well average 100 sheep a day at $3 or more each (1997 prices) all without keeping books, receipts, inventories or expensive equipment. Hard work it indeed is, but to quote the sage John Seymour, "Try earning that with a degree in philosophy!"

When you're no longer able to employ the wool-less armpit area as your shears' entry point, you'll need to use a technique called *opening*. This simply means turning the blades until instead of lying flat on the skin, they're perpendicular to it. The back of the "sled runner" blade still glides along the skin, but the clippers merely part the wool, severing nothing from the hide. So anyway, "open" a path from the wool-less area at the sheep's right armpit to the wool-less area under the thigh of the sheep's same side. The cut simply follows along the point where belly meets side, with mostly dirty wool to your left, and clean wool to the right. Now you have a clean entry-point, and can shear across to where the dirty wool ends on the left.

The chest wool is the most matted, with the possible exception of the hind legs. Once past it, you're shearing the belly. Here you must use caution about the ram's penis or *pizzle*. If it's an ewe, you have to be very careful as you move into the lower abdomen, because the teats and udder seem to fold right up into the belly wool. Hold them with the left hand, if need be, away from the clippers' path.

Very carefully, taking your time, clip the wool off the udder. If a ram, carefully fleece his "family jewels."

Now comes probably the most difficult part of shearing: the belly side of the hind legs. At some point in this whole effort the sheep has probably struggled mightily in protest, and if you were smart you have rested your back and legs, merely balancing the sheep and let it tire itself until it quit. You may have kept its shorn wool away from the flailing hooves. Maybe not. Who cares about belly wool anyway? If you were a very lucky shearer the flock was a bunch of *rosellas*. A rosella, be it of whatever breeding, has practically no wool on legs or belly. Knowing whose flocks consist of these can really tally up the income for a hireling shearer.

But back to the legs. Don't attempt to do as I've seen many tyros, to wit: don't grab the legs. It not only makes the sheep struggle (its legs are considerably stronger than your arms) but puts you and it off balance, leading to the most dread occurance in shearing--the sheep finding its feet and exiting while dragging its fleece all about in the dung and mud! Forget about catching it still clean if this happens. The dragging wool scares it all the more into continuing its parade in the mire.

No, instead of grabbing the legs, hold the left fist between the flank and thigh; mashing this spot straightens the leg. The wool is matted here, but again you must glide the points in that narrow area

of untangled wool against the skin. On the left hind leg, start near the body and work towards the hoof. On the right hind leg, one begins at the hoof and clips toward the body. This is not only difficult, it's nerve wracking. The front of the leg is easy, but the clippers must follow the contour as the leg narrows; this isn't easy.

After the inside of the hind legs comes the rear left hock and thigh. Be very careful behind the hock, not to hamstring the sheep. Some shearers don't even bother clipping any of this wool unless it's loaded with dung or has briars wound up in it. Some simply lay the blades parallel to the leg, and keep snipping off a quarter-inch or so at a time until it looks smooth, like trimming a man's beard. Whatever.

Now the belly area is clean. Step with your right foot on the floor between the sheep's hind legs, placing its right front leg behind your right leg. This way you can hold the sheep using your thighs against its back and belly. You are ready to shear its neck.

With the left hand on its head, fully extend its neck to the left.

Some sheep rebel at this, so be gentle. You need to begin by using the "opening" technique described above. Enter the wool on the chest, and open it all the way to the wool-less place just beneath the

sheep's right ear. Always go slow and careful near the eyes and ears. Many sheep have been blinded or had their ears cut off by careless neophyte shearers.

Having opened the neck wool, clip the neck, blow after blow, from chest to chin and on to the sheep's left ear, up to the point of the sheep's *poll* (top of head). Don't clip the topknot at this time. If it's a Cotswold sheep, don't clip the topknot at all; you'll trim it later to make it look nice. Some folks clip the topknot before opening the neck wool, some after completing the whole of the left side. Suit yourself. It is shorn by entering the wool from the direction of the nose, and clipping towards the poll.

Continue taking blow after blow on the left side of the neck, not crossing the backbone, but rather coming to within an inch or two of it. Here's where fat sheep make judging the location of the backbone hard.

Now you've reached the front of the sheep's left shoulder. Most sheep don't like having the back of the clippers gliding up into their wool in this area. Too bad. Take narrow blows from the sheep's knee toward its back. If you pause to relax each time the sheep struggles, it will soon submit quietly and you can proceed. Use the left hand to draw the shorn area tight, so the wool stands cutting better.

As you get beyond the left front leg, its rib cage can be shorn as follows: Using the left elbow to restrain the sheep's head away from the action, your left wrist may lightly hoist the sheep's left leg upward, to allow the shear points entry behind it.

Now clip down the left side to the thigh, blow after blow, not crossing the backbone. At some point of shearing near the hind leg, you may lay the sheep down on the ground, on its right (not shorn) side. A nice trick is to place your right foot right up under the sheep's chest, under its right shoulder. Sheep don't seem to try to struggle when their shoulder is thus kept but an inch or two off the floor. Your left foot is tight by the sheep's back. Balance yourself if you need to by again mashing the left fist between the sheep's flank and thigh to straighten its hind leg to shear. Sheep usually object to the feel of the blade running up into the wool around their rear. It seems worse to the sides of the anus/vagina area than between it and the udder. Persevere.

It helps if you removed all roughness from the back of the blade that serves as the "runner."

Sometimes there is matted wool on the tail. Undocked tails are a royal pain in the drain. Shearers suffer almost as much as sheep for the misplaced "kindness" of leaving the tail undocked. Charge an extra $3 or more to shear undocked tails, and real "humaniacs" will gladly cough it up. Move your left foot away from the sheep's back to a place under the sheep's shoulder--your right foot is at the belly side of the shoulder, your left foot at the neck side. The left hand may be used to steady the sheep if needed. Shear the whole tail if you can, right over almost to the other hip.

Now comes the long blow, or "long, bloody blow" as the hand shearing Brits and Aussies used to call it.

With the shearer's left foot already under the sheep's shoulder, his right foot slides a little to the

right, to a point by the sheep's flank/thigh region. The left hand takes hold of the back of the sheep's neck, bringing it up and round towards the shearer's hip. The shearer's right leg appears to kneel on the sheep, but it applies only light pressure there. The clippers enter the unshorn wool near the tail, and a series of long blows are taken all the way up the length of the sheep's back and neck until about 2 inches past the backbone.

The next maneuver is tricky, but not too strenuous once learned. Your right foot may step over its back now, bunching the clipped wool under the sheep. Don't stand

on the wool. Bring the sheep's head towards yourself, and stand it on its rump again, the clipped wool hanging prettily to the side.

Now you can shear the sheep's right side, starting at the head and working from back to belly, blow after blow. As you get towards the rear, you again lay the sheep down with your feet under its shoulder to keep it from getting up.

While the sheep is still under your control, trim off any unsightly

fribs of wool. Move the wool away from the sheep before you let it up, and hand feed it some of the corn in your pocket. This way when you send it off, it'll forget that you were "trying to kill it."

General Hints

Some points to remember are: 1. Don't clip over the same place twice if you can avoid it. These are called "second cuts." They cause little "neps" to form in spun yarn. If you do make a second cut, grab it and toss it away from the fleece. 2. Nicks tend to be the result of either pulling on the wool while clipping, or allowing loose skin to bulge up in rolls. They heal fast, but a dab of pine tar will discourage flies. 3. Don't strike or fight a sheep. It doesn't help at all, even if it lets off some of your steam. It tires you out, and painful fatigue is the enemy of good shearing and fun times. 4. Most important of all, have your shears razor sharp. Never leave the "wire edge" (see "Sharpening" in Smithing chapter) after honing them. To get them fit again when they feel a little dull (probably still sharp enough to shave with--but not sharp enough!) make a couple of sharpening strokes with the wet sharpening stone--on the stone's fine side--running along the length of the edge. Close the shear without letting the edges touch each other. Now open them while pressing the edges together. Close and open them in this manner a couple more times. Check their sharpness with a piece of waste wool--they must never flop over, nor refuse to cut. This is a good way to learn how wide a blow you can reasonably take. It's rare that it'll ever be more than 1½ inches wide with hand shears. 5. Don't rush; do good work. With time comes speed. 6. Do a sheep or two a day, reflecting on your mistakes which slowed you down in the interim. 7. Watch professional shearers use their power clippers at a wool festival or at a local farm. Observe those shearers whose physique prevents them from relying on brute force to control the sheep. Study their actions. There are some differences in using "blades" (handshears) and power "clippers," but most of it is identical. Note the "holds" and ways of controlling the sheep. 8. Persevere.

FEEDING SHEEP

Sheep are content with grass alone, if they have enough and are moved often enough to thwart parasites. Fences need to be very tight with certain breeds, but the Cotswolds we have kept through the years respected the most laughable of fences, as long as their topknots weren't sheared off. Rams ought to be kept in pens with fences at least 4 feet tall. Make certain the fences are built to avoid your sheeps' heads getting stuck in them. Many sheep die because of such carelessness.

A good manger will cut hay consumption in half, and the sheep will be healthier too. I like hay mangers on runners so they can be moved. I have never come across a commercially made one which was worth buying. Many of them hold the hay up high--this will ruin your fleeces for hand processing. It fills 'em with hay chaff as they pull the hay out from above. Many store bought feeders hold the hay in such a way to allow the sheep to pull it out under their front feet--where they trample it. Or they can take a bite, back up, and eat the hay over the back of the sheep next to them, filling the back and shoulder wool (the best wool on a sheep) with chaff.

The best hay feeder I have made is an open-topped box 3 feet tall, 2 feet wide, and (mine is) about 7 feet long. The length is determined by how many sheep need to be fed and by how heavy a device you can move. The bottom of the box has a solid, crack-free floor and solid walls 10 inches high all around. This way it'll serve as a grain feeder too, and conserve the fine leaves of the hay, the most nutritious part.

Above the solid walled bottom, are sloping parallel slats 1 inch

SHEEP MANGER PLAN

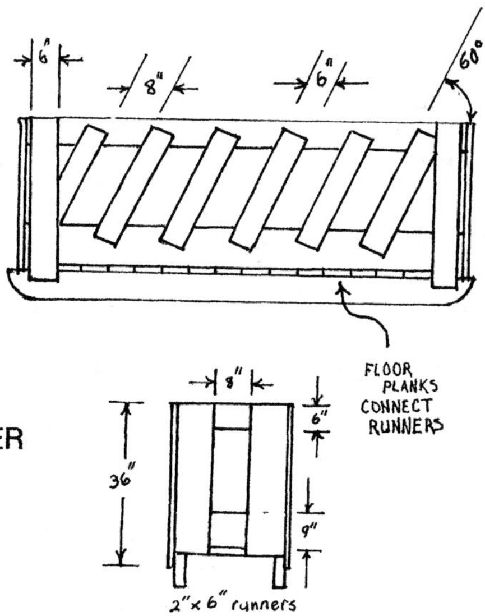

thick by 6 inch wide planks about 8" apart, set at about a 60° diagonal all around. These are for the the sheep to put their heads through. The slanting holes formed by the slats prevent them from backing up with each mouthful to eat over their neighbors' backs.

Being quasi-desert animals, sheep do really well on hay, especially nice leafy hay of at least 50% alfalfa content. Access to water at all times is important, though they can survive longer without water than say, a cow.

Lambs need a creep, which is a little shed made of packing pallets, old planks, or what have you, with openings big enough for the lambs to enter, but not the adults. Six-inch wide openings are about right. This should have cracked corn always present, and choice alfalfa hay. But especially the corn. Good clean oats or rye are also good, cracked or whole. Ground feed tends to pack hard in their tummies. In time, you may mix whole corn with the cracked, and if they start eating that, just feed whole. Any cracked corn left over from the previous day should be removed and fed to the ewes. Start with just a handful or two daily for very young lambs, and increase it to all they can clean up. Keep things clean, to avoid gut problems. Creep fed lambs make amazing gains.

Rams are typically slower, pickier eaters than ewes. This is another reason to not graze them together except in breeding season, because the rams may get skinny. They will normally take 1 lb. of grain per head daily, and all the good green hay they'll clean up. The hay is best if it is at least 50% alfalfa. If not, add a scant handful of soybean, sunflower, linseed or other high protein meal.

Shepherds set much store by alfalfa. It is the best sheep fodder, and really, the best fodder for probably all animals. Even sows have been wintered on primarily alfalfa hay! Horses get somewhat slobbery with it. But dairy cows do very well with it. It also has the advantage that if treated right, it'll out yield any other grass/hay crop. But there are substitutes. Clover, as has been said, affects the hormones of sheep so negatively as to be avoided at all costs during breeding season. Other times of the year it seems all right. Lespedeza (Japan clover) is about the equal of alfalfa, but not in yield. It will outyield all other clovers on acid ground in the states immediately to the South of Mason & Dixon's line, but is primarily good in July & August when most other pasture plants are lagging. Ordinary bluegrass and white clover feeds sheep well, but remember to rotate the pastures, because these grasses are particularly short, and that means more worms. Timothy hay if fed solo, or oat hay, will cause weak lambs at birth and for a week or so thereafter. Many sheep keepers look for excuses such as selenium deficiencies, preferring exotic ailments so their own management doesn't look bad. They dope 'em up, but their lambs are yet born weak. Even some vets spread such nonsense if it'll help 'em look smart to their clients. So ask around before settling on a family veterinarian.

During the last month of gestation, it is wise to begin feeding whole corn, oats or rye to the ewes. This is the time of greatest development of the unborn lambs. After lambing, "twinners" need a pound of corn daily during the first 2 to 3 months. Ewes with only one lamb should only be fed corn if they start really losing condition. Feed 'em too much and you'll need to vaccinate for "pulpy kidney," because the lambs will grow too fast.

Dry lot feeding, which means in a confined lot where they'll suffer no grass to grow, gets rid of worm. It is possible to eliminate worms over time using dry lot feeding. But it is more work, and sheep really need daily exercise.

Most sheep keepers turn the ram in with the ewes in early August, for January lambs. That way the lambs are well started before fly and worm season starts. Also, they are 7 months old and over 100 lbs. when marketed in July or August as choice lambs.

Non-commercial shepherds contrive to have their lambs come at the end of March to early April, and if they sell any lambs, they'll be under 100 lbs., and will bring just as much money as the fatter ones. This is because all those commercial lambs hit the market at once and the price drops. The lighter ones, if good looking, are snapped up at much higher prices than the slaughter lambs by feedlot owners to fatten as "fancy feeders." They sell them really fat in the winter when prices are at their highest.

Troubleshooting Sheep

Let's emphasize the point that sheep show very few symptoms of ailments until it's almost too late. Even long-time "sheep keepers" often utter things like "Sheep are always looking for an excuse to die." Even to novice shepherds such sentiments are absurd. They unwittingly divulge the real cause of the problem: lack of observation. Healthy sheep have good appetites, they don't lag when it's feeding time. Their dung is pelleted, or conglomerated pellets. White flecks in such blobs are usually remnants of parasitic worms. Healthy sheep don't limp, even a little; inspect their feet if they do. Healthy sheep don't stand around with hunched up backs, except when it's raining and they don't have shelter. Healthy sheep don't rub on things unless hot weather has come and you neglected to shear them. Healthy sheep don't have big globs of snot all around their noses. Healthy sheep aren't losing wool--wool loss is a symptom of fever. A normal flock is not nervously bunched in a corner of their pen, heads up and alert, tensely looking around. They act this way when predators have frightened them. Healthy sheep don't walk stiffly, they bend the knee lazily, or they happily lope to the feeder when called. A healthy sheep has a pinkish cast to its white skin. Black sheep too, but it's harder to become familiar with it. The whites of a healthy sheep's eye have small but plainly visible veins. Well kept sheep aren't bound by mats of brush, briars and burrs in their fleece. Allowing this situation obscures aberrant behavior. Look at each sheep for 30 seconds every day. It is a much longer time than you think. When you've become expert, 10 seconds will tell you if the sheep behaves heartily. Most problems revolve around lambs and lambing, so I'll only discuss those. Observing the sheep daily will make you more adept at describing other symptoms to your vet.

Lambing Problems

Birth problems. Now and then, especially with the big-headed breeds, a ewe has a hard time being delivered of her lambs. You'll need to help. The ewe is lying down, straining to push the lamb out. If the front hooves are both visible at the vulva, and the nose too, take hold of the hooves and as she strains, you pull. Never pull the head. At first you'll want to add only a pound or to to her efforts, increasing your pull if the first tries don't help. Pull only when she pushes.

If you don't see the head, it may be turned back. With very clean hands, reach in and find the head, turning it right. Be gentle, and remember you may need to push the lamb back in some to straighten it out. Similarly, one leg may be turned back. Find it and point it out the birth canal, helping pull if the ewe is exhausted from trying. If a ewe has been laboring for 24 hours or more, the lambs are almost certainly dead. Often the hind feet come out first; again, one leg may not be pointed right.

Other bad presentations include two lambs in the birth canal at once, deformed lambs, head and one leg turned back, lamb's body blocking the birth canal sideways inside, etc. In each case, the ewe has passed the lubricating "water bag," but nothing has come out for an hour or more afterward.

Prolapse

Some ewes evert the vagina. That is the vagina comes out of the vulva, in a meaty, sometimes bloody mess looking somewhat like a sock. There are various retainers which can be bought from vet supply houses, and homemade wire contraptions to hold the blob in after you've re-inserted it. Suffice to say you should get rid of any ewes to which this happens, and their offspring. This problem is inherited, and often recurrs despite your best efforts. Sometimes the entire uterus (lamb bed) prolapses, and such cases are usually fatal. If not from the condition itself, kind shepherds should mercifully and quickly dispatch the sheep.

Lambs

Some lambs are born really weak, so that they can't stand up on their front hooves. They pitifully walk on their knuckles. This is nearly always the result of not feeding enough legumes during pregnancy. A few day's milk from a well-fed mother brings 'em around.

Sometimes lambs get reversed eyelids. This often leads to infections when eyelashes rub on their eyes. Several times a day, you must check on the eyelids, and flip them back normally. If infected (they'll get a cloudy appearance) get some eye ointment from your vet. This problem isn't serious if you are a shepherd instead of a sheep keeper.

Some rare lambs are born with an allergy to milk. They show symptoms of this within 48 hours and die regardless of treatment. A similar ailment, tetanus, shows up within 2 weeks, and was contracted through the navel cord when it was still wet, from tetanus causing germs on the ground. Many shepherds dip or spray the navel with iodine when they have had tetanus on their place.

Soremouth is the name for painful blisters which develop all around the mouth of suckling lambs. This is quite contagious. If your flock doesn't have it, they won't get it unless you bring in new sheep whose flock had it. You can get it too, so exercise caution, if it's suspected. See your vet.

Scours are nothing more than diarrhea. Lambs get upset stomachs when: 1. They eat something that disagrees with their gut. It goes away, but now they've got a sticky rear end, a perfect host for *fly strike* (fleece maggots). Wash 'em off. 2. They have worms. 3. They have a gut infection; use antibiotics, like sulfamethazine given orally by syringe with needle removed. 4. Lush pastures. Feed a little hay before turning them in to such grass.

Sheep of any age can get bloat. This is caused by foam accumulating in the rumen (first stomach). It occurs mainly on legume (clover, alfalfa) pastures which are wet with dew. For some reason it occurs less with dry or rained-on pastures. If you're going to change pastures, and there is a lot of clover or alfalfa there, feed some grass first. The bloated sheep has a largely distended left side, just in front of the hip. It may not be able to get to its feet if it has been bloated for more than a half hour or so. A few tablespoons of vegetable oil mixed with a teaspoon of turpentine and carefully fed (to avoid getting it in the wrong place) with a syringe with no needle should subside the foam. If the sheep is too far gone, and is slipping into unconsciousness, it might be saved by stabbing the rumen with a sharp trocar, or even just a pocket knife blade, and holding the slit open while the stinky foam gushes all over your hand. Talk to other shepherds about this if you can, before you ever see bloat. I have kept sheep for many years and have never had a case of it other than indigestive bloat of lambs, which was treated by constant massage until the wee hours

when I was just too tired to continue. The next day the little lamb was fine.

Parasites

Sheep of all ages get foot rot, if you bring it onto your place. This is a parasitic fungus which cripples sheep and can live for long periods in the soil. Don't buy sheep from markets. Don't buy from farms where you see sheep limping unless you can ascertain that the limping is from a cause other than foot rot. The foot really stinks if they have this ailment, but often sheep feet stink anyway. Treat it with formaldehyde dip or (not approved, but it works) with copper-based hoof treatments sold for horses.

External parasites include mites and keds. These cause no end of discomfort to sheep. Keds are visible, especially at shearing time and look like large ticks. Mites are too small to see. The best course of action is to dip, spray or use a pour-on insecticide just after shearing, and again 2 or 3 weeks later. One benefit of Cotswold sheep is their wool parts along the backbone, so if treatment is necessary, it can be done any time using a pour-on instead of just at shearing time. Pour-on lice poisons are the easiest form to use safely.

Worms afflict sheep worse perhaps than any other domestic animal. There are different kinds of worms. Fortunately the modern broad spectrum wormers get rid of practically all species at one treatment. But worm eggs lie in wait on pastures, and sometimes in the sheep's gut. These form the seeds of the next cycle of torment, unless that cycle is broken by pasture rotation, or drylot husbandry. The tapeworm is killed only by a few available wormers. Some worms can be spread back into your pastures by deer traveling through. Deer are pretty, but not worth encouraging if you have livestock. Though it smacks of quackery, it is safe to assert that if your sheep are ill and you can find no cause, doctor for worms. Some further signs of worms are an absence of that pinkish color to the skin, or absence of healthy blood vessels in the whites of the sheep's eye. Another sign of worms--one which is the last sign before disaster--is *bottle jaw*. This is a baggy bulge which develops under the lower jaw. Even after treatment for worms, it takes several days for this swelling to disappear.

Fly-borne parasites include fleece maggots and nose bots. The fleece maggot, mentioned under Shearing, is really just the common blue blowfly, which lays its eggs either on dung tags around the sheep's rear or in wet, smelly wool. The eggs soon hatch and the maggots soon run out of dung to eat. They burrow into the fleece, consuming the skin, and then the meat and sometimes the rectum. The lambs are especially susceptible, so inspect them often if they walk funny. You'll have to catch them, and part the wool. The maggots are large, but blend in so well with the wool as to make 'em nearly invisible. Sheep plainly show discomfort by wiggling and twisting to try to reach the stricken place. Pure turpentine or kerosene will kill the maggots. Be vigilant, because severe damage takes long to heal.

Nose bots are the cause of "grub-in-the-head," a horrible affliction. The bot-fly hovers about the sheep's nose, never landing, but ejecting a deposit of just-hatching eggs into the sheep's nostril. These tiny larvae travel up into the sinuses, where they become large worms. They feed on the blood vessels there, causing untold discomfort, until the sheep sneezes out the pupated form of it, which emerges in the ground as a bot fly. Some injectible wormers kill nose bots. A good preventive when you see sheep with noses to the ground, obviously attempting to avoid the flies, then running away, and repeating this behavior, is to smear their noses with pine tar daily. An old time remedy was to mix their salt with pine tar in a salt feeder, which got the tar on their noses daily. Sheep learn to like pine tar.

15
RABBITS, ETC.

Every moving thing that liveth shall be meat for you; even as the green herb have I given you all things.
Genesis 9:3

I kept rabbits as a boy in Maryland. A roofed hutch, 3 ft. by 2½ ft. by 1½ ft. from floor to ceiling is big enough for a doe and her litter. The floor is ½ x 1 inch welded wire, the sides are 1 x 2 inch wire. An access door, 12 x 14 in. is made in front. The best pens have no wood in their construction, because rabbits chew them up. Three such pens are needed, and you ought to mount them so their floors are hip high, to cut down on the work of getting in and out of them, feeding, and cleaning the dung from beneath. They need no protection from the weather except wind protection from the north or northwest, and a nest box during kindling (parturition). The nest box is about 1 ft. by 1½ ft, by 8 in. tall. It should be closed in on floor, sides, but not top. In winter, insulation board under its floor helps the bunnies stay warm. Inside the box, place 2 inches of shavings or bark (or something similar) and a couple of inches of straw above this. Don't put the nest box in the pen until 29 days after breeding, or the doe will probably designate it as her toilet.

You have to be able to keep waterers from freezing in the winter, and *fines* (powdered feed) from clogging the feed holders.

The best feed for rabbits is alfalfa hay, fed in a manger on one side of the hutch, supplemented with corn, oats or barley, fed in a dispenser mounted at least 3 inches up one wall of the pen to keep dung out. Most folks just buy complete pelleted feed and a salt brick or spool, and be done with it. Feeding too much succulent feed (cabbage, lettuce, etc.) causes diarrhea.

Californian, American Chinchilla, Flemish Giants and New Zealands are the most useful breeds. In hot climates the lop-eared varieties keep cool better. Big rabbits mature late, and thus require more feed per pound of meat. You need at least two does to pay for a buck. The does are brought to the buck's pen, not vice versa, for breeding. One buck is good for at least 20 does.

It's best to have at least 8 bunnies per litter, but not over 10. Even this number will require a good-milking doe. She may be rebred 31 days after kindling. Careful husbandmen can get 6 litters/year per doe. That comes to a rabbit to eat every 4 days of the year with only 2 does and a buck! You'd have to be really crazy about rabbit for that, but I know some folks who are. With so many official impediments to buying and selling meat, it's hardly worth the effort to make a go of rabbit selling. Sadly, for all their police powers the authorities don't seem to ever prevent constant and widespread problems with food-borne illnesses.

Rabbit dung is the equal of poultry dung in my experience, and there were always plenty of worms for fish-bait under my childhood hutches.

Bullfrogs, chinchillas, ferrets, etc. are at time enthusiastically recommended by ardent devotees. All this stuff falls under the heading of agribusiness or hobby, not husbandry.

16
DRAFT BEASTS

Behold, we put bits in the horses' mouths, that they may obey us; and we turn about their whole body.
James 3:3

I bought a used two wheeled "walking tractor" before I took up residence in West Virginia. It was entirely gear driven, had heavy agricultural-tread tires, a roto-tiller, and sicklebar mower. After a brutally sluggish-paced experience with the roto-tiller on "channery loam" (a nice word for stony soil) I knew there had to be a better way. The ancient garden ground though fertile, just didn't yield more than two inches depth to the tiller. I tilled for days, and had little to show but blisters and poorly dug ground. I gave up, putting pigs in behind an electric fence. Now *there's* a way to till a garden. Moving them promptly, they loosened the ground enough for a bit more roto-tilling. Leaving them too long, they packed the soil worse than before. It's no use using them for tillage on hot, dry summer soil. But the roto-tiller was ineffective even for cultivating the spaces between rows of growing crops. Sure enough, a stone would catch the tines and violently bounce the tiller up and onto a crop plant.

I rapidly discovered the wheelhoe was less work and did a better job if used correctly--that is, if used weekly, before the soil had a chance to compact itself again.

Little by little, I saved pennies and bought a reversible plow blade and special hitch. The plow was only 6 inches wide, a tiny size by modern standards, and I'm sure it would have worked in better soil, but the little wheels, and meager

300 lbs. total weight of the "walking tractor" didn't really supply enough traction. I got blistered hands pushing to help.

So then I saved up and bought much larger diameter wheels, and heavy cast iron wheel weights. Bigger diameter makes more traction, and I could hardly hold back the machine, but it still barely had power enough to move the plow. I got a disk and one-row cultivator too. And a wagon which seemed highly susceptible to flat tires. All this junk didn't make work easier. It just changed its nature. I was sleeping well at night though, mainly because I was so totally exhausted!

Upon receiving a small inheritance, I quickly bought a small, used tractor. I was stunned at the tremendous difference in the traction of this machine and that of the "walking tractor." Oddly, it didn't seem to use a whole lot more gasoline, when one considers the work done. But it had a plow, a sicklebar mower, a trailer, a disk, a one row heavy duty cultivator, and a brush mower. It ran well, though parts seemed to break with disconcerting regularity, and little innocent chunks of metal cost up to as much as two fat hogs each! I did all sorts of repairs, but the worst part of the problem came with cold weather. I discovered the tractor was nigh impossible to get started when the temperatures went below zero. I had to use other vehicles to start it. Often I just dragged it up the road with the truck until it would finally catch. It had excellent traction, but I needed a tractor in the winter almost as much as the growing season. I finally decided to sell it. I lost about a thousand dollars in the deal, mainly because I shouldn't have bought it in the first place, I think. With the money from its sale, I bought a unbroken horse and some used harness.

The undisciplined horse cost me less than just one new replacement for a broken front axle spindle on the tractor. He wasn't a draft breed, but a fast moving Arabian. He knew nothing except he was taken away from his happy home by strangers. And I knew only a little more than he did. I had helped relatives in South Carolina break green horses when I was a child. I'm sure I wasn't doing much, but I observed a little, and enjoyed riding by myself in the fields and woods in those days.

I found the horse a willing partner, easily started in cold or heat. Responsive to voice command, and quiet, it was a pleasure to work with him. He didn't choke me with foul fumes, or compact the soil with a grating, grinding wheel. His strength and traction, though not near the equal of the tractor, was worlds ahead of the "walking tractor" and much faster. And he wasted no part of his feed. Horse dung is the most magnificent of manures. Even in a well-dunged field, the places where the horse paused to relieve himself in his work were plainly evident a month or two later in the crop. Blotches of lush green stood out among their good-but-not-great neighbors.

I discovered draft animals' feed was largely the product of sunshine. That is, once full grown, a horse or ox eats, excretes, and works. What it eats and what it excretes are about equal in terms of the land's fertility balance. The part converted to labor is almost totally carbohydrate, which is nothing more than the sunshine which fell on the field. All the protein (nitrogen compounds) and minerals are returned.

The tractor is not only noisy

and smelly, it is difficult to grow its feed, it will not repair itself if left to rest, and it won't go well in deep snow. When it breaks down in mid-field you can't unhitch it and put it away. There it sits until you fix it, snow, mud and all. There is only one reason I can see why a tractor is really handy, and that is in the hydraulic lifting most of them can be equipped for. But even so, it's cheaper for most husbandmen to hire this done--the once or twice a year it is needed--than to lay down the money to buy extra gear. A reliable tractor with a loader bucket will cost enough to pay for hundreds of hour's work by someone more experienced with it. If it breaks, that's his problem! Tractors are for big commercial operations, and even then, they have a hard time showing a better profit margin than a horse-powered farm.

Oxen and Horses

I believe if I were to start all over with draft animals, I'd choose oxen over horses. There are 6 reasons: 1. They're cheaper. You can buy a little week-old bull calf at a livestock auction or dairy farm real cheap, castrate it, feed it and teach it, and it'll be working hard for you inside of two years. It can do little things much sooner than that. 2. You don't need fancy craftsman-made harness. All a working steer's work tackle can and ought to be home made. 3. Oxen aren't near as exacting in their feeding requirements as horses. 4. They move more slowly, only about two-thirds as fast, so there's less skill needed at corners and narrow spots. 5. They're not as easily disquieted as horses. 6. Old oxen are more salable than old horses.

But horses have their advantages: They're better for wagons and carts because they *do* move faster. Their working lives are about twice that of oxen. They can go longer without water (but not a *whole lot* longer). Horses are better for riding; equipment is more available. Since I have no experience with mules I must offer the following second hand: They live longer than horses or oxen, are less likely to have digestive problems, are not quite as good for traction, and the good ones are costly.

Draft animals have one flaw: It takes common sense and compassion to do well with them. I have seen cripples and true half-wits driving tractors well. I've heard of the same with oxen, but only if they were already trained. Horses need understanding and reason from their drivers. Animals quickly adapt to routine, and eventually do their work almost without the drivers' interference.

Horses

One old method of getting a young, vigorous horse is to buy "washouts" from around the sulky racecourses. These used to be cheap, because there isn't much use for a pokey animal at a race. They've already been taught the use of the bit, and have been hitched to vehicles. But there are hazards, like health problems. Best get an old hand, or even a vet to help you.

Another place is on horse-powered farms. If the horses are purebred they'll likely set you back some real money. This is because nowadays merely "showing" animals has become quite a business! As ridiculous as it seems, an animal's looks or bloodlines have become financially more important than its more tangible worth.

Rather than get into a long treatise on evaluating horses, let me just say that old geezers who love messing with horses usually love to have the company of young folks they can share their knowledge with. Less aged enthusiasts often tend to treat newcomers like aliens. And older folks commonly have more time on their hands anyway.

As to breeds of horses, the really huge drafters, like Shires, Clydesdales, Percherons and Belgians were all originally bred for the purpose of hauling gigantic loads in cities. They alone have enough weight to start and stop heavy freight wagons on oft-times uneven pavement. They came into use on farms when the almost mythically large fields of the American midwest were brought under cultivation. These flat, rich lands cheapened the price of food so much that saving man-hours was the only way to compete. Thus, very large hitches of sometimes dozens of horses with one or two drivers got through the fields remarkably fast. The weight of these beasts insured enough traction to pull the ever enlarging equipment, until finally only tractors could do the work--it became customary to work 24 hours, with shifts minding the machines which never got a chance to cool off. This is the way money is now made in farming, like driving a taxi cab for a large cab company. The economies of scale cheapen food, by reducing the amount of human labor needed to grow it. And those once rural folk must find urban work so they can afford to eat that cheap food.

But the traditional horses of small farms of less than 20 acres (nowadays a "small farm" is one under 1000 acres) were of modest size. They stood not much more than 14 hands. (A hand is 4 inches, and is measured from the ground to where the neck meets the back--the *withers*). Smaller than 12 hands, and they don't have the weight for traction, no matter how strong they are. Small horses seem to be stronger, pound for pound, than big ones.

Flinging an 80 to 100 lb. harness up onto a high-backed drafter can be quite a job in itself.

Good small breeds include the Welsh Cob, particularly type C; old style Morgans; Haflingers; old style Connemaras; Fjords; Dales; Swedish Gotlands; and the like. In the horse drawn era of not so very long ago, *chunks* were the most demanded small farm horses. They were mixed breeding, hardier than most pure breeds, and fairly cheap. Pure bred horses tend to be less hardy than crossbreeds, despite their higher price tags.

Oxen Breeds

Mostly the same rules apply to oxen as to horses. Smaller oxen, like Jersey or Guernsey steers, will eat less than the big Holsteins and Brown Swiss, but also have less traction. Castrated cattle never cost as much as castrated or uncastrated horses. Small-ox yokes are lighter and don't need to be lifted as high onto their necks. Any calm natured steer, pure or cross bred can make a good drafter.

The Way of Horses

Horses not only have the most frail digestive system of all domestic animals, they are also the least intelligent. Well, maybe except sheep. Ninety-nine percent of the abuse inflicted on horses is due to the abuser's assumption that the horse is intelligent enough to "do it on purpose." We all too often interpret displays of memory as reason, of fright as ferocity, or uncertainty as rebellion. Horses can be made to remember very well; they are among the most teachable of animals. But they're easily frightened, and easily confused. It has been observed that they are completely non-agressive animals, except when stallions fight over mares. Punishment is never to be thought of as retribution for wickedness. With horses, undesirable behavior should in itself be so arranged as to punish the horse, through the use of what are known as *appliances*. These include various rope tricks, bits, etc. Also, desirable behavior must be made as pleasant as possible, to encourage the horse.

A horse's main defense is to run away from danger. And all things new are perceived as dangerous. Horses instinctively band together in herds to thwart predators and flies; some sleep while others watch. The only heirarchy appears to be that of the stallions, and even so, it's primary concern is breeding rights. Because horses are designed for constant movement, (only sleeping an average of 3 hours per day) they get bored when they have nothing to use up their energy, fidgeting in such ways as fence or building destruction, chewing on wooden parts of their pens (cribbing), striking (pawing with the front feet), etc. In this they are similar to dull-witted portions of human populations who, lacking work or entertainment, get into mischief. It is false kindness to over indulge such people or such horses.

Between each digestive organ in a horse lies a constricted orifice. For example, the stomach has plenty of room for digestion, but its exit opening is so small as to be easily clogged with pasty matter, such as ground or chewed corn or worse, wheat. Powdery feeds are therefore to be avoided at all costs. This is one reason why oats have always been the ideal grain; they are very high in fiber. So is wheat bran, but it lacks the oil and carbohydrate of the oats. Its main use is to add bulk to diets which of necessity must include corn or other low fiber ingredient.

As to roughages, bright timothy hay, free of dust or mold, has always been the fodder of choice. Oat hay, cut when the heads are just coming to the "milk" stage, or grass hay cut just before full flower, are about as good. Alfalfa is very good too, but requires more care in feeding, as does clover. These clover-family hays are far more likely to mold, and moldy hay is quite detrimental to horses. If the hay is dusty at all, it can be fed either in the rain or after hosing it off. Visibly moldy hay should not be fed to horses; pick moldy parts out. Also, clovers make horses slobber. Crimson clover cut at maturity has coarse hairy stems, and tends to clog a horse's gut. Most horsemen over feed their horses hay, to no detriment other than financial. A horse needs about 2% of its weight daily in hay, if there's no pasture. If idle, they need little or no grain; the more they work, the more grain and the less hay they need.

Here are the four most common afflictions to be careful for: Colic, Tetanus, Grain- or Grass-Founder, and Parasites. I present them in order of danger, because husbandmen must guard their animals' health as their own.

Colic is nothing more nor less than stopped up digestion. This is common, painful, and often fatal. It is caused by obstruction of the digestive tract. Changing feeds too abruptly, feeding too much corn, wheat, or ground feeds, or gaseous feeds, causes it. The horse sits like a dog, tries to roll, is not willing to eat, and shows obvious discomfort, sometimes a lot of sweat. Put your ear to his gut, and there's no healthy gurgling. Colic is your fault most of the time, but when it shows up, better get help from one of those aged horseman friends of yours, or a veterinarian. To prevent colic, don't feed beans unless they've been cooked. Soybean meal is heated in processing, so it's o.k., though it's sometimes powdery. If powdery feeds must be fed, have your miller mix wheat bran and molasses into it. (Molasses eliminates dust, which causes a pulmonary disorder called *heaves.)* Clean barley is the best horse feed, next to oats. Or you can use store-bought mixes. Or if you don't work your horse all that often, and not for long when you do, just don't bother with grain at all, except a handful or two as a treat.

Call the horse by name when you feed, especially when giving a treat of grain or a carrot. Sometimes just call, and then caress. This will get the horse to associate its name with goodness, and it'll always come at the sound of it.

Tetanus is caused by a microorganism found in the soil. A nail-prick or other wounds may give entry to the germs. A horse with tetanus is almost sure to die, unless it was vaccinated. Vaccines are usually combinations of immunizations all given in one injection. Injections are given most commonly in the heavy muscling of the upper half of the lower neck (not far from the shoulder). It is easier and less frightening than we'd assume.

Grass- or *grain-founder* is a condition brought on very commonly by the horse having engorged itself on overly rich feed. Horses that somehow gain entry to the corn or oat bin eat so much they founder. The congestion from too-rich blood causes the capillaries to burst in the toe and then the legs. The first symptom ordinarily is inhibited ability to move around. As the lameness worsens the horse's pulse accelerates, becoming full and hard. Temperature rises several degrees above normal (100.5°F, ±1°F, rectal) sometimes even reaching 106°F. Breathing is rapid and panting, nostrils widely opened, and snotty. The horse's facial expression is worried and showing acute suffering; the whole body is moist with sweat. Sooner or later diarrhea appears. Urine is darker, and slightly thicker. Appetite is depressed or gone, but thirst greatly increased. Affected feet are hot and dry, and the horse tries not to put weight on them. Any tapping on them or bearing weight causes dreadful pain. The horse may die before help arrives, or permanent damage may ensue in any case. This disorder also is quite common, especially in ponies, when the animal is turned out to the super-rich grass of early spring. Much better to make hay of the early grass growth, pasturing the *aftermath* once the first cutting is off. Make absolutely sure your horse positively cannot gain entry to any grain bin. Neighbors of mine have gotten founder troubles from both grain and grass. Even if the horse recovers, its feet may be damaged. "No feet, no horse," the old timers used to say.

Most country horses have parasites. This is because their owners don't understand the nature of parasites. There are a number worms which plague horses, but I suppose the common horse bot is the worst. *Bots* are the larvae of the gad-fly. This fly has no provisions for feeding or even landing on a horse! It deposits eggs while flying close, with a tail shaped like the letter "j." These eggs are visible, mostly in autumn and winter, as tiny white blobs on individual hairs, usually most abundant on the front legs and shoulders. The horse ingests these while scratching itself with its teeth. Once swallowed, the eggs hatch, remaining in the gut as larvae until springtime. These are *bots*. The larvae are passed out with the dung in warmer weather, and pupate in areas of accumulated dung, and in the soil. In late summer the flies begin emerging, to start the cycle again. Pupae in the ground and dung don't survive hard freezes, so the way to rid the horse of this pest is to administer a boticide after the first couple of freezes. Some bot-killers also kill a lot of other worms too. Gadflies migrate up to a few hundred yards, but will usually not bother a horse unless it's kept within range of other folks' horses not treated for bots. The bot eggs themselves can be combed off with special *bot combs*. Or they may be "neutralized" by wiping them lightly with a kerosene-soaked rag each day.

Other internal parasites are mainly the result of grazing infested pastures. Rotations are in order, and/or periodic worming with store-bought concoctions. Or you may feed hay the year round in a closed exercise paddock.

Water should be available to the horse at all times. If not, it needs to be supplied at least 3 times a day, much more often if the horse is working. Too, if the water isn't on hand at all times, it should be warmed to just cool or even lukewarm in winter--the horse may drink it too fast and get spasmodic colic.

One well kept horse ought to labor for about 20 years or more, sometimes much more.

Training A Horse

Most horses aren't wild and unruly, and can be trained by simply giving a new lesson each day, repeating it a few times, and then repeating all they've learned up to that point. A few rules are in order: 1. Don't beat a horse, especially around the front of the body. The only (rare) exception to this is lightly tapping your finger pads against the horse's forehead to get its attention. 2. Don't expect so unintelligent an animal to learn too much too fast. It's cruel to treat a horse as if it were smart enough to "know better." 3. Don't punish; rather contrive to have unwanted behavior punish itself. This isn't always completely possible, but if you at least try, you won't go far wrong. 4. Learn the dangers of the horse, and how to thwart them. 5. Seek to attain the horse's confidence in all commands. 6. Use widely accepted commands, don't improvise. Never change the meaning of a command--you'll only confuse the horse. Your children may one day need to control the horse, or you may sell. 7. Talk to the horse in a calm, gentle voice. 8. Work the horse as often as possible, to strengthen its memory and obedience. 9. Never teach a horse to play, like a dog. Horses must positively know who is boss. We are not equals. I have an acquaintance whose wife was killed by a horse he taught to be playful. There was no malice involved, but she still died while the horse was innocently playing.

A horse will grow fond of a kind master, increasingly seeking to please him. Horses cannot be frightened into reliable obedience, nor bribed. But both unpleasurable and pleasurable sensations are useful to intensify an experience. After this, repetition must be used to "fix" the activity in the horse's mind.

Four goals must be met in a well trained horse: 1. A calm horse, regardless of the situation. 2. Reliable and prompt obedience. 3. The habit of putting forth his full effort. 4. To correct faults, whims or vices.

The following directions are not meant to be infallible, nor are they useful on wild, unruly horses. Many horses resist or become frightened, and will eventually submit to rule, but the "monsters" must be trained by other methods, which I may elaborate some day. No one new to this should train a horse unless it has been at least haltered, and made to feel comfortable around people.

It is both a blessing and a trouble that horses forget even intense lessons. If a situation gets out of hand and the horse gets badly frightened, it will forget the whole problem with the passage of time and other, more pleasant experiences. A repeated bad episode is worse, but still can be overcome with time. These are blessings.

If a good experience fixes in the horse a favorable pattern of behavior, it still must be repeated daily (even several lessons a day) for some time before it is concretely retained. This is the trouble. A well-trained horse is worth the trouble.

Age to Teach

Young horses generally learn better than older ones. Day one is advocated by many wise trainers, but since it is unwise to actually demand much labor from a horse until at least two years old, one will be training for quite some time before the fruits of the work appear. Yet there is no better trained horse than one conditioned from birth, before any bad habits are learned. Also, young foals very quickly accept the leadership of the master.

Many experts suggest two years as the right time to train. Some say 4 years, because the horse is not only mature enough for full work on a daily basis, but has left behind most of its "foalish" playfulness. I agree with the latter, because in my experience, 4 to 5 year olds are much more coöperative. I don't think any age is too old, unless the poor thing is decrepit. But I don't like the idea of fooling with horses "trained" by those who refuse to teach standard commands, and who beat the horse into submission. Such animals eventually may choose between whether to obey a command or bolt from something frightening. Horses cannot trust a brutal master.

Safety

The safest place to be is close to the animal's shoulder, holding the halter or bridle close to its mouth with your closest hand. If something really drastic happens stay near the shoulder or get entirely out of reach. But letting go teaches the horse, too. Don't, if you can possibly avoid it. Many accidents and injuries have resulted trying to lead a green colt at the end of a lead rope or strap. Keeping the horse's head up prevents him kicking with the hind legs. Keeping its head down prevents rearing up on the hind legs. To put it another way, if a horse wants to rear up, it must fling its neck upward to get up on the hind legs. To kick out with both hind legs, it must drop its head to lift its back. A few thwarted tries to do either will eliminate it as an option in the horse's mind.

Handle the horse a good bit before training. The old and tried *third hand* is the safest and best procedure. This isn't anything more than a light pole or walking cane, which you use to caress the animal.

Begin at the base of the mane along the top of the neck. The dandruff here itches the horse, and he can't scratch it; he enjoys having it rubbed or scratched. Watch two horses scratching each other's necks, you'll see.

Now work the *third hand* across the withers to the back. If he tries to get away, start again in familiar territory. I do all this while the horse is eating hay. Little by little, and sometimes over a period of many days in the hyper-sensitive horse, I work the *third hand* to every place on the body until the horse will permit me to use my hands there. The most sensitive places are along the back legs, under the tail, and in the crotch area.

For safety's sake, a good strong halter should now be fitted. It should be loose enough to allow eating in comfort, but not loose enough to be pulled off easily. You can place it on him by sticking your hand through the nose loop with a handful of grain, or a carrot, and while the horse is eating, slide the halter up and over the head, fastening the latch smoothly and without sudden motions. Leave it on him in the stall or pen until he no longer notices it.

Training to Lead

After handling and haltering, the horse needs to be taught to lead. Most horses resist a pull on the halter with all their will power. So our first "appliance" is used: the *quarters hitch* or *butt-rope*.

Get a piece of sash-cord, or small light rope, tying a large non-slipping loop in one end, say 3 feet or so in diameter. Rub the horse with this light rope to alleviate the natural fear of it. With the horse secured by the halter to a tied lead-rope, ease this loop onto his back, so it easily falls down around the hindquarters perhaps up to a foot below the tail. The loose end is now passed along the back, past one side of the neck, and through the lowest, foremost loop on the halter.

You may now take the horse away from the hitching post. Give a tug on the halter rope. The horse will almost certainly resist. A sharp little jerk on the butt rope will bring him forward hastily. Caress until he quiets. When he's confident again, repeat it all again, using the butt rope as needed. Very soon the animal will follow where led, and the butt rope may be discarded.

Train a Loose Horse to Come

It is very useful, after several days of confident leading, to teach the horse to follow without a lead rope. It'll save much frustrating time trying to catch the loose horse at pasture, etc.

This method only works on gentle, teachable horses--the only kind you should consider for your first training attempts. As soon as you've taught him to lead, get or make a whip (a narrow lightweight wand of springy, tough wood like hickory, or bamboo, with a 10-inch shoelace or leather thong tied into a hole in the lightest end will do) and face the horse, standing just in front and slightly to his left.

Gently caress him with the whip on his back and rump. When ready to start, give the command "come here" followed forthwith by giving a light whip-stroke on his rump. This will make him walk forward. You walk backwards at the same rate, giving the command "come here," to stir him to follow. When he'll follow at command, you may cease to employ the whip. With time, gentle horses learn to enjoy following the master. But be aware it takes time, so be patient. And no more than 15 minutes of this a day. Reward him with a handful of oats or a carrot. (Keep some in your pocket.)

Bitting

Before putting a bit in a horse's mouth, examine the *bars* for redness, which indicates soreness. The bars are the toothless portion of the horse's mouth between the *molars* (back teeth) and the front teeth. There is a flesh-covered bar of bone here, right about where the corner of the mouth is. With the horse tied safely, you may slip the left thumb into the corner of the mouth at the "bars" without fear it'll be bitten. Reach across the horse's velvety soft tongue, and pull it to the left with the thumb. Now look at the right bar in the sunlight. It's unlikely to have any sores, but look anyway. I always check the bars before putting a bit in a horse's mouth--you can't control a horse if the bit causes constant pain.

Use the right thumb to check the other side similarly. The horse will probably pull back, or resist, but he's not going anywhere once you've got that tongue under your thumb.

Place the bridle on his head, right over the halter if you like, with the bit unfastened on one side. Once the bridle is attached, the throat latch buckled loose enough to place your fingers under it, pinch the sides of the jaws, actually between the jaws, to get him to open his mouth. While doing this, the other hand takes up the loose end of the bit, and slips it into the now opened teeth, and on back to where it can be refastened in its place. Don't get it too tight, wrinkling the corners of the mouth. Don't get it too loose either, or he'll learn to get his tongue over it. The object is to get him to learn to chomp his mouth shut while the bit is in, and to accept its presence. Many trainers now let him wear the bit ½ hour at a time, for a week or two, eating with it in, and out, gradually lengthening the time he wears it. Some *training bits* have little keys to encourage chomping. Some have leather or rubber covered bits. The best bits are either a large-diameter straight bar bit or a similar, but jointed snaffle, with cheek pieces or rubber stoppers to keep 'em from pulling the rings into the mouth. Either way, it should be the same width as the mouth is. Too small a bit causes pain, too loose jerks back and forth in the mouth, in a sawing motion. You can see these bits in horse-equipment catalogs. Adjust the *headstall* (bit-holder) so the bit makes contact, but not too tight.

Bitting harness consists of a bridle (no blinders) with a large, smooth bit with cheek guards; a *surcingle* which is a strap fastened around the horse's middle just back of the front legs and has rings to fasten reins; and a *crupper strap;* the crupper strap connects the surcingle to the *crupper*--a piece of smooth, sometimes padded leather which fastens under the tail. The purpose of the crupper is to keep the horse's head from pulling the surcingle forward via the reins.

There are two kinds of reins on the bitting harness. The *check reins* attach to the bit, and pass along each side of the head up through a loop in the bridle near the ear, and on to a ring in the middle of the surcingle, on his back. Its purpose is to keep the horse from putting its head down (such as in eating grass) where it would be difficult to control with the hand reins.

The *side reins* simulate the feel of the driving reins, and are attached from the rings of the bit to loops on the side of the surcingle.

The horse may wear the bitting harness for up to a half-hour in his first lessons, but he may be given up to 3 lessons a day. Each day the reins may be gradually shortened, starting from quite loose, but never getting so tight as to hold the head uncomfortably or draw the bit so tight that the mouth gets sore. Some horses will be used to it in just a couple of lessons, while others need a week or more of training before they settle down and stop paying so much attention to it.

There are many other things useful to teach before driving in lines; most commonly folks teach a horse to be *hobbled.* They tie the front feet together with either leather hobbles made specially for this, or with a large diameter cotton rope. A little at a time the horse is made to stand longer, until it can stand for hours. Then it is *Scotch hobbled* which is to say the lead rope is attached from the halter and tied to one of the front feet. Horses thus trained can be "tied" without a hitching post.

Wild horses are trained with special surcingles and straps to involuntarily lay the horse down on its side. They are taught they are powerless against the master, and are subjected to banging on

pans, bells and firecrackers, rattling newspapers and umbrellas, etc., until they take it all calmly and cease struggling. Others whirl the horses around by tying the halter to the tail, the fighting horse whirls until it's dizzy, etc. Make sure your new green horse is tame enough that none of these more complicated and costly methods will be needed.

Driving

After your horse has accepted the bit, you may replace the side reins of the bitting harness with real reins. Incidentally, I have seen training surcingles improvised out of rope, old leather belts, automobile seat belts, etc. If you use a rope, use a soft, thick one about 15 feet long, folded in half to form a tie-off loop. The loose ends may be inserted through this loop, and tied back to themselves. A double rope surcingle doesn't dig in as bad as a single.

Driving reins may similarly be improvised out of smooth (preferably braided) rope.

The driving reins attach to the bit, and must be inserted through side rings in the surcingle. This is because you don't want them getting over the back if the horse tries to face you, which it probably will. If the reins remain behind his butt as he turns, a gentle but firm pressure will still have its effect; if not, the pull will merely pull him towards you. For this reason, always keep the reins low when turning.

In any case, pull the reins mildly at first, if at all. It is easy to fix the wrong impression, making him back up. Prevent this at all costs. Horses are also likely to lower their head at excessive pressure, which results in a jerk to the mouth when the forward motion takes up the slack. All this is done in the paddock where the horse is kept, so he's not in frightening surroundings. For a few minutes give him plenty of liberty, so he gets used to walking away from you. It lets you walk behind, too. This first lesson should be fairly short. The horse can be started with a light flick of the whip; some use a tap from the reins. Stop him when he's calmed to the task of walking to the lines (reins) by gentle pressure on the bit. Don't speak until he's ready to learn commands.

After a one or two such driving lessons, teach him to guide left and right. First, left. Slacken the right rein and tug firmly on the left, not jerking, but swinging outward somewhat. If he comes around too far, hold the whip nudging his left shoulder. Don't pull him back with the right rein, nor strike with the whip, because you'll confuse him. If he doesn't respond to the pull on the left rein, tap his right shoulder with the whip to get him to swerve left, so there's no time to fight the bit.

After turning left a few times, teach right turns the same way. Repeat all rein-work learned thus far a few times, striving to use less and less pressure as he learns to obey the bit.

Teaching "Whoa"

This is the first and most important command. It is used only when the horse must stop and stand. It is dangerous, as well as a colossal aggravation when a horse continues on when told "whoa." Likewise, his value is affected if he constantly paces about.

When the horse is fresh, the day after he is taught to drive to reins, drive him about the paddock a while until he's going along smoothly. While moving, give the command "whoa," followed immediately by a firm swinging pull on both reins at once. The instant he is stopped, slacken. Keeping pressure will only confuse, and he'll try to work against it, becoming a *bit lugger*. Caress him gently with your whip for a job well done before starting again.

It's not a bad idea to stop at the same spot two or three more times. He'll probably stop at the command alone by that time. Don't stop there too many times, or he'll think stopping is only for that place, and that place is only for stopping. Stop a few other places. You may need the reins again, but repeat until he'll stop without 'em.

Teaching "Get Up"

The day after he learns *whoa* he can learn *get up*. So far he's been started by a tap on the rump with whip or lines. Go around the paddock a few times, as yesterday, stopping at *whoa*. After one of these stops, pause a moment, utter the command *get up*, and after a slight pause, tap his rump with the whip. He'll start. Once he's moving along smoothly again, stop with *whoa*, and start again as before with *get up*. Repeat until only the command is needed.

I like to see these commands repeated until he'll obey them even when he's loose and you're over the fence from him.

Teaching "Back"

After learning *get up*, teach the horse to back up. With the horse standing still, tighten the left rein, command him *back*, immediately giving a smart swinging tug on the right. He'll step backward, whereupon you repeat the command *back* while sort of see-sawing the pull between the reins. Stop on the command *whoa* the same as going forward. Don't wear him out by continually backing for too long a lesson period. Now you can give lessons with *get up*, *whoa*, and *back*, until his response is near automatic.

Preparing for the Harness

Harness training is more important than saddle training. Horses may pull with harness many times what they could pack on their backs. In most situations a horseman is far more comfortable and safe in a vehicle than riding. A single horse may easily pull four to eight passengers (depending on size and road conditions) where two are burdensome on most horses' backs. Besides, a single harness trained horse may prepare, cultivate, harvest, and transport crops too large for half a dozen strong and wise human laborers to cope with. And he'll do it for a small share of the crop, not wages.

But a driven horse is connected to the driver only through the two reins, sometimes not even that. There is no rider to convey commands through the legs and body inclination. Vocal commands are much more important.

Many horses which behaved well in training up to harnessing may become confused afterward, being disobedient or even uncontrollable. This is the natural consequence of inadequate preparation.

Pressure against a horse's chest, shoulders and/or neck induces backing up. Pressure against his hind end provokes him to start forward. Common horse harness has both a *collar* or *breast-collar* (either of which is pushed by the horse's shoulders and/or chest in moving the load forward) or a *breeching* (pronounced "britchin" by most drivers) which is a strap around the horse's hindquarters used in backing vehicles, or holding back a vehicle when going downhill.

If a harnessed horse is hitched to a load without due training, and started forward with the command *get up,* the unfamiliar pressure on the front induces him to reverse direction, backing right into the breeching. This makes him try forward again. Confusion mounts as he moves forward and back, getting too excited to heed a driver's commands. I have seen so-called "horse trainers" get agitated at this, believing the horse is smart enough to "do it on purpose." So a sharp whip-crack on the behind follows, frightening an already anxious beast into a forward bound. The "trainer" jerks the lines in the horse's mouth. The horse has now been punished front and rear, and his pitiable bewilderment is complete. A very good natured horse will now stand still during the whipping which nearly always follows, taking the undeserved punishment with humbled demeanor. I find it heartbreaking to watch. A more high-strung animal will rear, kick or bolt off, willing to try anything to free himself.

A real horseman will patiently labor to see the horse familiarized with harness pressures and noises before hitching to a load. And after he's confidently pulling things with chains, will introduce him to the even more confining feel of the shafts at his sides when pulling wagon or cart.

Before ever harnessing with anything more than a lightweight bitting harness, "poling" is in order. This is just like the *third hand* mentioned previously, but a 5 to 6 foot long smooth pole is used. Start with the horse in his bitting harness, check reins on, but no side reins. Holding his headstall, let him smell and feel the pole. Now gently rub it over the nose, up the side of the face to the poll of his head. Rub there a bit, then along the mane to the withers, then down along one shoulder. Continue rubbing, outside of foreleg, then the inside of same, back up the shoulder, along the back to the hindquarters, then down the outside of the hind leg, then inside it. Repeat the process on the horse's other side, then back to that side, then the other, now rubbing every part of the body.

If the horse becomes excited, refusing at any time to stand, allow him to touch the pole with his nose again, gradually working back over him again. Once he's perfectly submissive to pressure all over his body, it's time to work on the tail.

Most horses dislike abrupt tail grabbing, and will kick if not treated gently. But they have to learn to receive the harness crupper, which has more pressure than a bitting-harness crupper.

Stand to the left side of the horse's hind quarter. Touch both hind legs with the pole a little above the hocks. Holding the pole in the right hand, use the left hand to carefully raise the tail a little. Gently work the pole upward a few inches; pause, and repeat until the pole rests directly under the tail, lying across his buttocks. After a moment, raise his tail carefully and gently remove the pole, so as not to alarm him. Several repetitions of all this will render cruppering very easy.

Plenty of poling is always desirable, especially with sensitive horses. These are the best horses in harness because they notice everything and are often aware of approaching movement before it has a chance to startle them into a panic. In the old days these horses were admiringly pronounced "nervous."

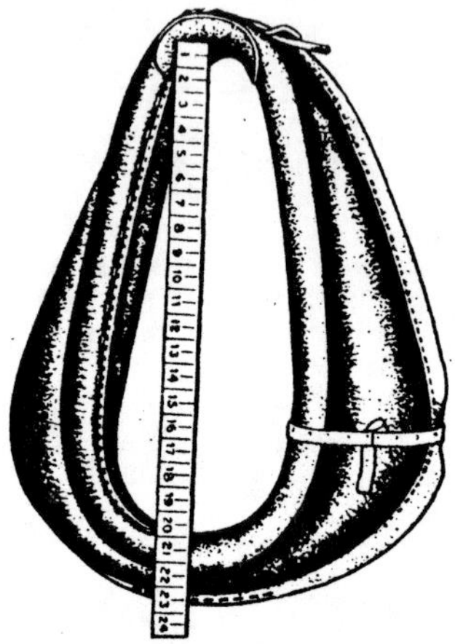

The Collar

The bridle is the first part of the harness, since it connects to the rest of the harness through the check reins. But since it has already been discussed enough, we

proceed to the collar. The collar is a hollow leather ring (sometimes faced with cloth) stuffed tightly with straw padding, which fits around a horse's neck. It isn't part of the harness proper, and is sold separately, because one size definitely doesn't fit all. The collar must be fitted to the horse, else sore shoulders will result. The collar transfers the entire weight of the load to the horse's shoulders.

To put on the collar, first let the horse smell it, then rub it all over him. If it has got a latch at the top, unlatch it and slip it up onto his neck. If not, you must first turn it so the wide end (bottom) is up, slipping it over his head. The way to do this is put the collar on say, your right arm. With the right hand, grasp his halter, now the left hand may pick up the collar and slide it over his head to his neck. When just past the ears, turn the collar right side up and seat it on his neck and shoulders. It may take a couple of tries with a new horse. Then give him a handful of the oats you had all along in your pocket. Take it back off and put it on a few times until it's the oats he's thinking of and no longer the collar. Lead him around a bit while he's wearing the collar.

Hames and Traces

The narrow side of the collar has a bulge called the *rim* or *wales*. Against this part two rigid bars are clamped. These bars are called *hames* and they are attached to the *traces*. The traces reach all the way back behind the horse, bearing the full strain of the pull against the load. There are large metal loops near the top of the hames; the driving reins run through these on their way back to your hands.

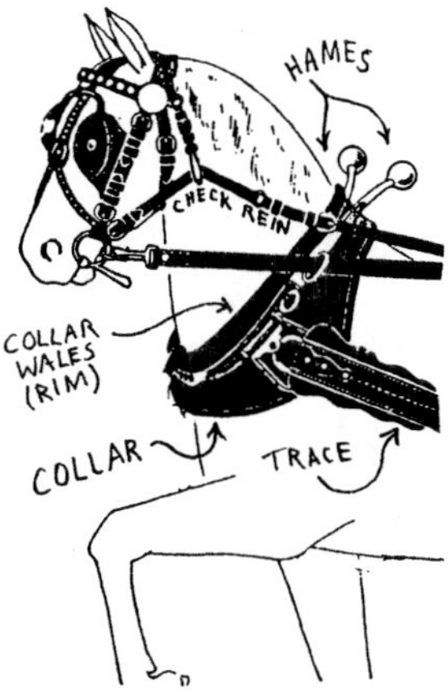

Crupper & Backstraps

Unlike on the bitting harness, the crupper doesn't attach to a surcingle. It attaches to the *hip safe* which is then attached to one or two *back straps*. The other ends of the back straps go to the collar, holding it in place.

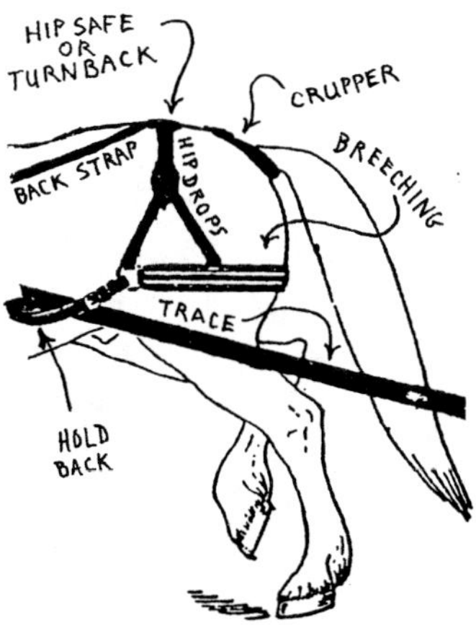

Saddle and Belly Band

On their way from hip safe to hames, the back straps pass through two loops on a wide leather pad across the back called the *saddle*. Unlike a riding saddle, the only real weight this saddle bears is that of traces, and the shafts if the horse is hooked to a vehicle. To keep this saddle from bouncing or rolling about on the horse, a *belly band* attaches just behind the front legs. It isn't really tight like a riding saddle's girth strap.

The check reins attach to a ring or hook on the center of the harness saddle.

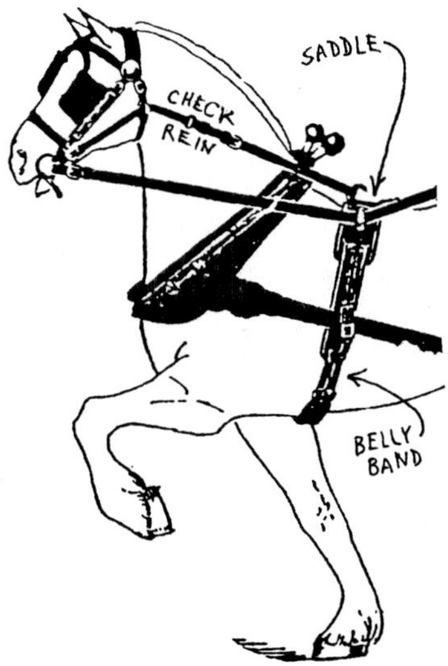

Hip Drops & Breeching

As previously stated, a breeching is a stout leather strap which rests behind the horse's hindquarters. It is carried by *hip drops* from the hip safe.

Plow and logging harness has no breeching at all, just hip drops to the traces to keep them from flopping too much. If there is a breeching, it attaches to the traces at the same place as the hip drops would on a plow harness by means of *lazy straps*. It is important that these not be too tight, otherwise too much pressure will be applied on the hip safe.

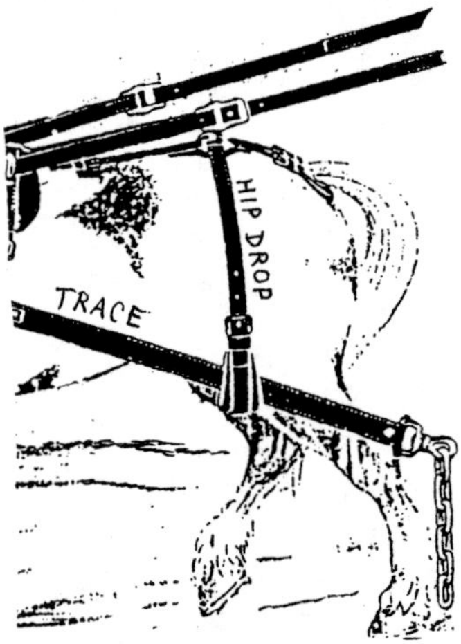

When hitched to a vehicle, the breeching ends tie to the vehicle's shafts by means of *holdbacks*. The vehicle shafts are also held by loops on the saddle sides called *tugs*.

Breast Collar

This is only for light vehicles, where the horse will have to pull less than a total weight of 500 lbs. Instead of a collar and hames the (much lighter weight) traces attach to the *breast collar body*, a wide strap which runs around the horse's chest. A *shoulder strap* runs from near the front of the breast collar body on both sides up over the back just behind the neck to keep the collar body from sagging. Breast harness is cheaper by far than work harness, but is only suitable for light driving, not heavy work. A new work harness often costs as much or even more than the horse. It can be bought used for a fraction of this, but be careful it isn't dry rotted or torn up. Well taken care of, a harness which was bought in used condition can be sold for about what was paid for it, but don't let it wear so thin as to be dangerous.

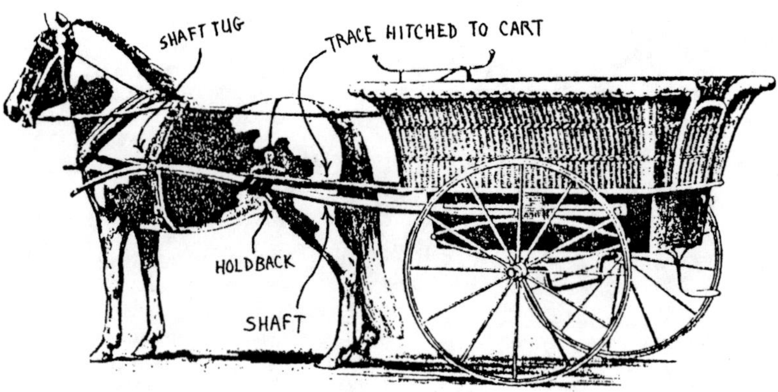

Though costly, lots of work and leather goes into a harness, and it's hard to see how harnessmakers make money.

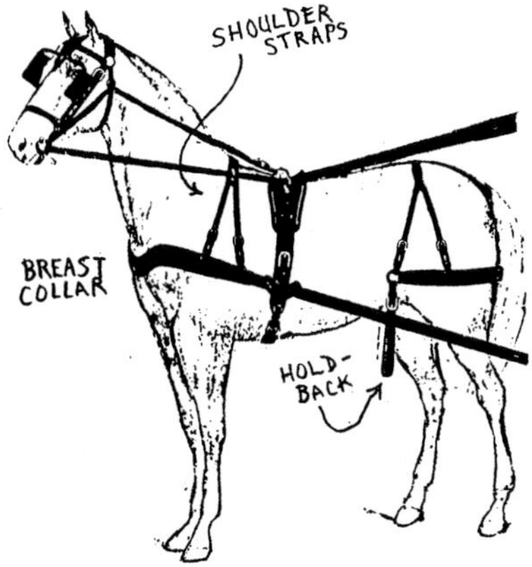

Singletrees

Though the singletree is not part of the harness, it is employed whenever the harness is used. So we might as well get to thinking about it as if it were part of the harness.

The singletree is a stout wooden or metal bar from 2 to 3 feet long, with a forward-facing hook at each end, and a backward facing swivel in the middle. The forward-facing hooks are attached to the ends of the traces; the middle swivel is attached to the load by means of a *split link* or *clevis*. The singletree keeps an even pressure on both the horse's shoulders at the collar. Metal singletrees are more convenient most of the time, but they sometimes unnerve new work horses by their rattling noises. The width of the singletree is dependent on the size of the horse. The traces ought to be roughly parallel with a correctly sized singletree, but as long as it isn't too narrow (confining the horse too much) it doesn't matter that much unless the horse is constantly worked.

Working the Horse

I'm not going to go into driving the horse with carts and wagons in this book. Mostly carts and wagons are recreational nowadays, and alternative transportation is getting a bit beyond mere husbandry.

Harnessing

The first time a horse is harnessed, I like to let him smell the harness, rubbing it with his nose. Put on his collar and bridle first. Take the harness from its hooks. (An assembled harness can be hung up on hooks at the hames, saddle and hip safe. It's best to hang it with the hame end on the left. You can lift it all with one arm by getting your right forearm under the hip safe, lift it off; then under the saddle, lift it off; lastly, under the hames, which you may hold in your right hand. The left arm may be used to brace the right under this rather heavy load.)

Now you are ready to harness from the horse's left or *near* side. Let him sniff the whole thing first. Then rub it on his neck, working down to his shoulder. The main thing is that he hear its creaking and jingling as it moves along, because with the blinders on the draft bridle, he can't really see much back there, even if he turns his head. Being tied with his lead rope, he can really only move away sideways, not front or back.

Trying to keep him from getting too excited, smoothly lift the whole mess gently upon his back and shoulders, separating the hames and easing them down on either side of his neck. Get the belly band buckles on their respective sides. Last, get the hip drops onto their correct sides. Now remove the whole thing and repeat.

This time, line up the hames on the collar, fastening them comfortably snug between rim and collar body. Carefully and quietly reach under his belly and get hold of the belly band, which is hanging on the *off* (right) side. Bring it right under, and buckle it, not too tight.

Ease back to his hindquarter, caressing the horse gently, and talking calmly to him. Pull, don't jerk, the hip safe firmly back, seating the hames and collar against his shoulder. Now attach the crupper.

Most harness has some way to hang the rear ends of the traces to the ring on the hip safe. This way, they won't drag on the ground, catching on roots and rocks, and scaring the daylights out of the horse. But they'll still rattle, and slap the horse's hindquarters somewhat. Old harness is best, if in good repair, because it doesn't smell so strongly of leather and because it won't rattle and squeak so much. Decrepit harness should absolutely not be used with a new workhorse.

Only horses which are quite used to harness will tolerate having the whole thing simply flung onto their backs in a careless manner. So don't do it. Even a horse which will tolerate this will also tolerate putting it on gently.

An unworked horse's flesh is very tender, so inspect for abrasions. The collar sometimes rubs abrasions on the shoulders. The saddle abrades the back; the crupper, under the tail. If abrasions begin to appear, training must temporarily cease. Wash the sores with soap and water, then apply mild salt water to harden and heal. As soon as the soreness has faded, resume training.

Driving in Harness

Begin in his pen, as before. If your harness has a breeching, tie it to the traces in such a way as to apply very slight pressure on the breeching and the collar. Start him at the command *get up;* stop at *whoa;* back saying *back*. Turn him right and left using the reins. When he appears confident and smooth, he may be taken about a larger field. At first, because of the openness of his surroundings, he may act strangely, but go through all the commands he knows. Now he's ready to begin pulling.

Hitching to a Load

Don't hook him right up to a stone boat or plow. These require a very large effort, and will certainly discourage if not terrify him. Instead, just pick smooth ground, like grass or cultivated fallow, and let the traces drag. After that seems comfortable to the horse, attach the singletree. When he's used to that, you may attach a hand line to the singletree, and have an assistant pull very lightly backwards on it while you go through the commands. If no assistant is available, tie a smooth wooden fence post to the singletree, and have him drag that. Don't take him where the wood will start bouncing up and down on rocks, or on a steep grade where it will ride up on his heels. Keep everything easy and smooth.

A few days with the fence post will convince him pulling is fun. You can then tie two fence posts, and as soon as he's comfortable with it, add another. Then an old car hood with say, 200-300 lbs. of dirt, rocks, etc. Don't have him go over rough ground until he's confident dragging a heavy load, say 500-600 lbs. Also, when the loads start getting really heavy, loosen his check reins, because it is very hard for him to pull with his head up too high.

You may hitch to a one-horse disk cultivator. This will make lots of noise, and when adjusted for maximum penetration (the disks are set and the sharpest angle) it will make less noise and won't ride up close to him if he should jerk. Once he's used to this, lessen the angle until he'll draw it without jerking, even when the disks are straight out (least drag).

If you have no disk, various types of harrows may be substituted, varying the penetration similarly. If you have no harrows, try having him pull bigger logs until he's pulling about his own weight or more.

The more you work him, the better he will get at it, until he'll know exactly what to do in every field with every implement, and your work becomes very easy indeed.

Plowing

Everyone seems to want his horse to plow. Plowing is very hard work. A very large horse might pull up to an 8 inch plow in ground with no grass or roots. Most horses are doing what they can with 6 inches. A plow does its best work when it is running 70% as deep as it is wide. Plowing is such a difficult thing for both horse and driver to get right

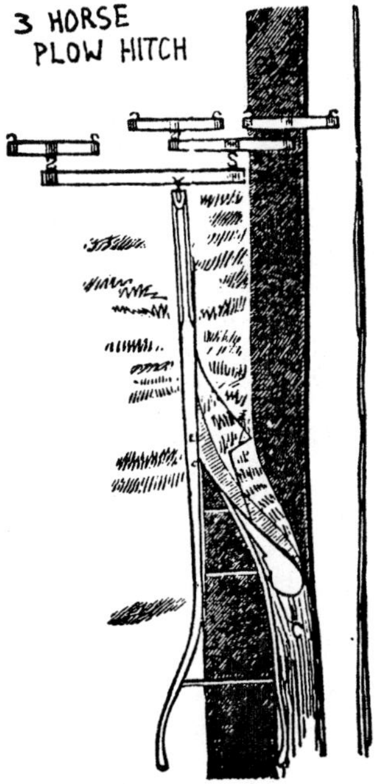

that I must say you should attend a plowing contest, or seek advice from old hands at it before you attempt it. Even then, the horse

ought to be very used to working first. The hitch must be just so, and the way of plows isn't easily taught in books. In any case it would take a much bigger book than this one to do any justice at all to it.

But plowing is the least useful of the horse's services. Most ground needs plowing but very rarely. I am tempted to say plowing is never needed, but then I'd be flying in the face of authorities far more established than myself.

It's cheaper to just hire a field plowed once, and then you'll never need to plow it again with anything more than a *bull-tongue plow,* which is hardly a plow at all. It is just a large, pointed cultivator shovel on the bottom of an ordinary but light plow beam. I never even use one of these! A set of spring-tine cultivators, with good sharp points will cultivate the must defiant ground, even stony ground,

down to a seed bed fine enough for any crop except the small seeded ones like carrots. Even those require only a small bit more tillage using a spike-tooth harrow. In good well-drained soil, a horse can be out working the day after the rain stops. A tractor must wait until it's quite dry or you'll risk packing, puddling and in general ruining the seedbed until next year.

Walk Fast

Train your horse to walk fast. A pokey horse will waste your time when you want to finish up a bit of cultivating. The difference is at least 50% of time saved, and less work for the horse, really. A pokey horse will break into a trot just to go as fast as a quick horse can walk. At the trot he's using half again as much energy, and tires too fast. The trot is for road transport. Faster than a trot is only useful for riding horses, and won't be discussed here.

Breaking the Runaway

If you have a horse that just will not cease breaking into a run, you must employ a *foot rope* for a while to break him of the vice. Attach a foot strap with a strong ring in it to his fetlock on either the right or left side, it doesn't matter. Tie a rope to the center ring of his belly band. Run the rope down through the ring in the foot strap, and back up through one of the rings in that side of the saddle. You hold onto this rope like a "third rein." Give him enough slack to use his legs as normal. When he begins to bolt, pull back as hard as you can on the foot rope. He's very unlikely to fall, but he certainly will stop. When he's back to manageable, give him his foot and try again. It only takes one or two lessons to make him quit bolting. Incidentally, a horse with one front foot raised cannot lift his other feet to kick you; hooking the front foot strap to the surcingle so his foot stays raised enables you to handle the hind legs of even confirmed kickers. But to lift a hind foot of such a horse requires the use of a twitch, which any farrier can explain to you; the foot strap and surcingle are only for handling, not working. This *"single safety"* was introduced by the great John S.

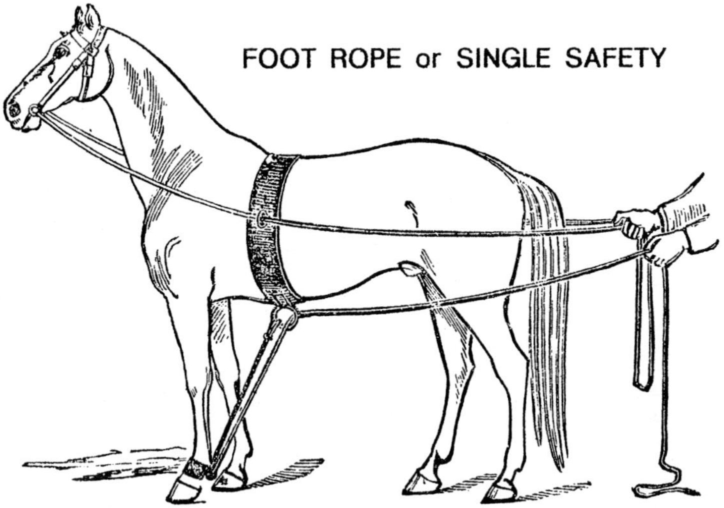

Rarey (1828-1866), the most famous horse trainer in modern history. Horses he trained were said to be "Rareyfied." The "running W" is a very harsh punishment, and consists of a foot rope on both front feet. It brings the horse crashing to his knees when pulled. It is very humiliating to the horse, and is so severe that if not used just exactly right, has been known to cripple and ruin good horses. I've never used one.

Using Reins

Though one never sees this in photos or instruction books, the reins are really only correctly handled in the left hand. One rein passes between the index and middle finger, and the other passes between thumb and index finger. One hand thus grips them both and keeps them separate. The right rein is given more slack than the left, and is passed over the left handle of the implement. The right hand holds the right implement handle. Thus, when the left hand is pulled straight back, the left side of the bit alone receives pressure. When the left hand is swung leftward, the right side of the bit is pulled. When the left hand is swung diagonally back, both sides of the bit receive pressure. With but very little practice, all drivers can master this.

It is disappointing to see an expert plowman, who can guide horses to plow arrow-straight furrows, with the reins tied together and wrapped behind his back. The horse's work is already very hard, and pulling the plowman with his mouth is even more fatiguing than the traces. Just because the horses will do it doesn't mean their mouths aren't being toughened so that only such methods will control them. Many become "luggers." These are a misery to work, because you work against them and are as tired as they are at the end of the day.

Grooming

The horse, after he's done his work, needs to be curried, at least a little. If stall kept, where they have no exercise paddock, you must curry them with curry comb until their sweat dries. The best curry combs now made, in my opinion, are the ones with long conical fingers of rubber.

Rub in a circular motion, removing hair from the curry comb as it accumulates. The place the horse needs rubbing the most is on the legs and shoulders. Lazy or incompetent grooms only comb the body. If you must neglect any part, the body would be it. The circulation is greatly stimulated by currying the legs, and the next day you'll see the benefits of spending your time there. Steel "curry" combs are not really for currying. They are mostly for cleaning brushes and touch-up work.

Teaching *Gee, Haw, Yea* and *Steady*

Gee means turn right. *Haw* means turn left. *Yea* means you've turned far enough, now go straight. These commands have practically no use for a single horse. They are used with large teams where many horses must follow a leader, and there aren't enough reins for them all. Since large teams represent farming as a business or just show-ring fanciers, I won't waste time on it here.

"Steady" is a command nearly always given in a low, smooth tone of voice. It is used for warning the horse of an upcoming obstacle or situation which may startle him. If you're approaching a visible rock or root with your spring-tine harrow, and you know there's going to be increased resistance, say "steady" to your horse. If you're dragging a sled or some logs and you're coming to a rough, rocky spot where there's going to be some bouncing, say "steady." Once he learns this one, and it takes a good while of working for him to really catch on, it can be used to calm him in almost any frightening situation.

The six standard commands, "whoa, get up, back, gee, haw," and "steady" all sound unique enough for horses to easily differentiate between them. Don't invent whimsical commands for a work horse.

Picketing the Horse

Horsemen refer to tethering horses as *picketing.* The spike to which the tether rope is attached is called the *picket pin.* I like these to be made of at least half-inch diameter iron rod, about 2 feet long. The head should be smooth on the underside, and at least 1½ inches in diameter, so a non-slipping loop in the picket rope will easily pivot. A tapering point makes driving it into the ground very easy. And it should be driven all the way down, with the picket rope just loose enough to spin round it.

The picket rope is snapped to the ring at the bottom of the halter. I like it about 10 to 12 feet long. It is best made of braided, soft, smooth rope, at least ⅝ inch diameter. This is because the first two days a horse is picketed, he will get rope burned, mainly behind his hind pasterns. So check on him frequently, and untangle him if necessary. Treat the rope burns with pine tar. Just daub it on using a long stick with wool wrapped around the tip. He won't like this, but it keeps the flies off and helps heal it quickly. If the rope burns are pretty bad, don't picket him for a couple of days. Next time, watch again, and treat if he rope burns again. Two or three lessons picketing, and horses learn not to tangle in the rope. Move them periodically, and don't leave 'em in the hot sun with the flies. A horse which has learned to picket is very easily led with any sort of lead rope.

Examine the ground before picketing to avoid nails, glass, groundhog holes or hornet's nests.

Remember This
To Tell the Age of Horses

To tell the age of any horse,
Inspect the lower jaw, or course.
The sixth front tooth the tale will tell,
And every doubt and fear dispel.

Two middle "nippers" you behold
Before the colt is 2 weeks old.
Before 8 weeks 2 more will come;
Eight months, the "corners" cut the gum.

Two outside grooves will disappear
From middle 2 in just one year.
In 2 years from the second pair;
In 3 the corners, too are bare.

At 2, the middle "nippers" drop;
At 3, the second pair can't stop.
When 4 years old, the third pair goes.
At 5 a full new set he shows.

The deep black spots will pass from view.
At 6 years, from the middle two.
The second pair at 7 years;
At 8, the spot each "corner" clears.
From the middle "nippers," upper jaw,
At 9, the black spots will withdraw.
The second pair at 10 are white;
Eleven finds the "corners" light.

As time goes on, the horsemen know,
The oval teeth, 3-sided grow;
The longer get, project before,
Till 20, which we know no more.

Prof. Oscar R. Gleason
(Circa 1900)

17
OXEN

Where no oxen are, *the crib* is *clean: but much increase* is *by the strength of the ox. Proverbs 14:4*

I have never trained oxen myself, and have only participated in it as an assistant to the trainer. There are two methods of training them. The first is slow, but easy even for children, the second is quick, and takes somewhat more skill.

The first method is to start with a calf. The Holstein or its crosses are the most popular in my part of West Virginia, because they're so big at maturity and thus exert tremendous traction. With a very young calf a small yoke is fashioned of basswood. Basswood has ever been the proper wood for ox yokes, because it is so very light, so easily wrought, and as very old man once told me "basswood's tuuuuff!" They used to use it to make 15 foot handles for hay forks for building tall hay-stacks. The common ash or hickory handle would be far too heavy if that long.

The midget single yoke for a newborn calf is made of 3 inch square stuff about 14 to 16 inches

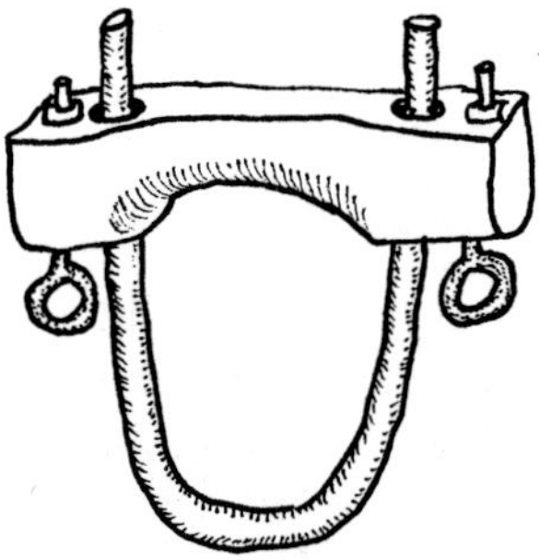

long. It has a hickory bow of perhaps 5 to 6 inches' width fitted into holes just as far apart, long enough to slide to where it will fit the calf snugly when slid towards the shoulder. Hickory is always used for the bows, because it can be bent very well without breaking. Just heat the bowed part with boiling water until it becomes more flexible, and bend it around a log or tree stump of the desired size. Clamp the yoke onto its ends and let it cool until it sets. The two draft lines attach at the outside ends of the yoke. Where the yoke crosses the neck, a small curve is made, and well smoothed and rounded, so it won't be too uncomfortable.

A little halter is fitted to the calf, and a lead rope is attached. A straight buggy whip is held in one hand, the lead rope in the other. The calf is taught to lead in much the same way as the horse, except the whip may be used instead of the quarters hitch or butt rope. When the calf can be led well, he may be fitted with his single yoke, with an assistant to draw on the pulling ropes (traces) just enough to keep the yoke from falling forward when the calf's head is lowered.

If training a team of small calves, the first is taught to lead, then the other. When both can be led well without resistance, they

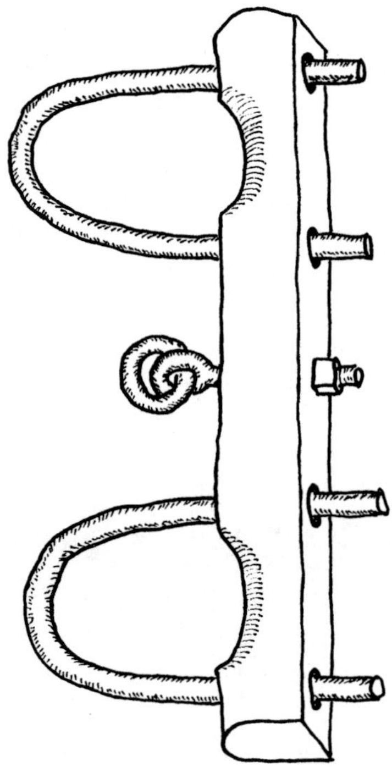

can have a double yoke fitted. Its dimensions are the same as the single yoke, but it's longer and has two hickory bows about 12 to 14 inches away from one another. Such a yoke can be fashioned in an afternoon. It has only one draft line, which is attached in the middle to an iron ring bolted all the way through the yoke from top to bottom, for strength. Ox yokes must be large enough for the animals, so several will be needed by the time the animals are big enough to pull anything substantial. The bigger the animals, the heavier ought to be the yoke. I have a yoke for mature oxen which is made of a basswood "six-by-eight." It is about 4 feet long. Its iron ring is hand forged of three-quarter inch diameter round rod.

Commonly two drivers train an ox team. One handles the trace line(s) or chain, and the other issues commands and enforces them using the whip. In the past, sharpened sticks, called *goads* served the same purpose, especially on reluctant oxen. Mostly only castrated animals are used for teams, but this wasn't always so. Old veterinary manuals recommended not castrating bulls until they were at least six months old, so they would have more muscle development. One even says wait until 3 to 4 years old! Either way it'd be a difficult job for average folks to undertake without help, so most ox drivers castrate as soon as they can.

Oxen need to learn the same commands as horses, viz., *whoa, get up, back, gee haw, yea,* and *steady*. In fact, they need it even more, because they don't have a bit in their mouths. Luckily, cattle are far less prone to run away or startle at trifles. And they work more slowly, about two-thirds the speed of horses.

When training from very early age, the trainer may guide the calf to obey through simple maneuvering tugs and shoves in the right direction until the calf learns it. I know a number of old men who taught calves this way before they themselves were ten years old! The calves grow quickly, and if trained regularly (best every day) they'll never know they've gotten too big for you to boss around. They make the best of all draft animals when thus treated.

A further advantage to oxen is they don't forget or get fuzzy on what is expected. Ox teams not worked for a year will be obedient as if their last work day was yesterday. Not so with horses, which, though they don't forget too badly, get rusty at it.

As to training older steers, I will offer the following quote from *Williams' New System of Handling & Educating the Horse, 1878.*

TRAINING STEERS AS TAUGHT BY A. H. ROCKWELL

How to Teach Steers To Obey the Whip

First get your steer into a room, or small yard, so that he cannot run from you, then approach him gently and slowly, and if he runs do not be in a hurry; wait till he gets to the end of the enclosure, then approach him slowly, as before. A steer will often run from you in this way 6 or 8 times. Do not try to stop him with your whip, or force him to think he will be at all injured, until he will stand and suffer you to approach him. As soon as this is accomplished, gently tie a rope around his body, near the shoulders, rather loosely. Then take another rope, or strap, and gently fasten one end to the near forward foot, then pass the other end over the rope, or surcingle beneath the body. This rope should be sufficiently long to allow him to run to the end of the enclosure without your moving, at the same time you holding the rope firm to compel him to move on 3 legs. Then approach him again, quietly; he will not run off in this way but a few times before he fully comprehends that he will not be injured, when he will suffer you to approach and handle him just as you please. Now take a short hold of the strap with your left hand, your whip in your right, which pass over his shoulders, and quietly touch him upon the off side of his head, at the same time saying "haw," and continue this until he moves his head a little towards you. Then stop and caress him upon and about his neck and head. Repeat this until he haws around towards you. If he attempts to run from you, pull upon the strap, saying "whoa" and at the same time hitting him lightly upon the head with the whip. As soon as he learns in this way to stop at the word of command, and comes toward you readily, take off the strap and surcingle, and turn him out. Then take the mate, and give him the same course of training until you can accomplish with him a like result. Then turn him out. By this time the first steer is cool, and rested, ready to receive another lesson. Now drive him into the inclosure, and repeat his lessons with the whip. Then quietly touch him gently upon the near side of the head, at the same time saying "gee," until he will step around from you, then caress and repeat until he will gee or haw readily. Then repeat the same lesson with his mate, which is all that is necessary to be done the first half day. Then take both together in the same room, or yard, and repeat these lessons until they have a thorough understanding of what you have done. Then take one of them near the wall, standing by his side hitting with your whip gently over the head, at the same time saying "back," until he will step back, for which caress. Repeat this until he will go back at the word readily. Then give the other steer the same lesson. You will find this course of training will make your steers quite obedient, and willing to haw, gee, or back, which will be all that you should require in one day's time. The next morning, or when you have time, repeat this lesson with both together. Then place the yoke upon them, and let them go for an hour or two, or sufficiently long to become fully reconciled to the restraint of the yoke. Then repeat your lessons in open yard, until they fully understand what you require of them yoked together.

This course of training does not excite, while it conveys the idea most clearly of what is required, and will enable the trainer to accomplish more in a given time than has yet been done by any other system of managing steers. If your steers have learned to run away from you, which is a common result of the ordinary method of training, put on the rope and strap, to the foot. If hitched to a wagon, or stone load, let your man hold the foot strap, which runs back between the steers, and the moment they attempt to run, let him pull up the feet, while you whip over the head, which will stop them immediately, and will very soon break up the habit.

The training paddock ought to be not much bigger than 20 to 25 feet square.

When working oxen, stop them at the end of long furrows, say 200 yards or more, and let them have a neck rub where the yoke pressed. This way you'll have their love and full effort. Oxen may be shod like horses if you take them on roads a lot. Like horses, hard ground wears their hooves too fast for health. An ox or horse used only in soft farm ground and woodland rarely needs any shoes at all.

18
DAIRY

Surely the churning of milk bringeth forth butter.
Proverbs 30:33

Milk is the perfect food for man, most beasts, and germs. Carefully handled to exclude the latter, it can be kept for many months safely in the form of cheese, or for weeks in the form of butter, yogurt, etc. This means not using anything for holding the milk except clean, smooth vessels. Aluminum vessels tend to pit, and also discolor cheese because of the lactic acid's reaction to it. Enamel is good, but be sure it has no cracks. The best is stainless steel, which is expensive but cheaper by far in the long run. When washing, start with warm water & soap or detergent. Rinse first with hot, then cold water. If you start with hot water, milk particles solidify on the inside of the bucket. Scraping this stuff off is then necessary and very troublesome. Don't let milk vessels sit around wet. Drain and dry them immediately after the cold rinse. Don't dry with a towel.

If despite your careful sanitation, the milk still tastes amiss, check for mastitis in the cow. If not that, observe what she's eating; nothing makes milk taste chronically sour like sweet vernal grass.

Milk is in its healthiest state fresh from the cow. Even cows with tuberculosis have never been shown to pass this through their milk. Those who insist the house-cow owner must pasteurize to be safe are oblivious to reality. Pasteurized milk is a dead product, homogenized to hide the blood and mastitis clots, which is why store-bought milk putrefies instead of ripening into cheese. We firmly believe the health risks of pasteurizing milk are just too great to justify doing it, so we don't.

Cow's milk is eminently useful because its cream rises to the top where it can be skimmed off to make butter or ice cream. The best way to *set* the milk (raise the cream) is to put it straight into tall jars or cans (or just leave it in the milking bucket) and cool by the fastest possible means. If you followed my advice way back in the chapter on water and installed a gravity-fed spring water system, you have the cheapest and best cooling system already available. The old *springhouse* had troughs to hold the *shotgun cans*. These were tall and narrow to encourage rapid cooling, and amazingly efficient. The cream rises for some reason more completely in these than the even older-fashioned shallow setting pans. Anyway, cream separators are for goat keepers and those who enjoy disassembling, washing and scalding, drying and re-assembling lots of parts twice a day. An added advantage of very rapid cooling is the milk keeps far longer; the shock of rapid temperature change always affects enzyme-producing germs in milk.

Butter

Before making butter, it is essential with most cows' cream to ripen it first. All this means is to let it sit at a moderate temperature until it expands a little in bulk, or has a very slightly sour taste. It'll still make butter if more sour, and many enjoy this (it is true sour cream, like they serve in Europe) but the light ripening is preferred by most.

At some time before churning, cream really needs to be cooled to below 50°F. But the best churning temperatures are between 55° and 66°F. I think 60-63°F is best.

The simplest way to make butter is to half fill a jar with cream and shake it. No need to get too vigorous, just keep the cream sloshing. After a minute or so, loosen the lid enough to let out the gas, and re-tighten. Continue shaking it until the butter "comes." Little blobs of butter the size of course sand or bird shot appear first. These soon grow to the size of BBs. Somewhere between here and the size of a kernel of wheat, stop shaking and drain off the butter milk; save it for baking, cheesemaking, pouring on the hog's back to kill lice, or just drinking. Now, wash the butter by putting some cold water in the jar with the butter kernels, and shake it a little. Pour it off. Wash the butter a time or two more until the water comes off clear. Now haul it out of the jar, and mash it all together, getting as much air and water out of it as possible by just working it in a large bowl of water with your fingers until you're sure it's all air-free. Then make it into blocks, balls or forms with a *butter mold*. If the latter, warm the mold first with warm water. You can press it with a wet butter mold; water and butter don't really mix once the air-pockets are all removed.

Many churns are on the market, both manual and electric, and most of 'em work fairly well. My favorite isn't sold any more, but it is essentially just a water-tight can or jar suspended from a frame or just by a couple of ropes attached to the ceiling. All you do is push and pull it to slosh the cream.

Occasionally, when churning fresh, unripened cream, the butter "goes to sleep." This means it gets almost to the point of "breaking" and just won't break. If after say, 10 minutes the cool sweet cream doesn't break, "shock" it with ½-1 cup of very warm water and start churning again. It should break almost instantly. Now handle as before described.

Most butter makers salt their product with about a tablespoon of salt per pound of butter. Just work it in while working the air out. We don't salt it at all. Butter keeps even in summer if held in water so it's not exposed to air. Change the water daily. Butter keeps indefinitely frozen, even without salt. Rancidity is the product of air, heat and sunshine.

Speaking of which, butter is almost pure sunshine. Selling it (and there is a great demand at all times for country butter) removes practically no fertility from the soil. Milk is another matter entirely, being so chock-full of vitamins and minerals.

Germs, Good & Bad

If you don't cook (pasteurize) your milk, there is no need to buy all those laboratory products to make truly sublime cheese. By now all cheese making recipes have been affected by the pasteurized-milk propagandists, to the hurt of both health and purse. They all say to buy *starters,* which are actually the very things you killed by pasteurizing! We have made outrageously good cheese consistently without any bought products, except a bit of salt, and yes, rennet.

Contained in untreated milk are two types of ferment-producing bacteria. One type can stand temperatures around 85°F and one survives temperatures around 120°F. There being other germs in the milk (it can't be helped) one must heat the milk to 85° or 120° respectively, before making cheese, to give the advantage to the desired fermenting germs. These will now be vastly in the majority, and they rapidly crowd and suffocate undesirable germs remaining in the milk. If you aren't careful in keeping the milk reasonably clean before you begin, your cheese will become foul and gassy. It will be pig fodder. But take heart, if you start with clean utensils, and wash and dry the cow's udder, and keep dung and cats from the bucket, you'll almost certainly prevail.

Cheese is basically milk with the water removed, and fermented and aged to bring out desirable flavors. To remove water, *rennet* is added to the milk, which causes the solids to adhere to each other, expelling the liquids, which are called *whey.* The solids are called *curds.* Milk may be curdled with a number of substances. A quarter cup of vinegar will curdle a gallon of milk. So will a sixth of a rennet tablet dissolved in ¼ cup of tepid water. So will the juice of a plant called "Lady's Bedstraw" *(Galium verum).* There are many ways to curdle milk, you see, including just letting it sit until enough milk sugar *(lactose)* is converted by fermentation into *lactic acid.*

Rennet has been the standard curdling agent since before recorded history. It is made of the stomach of suckling calves, lambs, piglets, goat kids, etc. The animal, at several weeks old, is slaughtered for veal, roast suckling pig, etc., shortly after having been allowed to suckle. The milk-swollen stomach is removed, cleaned without water, and dried as follows. The opening is salted and tied shut, after inflating it like a little balloon. Let it dry, hanging like this. These dried stomachs, or "vells" keep indefinitely. The dried stomach is placed in a jar with 2 handfuls of salt. Pour three quarts of cold water over this. Let it stand for 5 days and it's ready for use. An ounce will curdle up to 6 gallons. We never made rennet this way because it was never convenient.

Making Cheese

The way Eva makes her *Edam* cheese (which I find mild and very delicious) is as follows:

1. Let the evening milk stand overnight unrefrigerated. Skim.
2. Add the morning milk, mix. Now we have about 4 hours in our enamel canning pot.
3. Warm to 85°F and curdle with enough rennet to coagulate in 40 minutes. If you've made cheese before, it ought to be a little softer than the curd on other cheeses. If you haven't made cheese before, try poking your very clean finger into the curd at an angle. Lift it up through the curd, rupturing what curd is in the way. A clean break is what is used for most cheeses, but this one will be a little softer.
4. Cut the curd like this: With a long thin knife, cut vertically into the curd in parallel slices 1 inch apart. Then turn the pot so the slices are crossing your path, and cut again in parallel 1 inch vertical slices, so it looks cross-hatched. Now you must slip the knife in diagonally for the next set of parallel cuts. Then diagonally from the opposite side, so all four sets of parallel cuts form final chunk sizes of about an inch. Precision isn't important. There are special tools for this somewhere out there, but we don't own any. The first cuts are gently made. As you proceed, the stuff shrinks, and you can get more vigorous, finally breaking the curd into finer and finer pieces with your very clean hand.
5. As the whey continues to separate, a portion is drawn off and heated on a stove. Meanwhile, the curd and remaining whey is kept in constant agitation.
6. The whey on the stove is heated until when added back to the whole mass it will bring the whole contents of the tub back up to 85°F. This heating of a portion and returning it to the main vessel may have to be repeated several times. This is the best way, because sudden temperature changes shock the good germs, perhaps harming the outcome. Three minutes per degree raised is about right.
7. This temperature (85°F)--and the constant agitation--are kept up for a full hour, at which time the particles of curd ought to be about the size of kernels of wheat, firm and hard, but still entirely sweet.
8. Let the curd settle on the bottom of the tub.
9. When it has matted enough to hang together, fold it in half. That is, double it up at one side of the tub, and elevate that side a bit.

10. Dip off all the whey, then put a weighted (clean) plank on the curd to mash out and drain the remaining whey.
11. When you've removed all the whey that will run off readily, break the curd with the hands, and pack into a mold firmly with hand pressure until it will hold its shape. In northern Holland special wooden molds are used to make cheeses as round as cannonballs.

We get by with plastic cylinders cut from old vinegar jugs about 5 inches in diameter. Wooden disks are cut to fit each end. Our cheeses aren't round like they should be, but cylindrical, maybe 5 inches tall. With 4 gallons of milk to start, we now have about 4 lbs. of cheese.
12. When the curd-ball will hold its shape, it is removed, wrapped in clean cotton cloth, and replaced in the mold. There it remains for 3 hours, clamped with a C-clamp and laid on its side to drain.
13. Remove the bandages, cover the cheese with salt, turning & rubbing salt in occasionally for 3-4 days. Set the cheese on one of the wooden disks from the improvised cheese press, to help with drainage. Keep salted as long as any whey continues being drawn out.
14. When the whey is out, remove the surface salt, leave on the wooden disk, and cure for 3 to 4 weeks. Don't let it stay where the temperature will rise above 65°F. In a refrigerator is fine. Turn it daily.
15. When ready, in 3 to 4 weeks, the cheese will be hard and firm, and covered with white moldy stuff. Wash the cheese now in whey, scrubbing off all the mold, dry and grease with vegetable oil. It can now be wrapped in wax paper and kept in a cool, humid mouse-proof place for 6 to 8 months to mature. At 3 or 4 weeks it is tough and stretchy with practically no flavor. At 8 months, it's perfect.

The best way to make these cheeses is to make one at least once a week during the milk glut. If you vary the recipe in any way, always keep a note with it so if at maturity it is superior, you know what you did differently, and can improve your recipe. Cheese that won't keep isn't worth discussing here, because your main reason for cheese is to have dairy products in that inevitable dry period just before the next calving. Believe me when you've had all the milk you wanted for 10 months, it is hard behaving normal without it.

Yogurt

This is easy. Just get some yogurt at the store, a plain one with no additives, add a spoonful to any amount of fresh warm milk, cover, and keep at body temperature for several hours. Once it has set firm, keep it cool, or just eat it. Mom used to make hers in a gas clothes drier--its pilot burner kept it warm--but make sure to put a big sign on the drier, or you know what's going to happen.

Ice Cream

There's nothing like having practically unlimited supplies of this--especially since it will be better than the best gourmet stuff if you do it right. I have made it in a deep freeze, but a hand-crank freezer is better.

When you skim the cream (if you have a Jersey or Guernsey cow) you'll notice some of the cream is extra thick. This cannot be bought in the store. Store-bought "heavy" cream is as thick as the thin stuff from a good cow. The thick stuff is called *double cream.* It's really too thick for ice cream unless you make fruit flavored stuff. Then the water in the fruit will dilute it. John Seymour gave a really successful recipe in his excellent book *The Guide to Self-Sufficiency* (London, Dorling Kindersley, ©1976) which went like this:

1 lb. strawberries
½ cup sugar
½ cup water
2½ cups single cream or 1¼ cups double cream

Make syrup out of the sugar and water. Mash the strawberries and (he says) strain the pips out (we don't). Pour the cooled syrup into the strawberry mash. Add the single cream as is, or double cream beaten first. Freeze. If you freeze it in the deep freeze, remove it every 10 to 15 minutes and stir it. Stir it every other minute when it just starts to freeze, or you'll get too many ice crystals forming in it. In practice, I find if you beat the double cream, you wind up getting butterfat lingering on your tongue.

My favorite ice cream recipe uses lots of eggs. I like about a dozen beaten eggs added per gallon of ice cream. Don't use dirty eggs (unless they're really fresh) or store-bought eggs. These really ought to be cooked to be safe. Here's the recipe:

6 egg whites
6 egg yolks
¼ teaspoon salt
1 quart single cream
1 quart whole milk
3 ounces unsweetened chocolate

Mix 1 cup sugar, egg yolks, and salt until well blended. Scald milk to 160-180°F, stirring to avoid sticking. Add chocolate, stirring until dissolved. While cooling, stir into egg yolk/salt/sugar mixture. Set aside. Beat egg whites until thick, add ½ cup sugar; beat until mix forms peaks. Fold egg whites into cooled chocolate mixture, add cream, chill. When chilled freeze in hand crank freezer.

This is supposed to keep for 6 months in the freezer. Mine never keeps more than a hour or two, especially when children are around.

19
POULTRY

Also fowls were prepared for me.
Nehemiah 5:18

I tried one method of slaughtering chickens after another, all gotten from books and pamphlets on the subject. Each was either revoltingly cruel, or appeared so, or too messy, or too much work to be worth doing.

I, like most, find killing animals very unpleasant. I wanted to rather make the whole process easy and painless for myself and beast.

Yes, I have thought through the argument that we ought not do it at all, which sounds righteous enough at first. But we must note that all vegetarian animals compete with vegetarian human beings for food, or would if their procreation were not interfered with. But it means that either we must (1) Never interfere with disease, mass starvation due to overpopulation, or pestilence and predation, or (2) We must lock up all herbivore animals in controlled conditions so they don't unmanageably procreate. Then, if we did that, we'd have to lock up all the predators for the same reason. The whole world will have then become civilized, from a kindness-to-animals point of view. I think such a scenario is ugly, and would rather see my canine teeth for what they are: God's gift to help me eat some of my favorite foods, which include meat.

But it goes further than that. As barbaric as meat eaters are depicted by aggressive vegetarians, are they as violent as the vegetarians? I think not. Vegetarians have beaten up ladies in fur or leather coats. They have firebombed law-abiding shop keepers. They have assassinated physicians and scientists. They often make statements equating human life with animal life; some even put animals first! Adolf Hitler, who gave the orders to murder some 6 million Jewish people, was himself a strict vegetarian, sparing animals the pain and indignity of laboratory research by substituting men, women, and yes, children. May God help me to never excuse violence and coercion of my fellow man in the name of animal "rights!"

I know one adult vegetarian who simply doesn't like the taste of meat. I applaud her for not letting meat eating zealots (if there are any) tell her what she must eat. She cooks very tasty meat dishes for her husband though.

But my views on morality don't make chicken butchering any easier. Let me briefly explain the methods I rejected, and why, to spare you the same unpleasantnesses.

I didn't like killing chickens by dislocating the neck; too much blood accumulates in the neck. Nor simply chopping off the head; a flapping, running, blood spewing headless chicken is revolting to see and clean up. Same goes for wringing the neck.

As for plucking, most authorities argue for dry-plucking the feathers. In order to be successful, this must commence while the bird is still flapping and must be nearly complete by the time it stops. This is not only fiddly, but very un-esthetic. It once took me 4 hours to dry pluck a goose. Never again.

How We Kill a Chicken

I slay fowl like this: First I tie a bit of hay-string around the bird's neck, just barely tight enough not to slide off. Then I insert the whole thing into a "killing cone." These are available at all complete poultry supply dealers, but I made my own of sheet metal

Butchering Cone

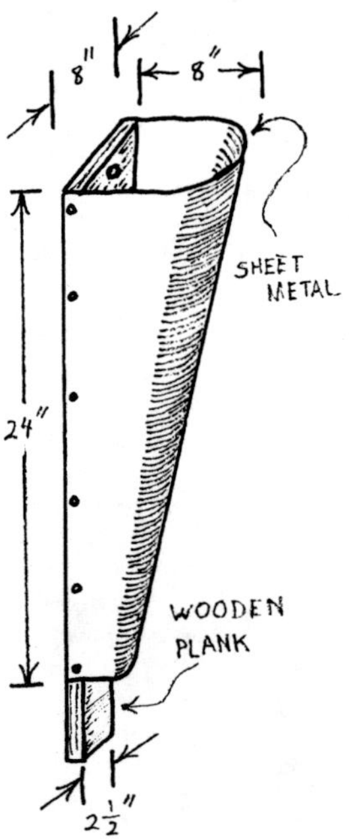

tacked to a 2 foot long tapering plank. This flat-backed cone has an opening at the bottom of about 2½ inches, and the top opening is 8 inches. The plank has a hole at the top to hang on a nail. I place the bird, head down, in this cone, the string hanging out the bottom. I now tug on the string until I can grab the bird's head, and with a very sharp knife, I cut off the head in one stroke. A bucket under the cone catches all the blood, and I drop the head in too. The shape of the cone keeps the bird from popping out and flopping all over, but allows a complete bleed. Squeamish types need not even watch anything but the actual knife stroke.

Scalding & Plucking

When the bird is dead calm, if you'll pardon the expression, it can be scalded. We just slosh it up and down for 30 seconds in a large pot of water at 126°F. The feathers loosen right up and the whole plucking process takes maybe a couple of minutes. Too hot water (140°F) causes the outer skin to come off; way too hot (160°+) starts it cooking! For geese, we use a very large pot, and a large handful of hardwood ashes (some use detergent) to cut through the plumage oils.

Pluck the birds by just grabbing feathers and pulling, not against the skin, for it easily tears. The hardest to pull are the wing and tail feathers. An incomplete scald makes it hard to get the roots of the big feathers out. So does letting it cool too much after scalding. You may have to rescald. Get the pinfeathers out with a rubbing motion under a stream of cold water. The difficult ones may require pliers. The little "hairs" left can be gotten rid of by singeing with a propane torch or a candle. Or if you don't care, then just leave them.

Drawing

Save a lot of time and trouble by taking the guts out of the back. First cut out the oil-sac at the base of the tail on the back. It makes the meat taste yucky. Second, slit the back of the neck skin to the back. Next, cut the neck-bone off as close to the body as you can. Peel the crop (the place where the grain piles up before digestion) loose from the neck cavity and cut it off. Remove the feet. Now, flip the bird onto its breast and with a pair of heavy kitchen shears, carefully cut the back in half. Avoid cutting the guts by keeping the lower blade tip close to the back bone. Haul out the guts in one mass.

Carefully remove the dark-green gall bladder from the liver. You can do this by lifting on the lobes of the liver, and letting the intestines pull the gall bladder downward. Cut it out of the liver. Don't cut the gall bladder itself because it will render everything it touches extremely bitter. Better to lose a bit of liver than all the giblets. Remove the gizzard, cool, cut, empty and peel it. Remove the heart. Feed the guts to the pigs, rinse the giblets and body cavity, and you're done.

Older and/or bigger fowl are more difficult. You cannot split the back as easily. Loosen crop, gullet and windpipe. Gently pull them away from the skin and out of the body. Cut them off where they enter the body. Stick a finger or two into the neck opening, and encircle the cavity, breaking all the little strings which hold the guts.

Make as small an incision as you can from a half inch away from the breast bone to about a half inch from the vent. Practiced poultry dressers need an astonishingly small opening. The cut needs to go through the fat but don't cut the intestines. Insert your forefinger to lift up the intestine leading to the vent. Cut right around the vent hole, avoiding cutting into the intestines. I'd rather lose a little meat than risk cutting them.

Work two fingers into the body, and haul out the guts. Don't break them; they'll ruin the meat. Now cut the gonads from the either side of the center backbone. Also, remove the lungs, which look like pink sponges, if they didn't come out with the guts. Rinse the gut cavity.

The drumsticks may be held close by sticking them under the bit of skin between the vent hole and the gut removal hole. Fold the wings behind the back to hold the neck-skin flap shut. You're done!

Put the feathers in your compost pile, or use them for pillow stuffings. If a goose, save the down for winter clothing.

Let the fowl sit overnight in a cool, varmint-proof place, away from the sun. In hot weather it may be submerged in buttermilk. The aging tenderizes it. Old hens and roosters make the finest soups because of their richer flavor, which is a good thing, because they're too tough to cook by frying, roasting or barbequeing. But if you do cook 'em by one of these methods they can be chopped up for the finest of chicken salad.

With practice, a chicken is easily killed and cleaned in no more than 10 minutes.

20
PORK

For every creature of God is *good, and nothing to be refused, if it be received with thanksgiving.*
1 Timothy 4:4

I was very fortunate in my upbringing to have spent a great many childhood days in the company of rural family members. I witnessed the butchering of hogs since I was very small. We had juicy country ham, smoked bacon, cured sausages, and Southern barbeque a-plenty.

Some people justify their squeamishness with unfamiliar doings, like killing animals, by asserting such things are somehow wrong. I have actually heard grown men say "I could never kill anything" as they stare blankly through the bug stained windshield of their big automobiles. Who knows how many dead animals were left in its wake, for no other reason than that they "got in the way!"

I never thought it was "fun" to end an animal's life. But I must say I often feel a similar pang at killing a plant I have been acquainted with from seedling. Loving labors in the field, tilling, manuring, planting, watching for the first hopeful seedlings, all have a purpose, I know. Thinning lessens my sum total of plants, but I know I must, to assure a healthy harvest. Without all those extra plants, I must protect the remaining little inanimate innocents from all manner of predation, disease and weed competition. I must feed them in times of drought, and sometimes prune them to help the remaining fruits grow. How is it so easy for some people to kill such things?

Some zealots even declare "Don't kill anything with a face!" It makes one wonder if they ever ate pumpkin pie, or at least if they ever had fun with Jack-o'-lanterns?

The zealots have now had their way with the law. John Seymour, that great British dean of self sufficient agriculture, sometime lamented that where once livestock was walked along the road a mile or so to a local butcher, spending a quiet night in his enclosure to be slaughtered the next day, today they must be hauled in terrifying trailers and trucks, jammed in with dozens of others to a killing factory reeking of blood and guts, all in the name of progress. I say rather it is all in the name of "humaneness" or "sanitation."

In the days of local butchers one rarely heard of the kind of salmonella and coliform infections so ubiquitous today. The answer seems to be ever more of the kind of expensive regulation which forces small abattoirs to fold for want of capital. The mom and pop business always had to keep clean and safe. One malign rumor would end their livelihood. Today's meat packers can always fall back on the excuse that it was the government inspector's job to catch such problems, since inspection is mandatory nowadays.

These are some reasons I prefer

to kill my own hogs, despite the fact that I'd prefer not to do it. It is a matter of commitment versus convenience.

Killing My First Hog

It was a cold, raw February day. I had read that it only took about 1½ hours to kill and clean a hog, so I had plenty of time to get the job done before dark.

I had the great barrow penned up overnight in a horse stall. Everything was in readiness, or so I thought. I placed some corn on the ground and led him to it. While he was merrily eating, I drew imaginary crossing lines on its forehead from eyes to ears. I then shot him at that spot, and down he instantly went: Dead in a twinkling of the eye. Leaping to his back, I slipped the razor-sharp boning knife into his throat, halfway between the breastbone and the *throat latch,* which is the creased skin just back of its jawbone. The knife tip was pointing directly at his tail. Holding the knife tip stationary, the handle was brought toward the breast bone, and withdrawn.

There was a gusher of blood as the final nervous convulsions commenced, pumping the rest of it out of the meat. In a few moments, which seemed eternal, all was still.

The supposedly 200 lb. hog was now found to be considerably larger. Looking back to many hogs butchered over the years, that one must have been about 450 lbs.! I had to use a winch to drag the big hog into the stall I had prepared. No way could I hoist it up into a steaming vat of scalding water. So I had to dribble 145°F water on the hog a little at a time to loosen the hair and scurf. Eva dragged 5 gallon jugs of hot water all night to the job. I poured, scraped, poured and scraped. Finally, the pig was clean. I hoisted it to a barn beam, gutted it, sloshed it with cold water, and went to bed at 2 a.m., thoroughly exhausted.

I learned a lot from that event, and the next year the butchering went far smoother. For one thing, I made sure to not wait until the pig was so huge before carrying out my butchering duty.

In subsequent years I found about one hog in ten killed by gunshot will have flecks of blood in the meat, no matter how well you stick it. Another thing I learned was that shooting in the brain isn't always as quick and painless as enthusiasts claim. I now favor killing without stunning, by the old-timer's ancient method of sticking. Done right, there's less chance of undue suffering than by shooting.

As to cleaning the carcass, I have tried many methods. I have scorched the hogs, like they do in many parts of Europe, with burning straw, followed by scraping, dousing and brushing. I have skinned hogs. I have simply washed 'em and brushed 'em clean. None of these methods was satisfactory, and I'm back to scalding. Scorching often winds up cooking the skin too done, and scraping may then take chunks of skin and fat with it. Skinning wastes the very edible skin, unless you tan it into leather. This is very laborious with pigskin. Just cleaning them up and brushing is no good, because the skin isn't edible, and you wind up having to skin it anyway. Except it is even more difficult after the pig has been gutted.

Slaughtering Hogs

There are a few tools you need

to do a good, relatively easy job of it: 1. A tripod, overhead beam, tractor lift, or tree branch about 9 to 12 feet off the ground, to hang the carcass. 2. A hoist, come-along, or block and tackle. 3. A gambrel, which is a device used to hang the carcass. It looks like a giant coat hanger with hooks at the corners for the gambrel tendons. You can improvise this out of a draft horse singletree about 2½ feet long, or even just a 3-inch thick oak or hickory stick tapered to ¾-inch at the ends, with little notches to keep the gambrel-tendons from sliding off, and a non-slipping place to attach the hoist in the middle. I have found that removable hooks to attach to the ends of a singletree make handling much easier for one man. 4. A 15 to 20 inch length of chain with rings at the ends so it can form a running loop like a dog's choker collar, except it's for the pig's foot. Get chain heavy enough to hold even a 500 lb. animal for hoisting. Failing this, you can use a flexible rope tied into a loop with a double slipknot, or better, a lark's head knot. 5. A razor-sharp knife with a thin blade about 6 inches long. 6. A wash tub and a bucket, at least, to heat and pour water for scalding. If you're going to butcher a lot of hogs, I'd recommend a vat big enough to dunk the hog. An old bathtub is good; with the drain holes sealed shut. It needs a few wooden slats in the water at the bottom to keep the hog from burning if a fire is under it. 7. Dull knives, candlesticks, or factory-made bell scrapers to scrape off the hair and scurf. 8. Clean rags, old towels or gunny sacks. 9. A meat saw to help you split the carcass. I got by just fine for years with just a carpenter's crosscut saw, and could hardly see much difference. Some folks chop it in half with a cleaver or axe, I don't think this is wise for your first hog. 10. A thermometer to measure the scald water's temperature.

Your first hog ought to be butchered when the weather has gotten cool enough to preclude fly problems and also to chill the carcass reasonably fast. I like to wait until the ground has frozen solid at least a couple of times, and slaughter on a day which had a frosty morning.

Get the hog into a quiet pen by itself. Water but don't feed it the day before killing. This makes the job somewhat easier.

I start the fire under the scald water vessel first, then put a little feed on the ground. While the pig is eating, slowly move up behind it with plenty of slack on your hoist. Loop the foot chain on the hind leg, pulling it tight. Keeping tension on the hind leg, smoothly and not too quickly hoist the hog up, so his nose isn't on the ground. Some hogs struggle, some don't. If you fed plenty of wood ashes or limestone, there will never be a broken bone this way. Standing beside the hog, place one hand on its shoulder to steady it. The other hand inserts the very sharp knife, blade up, into the hog's throat halfway between the throat latch and the breast bone. The knife tip is pointing toward the pig's tail. Keep it in the center; you don't want to spoil any shoulder meat by cutting into it. Push it all the way in, and with the tip stationary, bring the handle up to the breast bone and withdraw it. The bleeding will be over very fast if you pierced the heart (which you really ought not do) but it will take up to ten minutes, and likely be more complete, if you don't. Never fear, immediately following the interruption of blood flow to the brain, the animal falls asleep. I have never had a hog show pain at the sharp knife going in, I don't know if they even have any nerve endings at the sticking point.

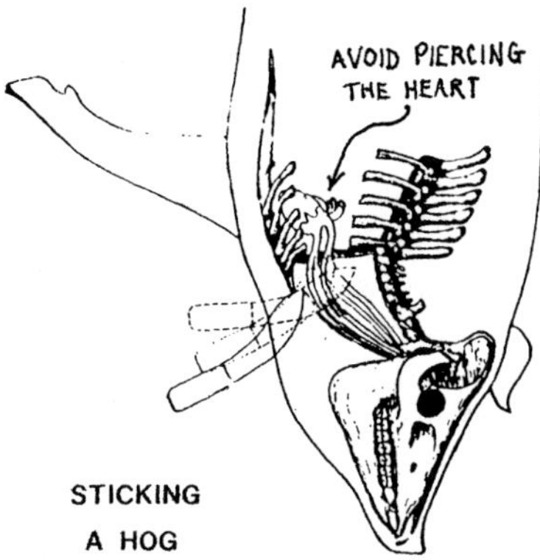

STICKING A HOG

If you feel you must shoot the hog first, that's o.k. I would use a .22 cal. pistol rather than a rifle. Over the years I have discovered that hogs are very prone to move the head at the last moment while you're pulling a rifle's trigger. This is because it looks like a stick in your hand. The .22 pistol just looks like your finger. Can you believe there are people who think outlawing pistols would be a good thing? As soon as the pig drops, cut its throat in the manner just described. It'll probably gush a big mess of blood. Many folks use this to make black pudding--blood sausage. We don't, but don't condemn others who do (Genesis 9:4, Acts 15:20,29). There are other uses, like feeding to chickens, compost starter, fish bait (congealed).

Scalding

Now the hog is bled. It needs to be scalded. In the "soft hair season" of February through April, 136°F is hot enough for your scald water, and it'll take about 4 minutes' scalding to make the hair slip. In the "hard hair season" of October thru January 143°F for 4½ minutes will do. A bit of wood ashes or slack lime added to the scald water will help take off the hair easier. Place the hog in the vat when the water is right, and don't let it get any hotter. Just a few degrees too hot will set the hairs, and you'll never get 'em out! Keep the hog moving in the vat so all parts will take some air at least once during the scald. Remember to keep the hog off the bottom where the fire is, or it will over heat, setting the hair and perhaps cooking some of the fat. A mat of wood slats in the bottom does the trick.

If you're not able to dunk the hog, use water heated to 150°F so it will cool to about the right temperature during handling. I like to work on some sort of platform or clean concrete or wooden floor. Wrap the hog fairly tight in gunny sacks all over, and dribble the hot water on it, keeping it soaked and hot for several minutes. Lift the sacking now and then to test for hair slippage. Alternatively you can just slowly dribble the heated water with a dipper on portions to be scraped, scrape them when the hair loosens, and move on. More than one person doing this makes the work go faster, and gives one time to refill the wash tub as needed. You'll need at least 10 gallons of hot water, I seemed to need about 30 to 40 gallons. Pour and scrape; pour and scrape.

The head is probably the hardest place to scrape, inside the ears, and along the snout and lower jaws; get it all. Each foot may be immersed in a bucket of the heated water and the hair removed by grabbing about the knee, and slide the hand down to the toes with a twisting motion. Scrape between the toes, and the outer skin will come off nicely. Yank off the toenails with a small hook made of heavy wire. Be careful scraping the soft belly skin.

No matter what color your hog was when you started it is now all white. A few hairs may remain. You can either shave or singe them off. As long as the scurf, or outer skin is off, it is clean.

Raise the gambrel tendon like this: Take a hind leg in hand and examine the back of the foot where the dew claws are. I like to make two parallel cuts about a half inch apart, equidistant from the center, between these dew claws, and about 4 or 5 inches long going in the direction of the ham. Right under the uncut center is the very strong gambrel tendon. I don't know how many books I have seen instructing folks to hang the hog by the Achilles tendon on the ham side of the hock. They're wrong and ought to be ashamed of themselves wasting so much meat.

Hook the gambrel into the gambrel tendon thus exposed. Now do the other leg the same way.

I cut the head off now. It's easier to gut with it off. Start with a sharp knife and cut way down the lower jaw, following the jawbone and trying to get as much cheek meat as you can to remain on the carcass, not the head. Keep cutting right up to behind the ears, until you get stopped by the backbone. The end of the backbone, called the Atlas bone, sticks into a socket in the skull. Cut as close as you can to the skull, right around this. Now grab both ears, and twist. After a full rotation the surprisingly heavy head will come off in your hands. You may need to free it with the knife.

Eviscerating

It's time to gut the pig. Professional slaughterhouse workers can completely eviscerate a hog in 10 to 20 seconds. It takes me almost that many minutes, and may well take longer for a first timer. With the hog lying on its back, score through the skin to the breast bone, all the way down to where you stuck it. Now you can slip the knife under the breast bone, being careful not to stab the organs inside, and lift, splitting the bone all the way in half. Or you can just cut it in half with the saw, again, not touching anything inside. Hoist the hog part way up and before it's too far, cut right round the anus and the vulva if it's a female. These organs' internal portions form tubes which pass through the center hole in the pelvis or "aitchbone" beneath the tail. The knife is inserted its whole length, but stay away from the organs and their attachment. To prick any part of the intestines, but most especially the rectum, large intestine, or urine-carrying portions will ruin an amazingly large part of the meat. As you complete the cut circling the anus (and vulva) grab the organ with your free hand to keep it from falling down into the body cavity. Take a bit of string and tie these organs tightly shut.

You can now score from the anus cut down to the split breast bone, right down the center, but not cutting through the inner belly wall, just the skin. If it is a male, dissect the pizzle and sheath, but leave it attached at its base, between the hams.

Make a short, shallow cut through the belly wall up near the crotch, and hold the handle of the knife inside. With the blade point sticking out of the belly, cut down from the inside all the way to the breast bone. Now the guts will fall down and outward through the opening. The reason we hold the knife this way is to prevent stabbing the guts.

Cut the hams apart from one another following the cartilage seam right to the aitch bone. Place the point of the knife into the center of the aitch bone and smack the end of the handle with the heel of your hand to split it in half.

Grab the guts with your left hand, supporting the intestines with your left forearm, sever the artery near the top of the liver on the right side to liberate the liver. Slip the hand behind the stomach; roll it forward and out. Cut through the gullet (esophagus) to loose the guts, which one hopes will flop right into the big washtub you have waiting beneath.

Carefully remove the pluck, which is the heart, liver and lungs, all bound up by the windpipe. In the liver is a dark green bag of horrible bitter bile--the gall bladder. Be careful to not cut or rupture this, but cut the liver holding it. I'd rather lose half the liver cutting around this than risk cutting into it or breaking it. It will simply ruin anything it touches. You can hang the pluck on a hook through the windpipe, though I prefer to get this stuff wrapped and into a varmint-proof place.

Finishing Touches

Slosh the inside with clean water, hoist the carcass up away from varmints and get some rest. Hang the head up too. If it's not too late in the day, you'll want to remove the stomach, slit it near the opening, about a 2½ to 3 inch cut, and get all the gorge out. Then turn it inside out and wash it thoroughly. Place it in scalding water until the inner lining can be easily removed. Scrape off all the inner lining and rinse the stomach in cold water. Pack down in salt until needed for stuffing with sausage or headcheese.

Small intestines are used for sausage casings. If at all possible, work them while still warm. Tie off the ends, and run them through the hands to untangle them. Peel off the ruffle fat. I now flush them by turning a garden hose into the end, and running water through. This only works if the intestines are not kinked, otherwise they'll burst from the pressure. I always turned them inside out by sliding them on a dowel. Then take the end and make a cuff by folding it back on itself. Slide this ever growing cuff back over the stick, continuing until the whole thing's inside out. Lay the inverted intestines on a board, and scrape with the back of a knife until all the mucus lining is off. Rinse, and repeat. Rinse and pack down in salt until needed. I don't fool with the large intestine because it just stinks too much for me to find it appetizing.

Some folks wash out the bladder and inflate it like a balloon or fill it with lard. I can't see much purpose in the latter, though urchins in the old days used to come a-running when they heard the sounds of butchering day. It wasn't morbid curiosity, they just wanted to play with the balloons.

If you should have to butcher in hot weather, there is a way to get by. I don't recommend this until you're familiar with the whole process.

Kill the pig early in the morning, before it's fully light, and have it gutted before the flies are fully active. Get a clean water tight barrel, and fill it about a third full of cold water. Put in a bunch of gallon-sized chunks of ice. Add 3 lbs. of salt. When the hog has been cleaned and gutted, split the carcass in half down the backbone, and then cut each half into three portions: a shoulder, a loin, and a ham. Place all the portions, including the head into this ice brine. It'll cool pretty fast to around 35°F, and won't absorb much, if any salt by the time you cut it all up. This is what folks often must do in warm climates where it never gets cold enough to cool the meat.

21
Charcuterie

Meats for the belly, and the belly for meats.
1 Corinthians 6:13

Charcuterie is French for pork-butchering. It is so important a subject as to have a word distinct from all other butchery. For pork is unique among the common meats in that it cures and keeps more agreeably than any other meat, and it provides lard, a peerless cooking and preserving grease. Moderns wrinkle their noses at fats, I know, but if they once again had any physically demanding labor to do, they would soon see the healthful advantages of including in their diet this gift of God to working men. I cannot address the gamut of charcuterie, but will cover the important points as I employ them.

Wholesale Cuts

Left hanging in the last chapter was the cooling pork carcass. I made brief mention of cutting it into chunks for cooling in brine. It is identical to cutting the already cooled carcass, to wit: First, look to your gambrel. Unless you have separate hooks holding the legs to the hooks on the singletree, you will find that when you cut the carcass in half lengthwise down the backbone, one half will be heavier than the other, and will deposit itself on the ground. As soon as that happens, the other half will deposit itself on the ground. You must either tie the gambrel to the legs, or think ahead, and use the hooks.

When the hog is secured from sliding off the gambrel, saw all the way down through the backbone. The hoist may have to be adjusted as you go, and your arm will get quite tired sawing. Take your time. Fresh pork just doesn't spoil as fast as store bought stuff, which after all, is a week or two old by

PORK CHART

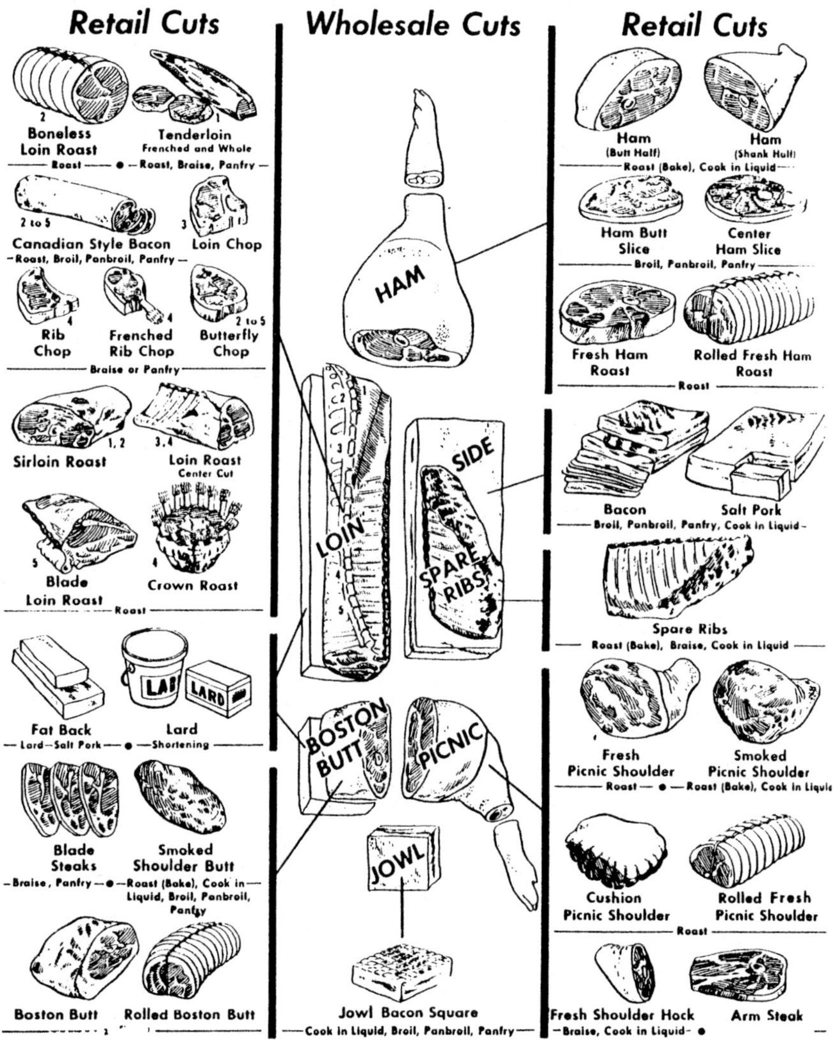
Retail Cuts
Wholesale Cuts
Retail Cuts
Boneless Loin Roast
Tenderloin Frenched and Whole
Roast — ● — Roast, Braise, Panfry
Canadian Style Bacon
Loin Chop
—Roast, Broil, Panbroil, Panfry—
Rib Chop
Frenched Rib Chop
Butterfly Chop
Braise or Panfry
Sirloin Roast
Loin Roast Center Cut
Blade Loin Roast
Crown Roast
Roast
Fat Back
Lard
— Lard—Salt Pork — ● —Shortening
Blade Steaks
Smoked Shoulder Butt
—Braise, Panfry—●—Roast (Bake), Cook in Liquid, Broil, Panbroil, Panfry
Boston Butt
Rolled Boston Butt
HAM
LOIN
SIDE
SPARE RIBS
BOSTON BUTT
PICNIC
JOWL
Jowl Bacon Square
—Cook in Liquid, Broil, Panbroil, Panfry—
Ham (Butt Half)
Ham (Shank Half)
—Roast (Bake), Cook in Liquid—
Ham Butt Slice
Center Ham Slice
—Broil, Panbroil, Panfry—
Fresh Ham Roast
Rolled Fresh Ham Roast
Roast
Bacon
Salt Pork
—Broil, Panbroil, Panfry, Cook in Liquid—
Spare Ribs
— Roast (Bake), Braise, Cook in Liquid —
Fresh Picnic Shoulder
Smoked Picnic Shoulder
—Roast— ● —Roast (Bake), Cook in Liquid
Cushion Picnic Shoulder
Rolled Fresh Picnic Shoulder
Roast
Fresh Shoulder Hock
Arm Steak
—Braise, Cook in Liquid— ●

the time it gets to the meat case. After cutting you have two *sides*.

Cut off one shoulder by finding the second and third ribs. Saw from between their tips straight across at right angles to the back. If you cut through the bones, saving the backfat and skin for last, you can put down your saw, cradle the whole shoulder and leg, and complete the cut with the knife, thereby avoiding the meat falling on the floor.

Though I prefer to leave the loin on the pig until I've finished working with the shoulder, I'll explain where to cut it loose from the ham, in case you need to chill it in brine. Find the part of the aitch bone you already split. Saw right through the backbone at a point 3 finger-widths below the aitch bone (on the loin side). Saw at right angles to the leg shank. You may cut part way through, finishing the cut with the knife to allow you to cradle the loin as you did when cutting the shoulder off. In fact, most charcutiers don't like making a perfectly straight cut, but when they get to the knife part, they angle upwards so as to leave the flat, thin meat with the loin rather than the ham. It gives you a little more bacon and a little less sausage.

The process of cutting these *wholesale cuts,* as they're known, is now complete.

Cutting the Shoulder

Now back to that shoulder. Bone out the neck bone with a sharp knife.

Remove the jowl, which is the relatively fat portion of the neck that used to attach to the head. Cut straight across with the knife parallel to the cut on the loin side of the shoulder. Many folks trim this up neat and square, and pound it flat with the side of a meat cleaver or butcher's axe. Then it's called the *bacon square.* If left as is and trimmed smooth, it's called the *bath chop,* a term hardly known any more. It can be cured and used the same as bacon, or skinned and ground into the sausage. I much prefer the former.

Saw the foot off ½ inch above the knee joint. Split the foot in half, starting between the toes with the knife. You may need to finish using the saw. I just use a little carpenter's hatchet as a meat cleaver. The feet are delicious pickled or are used in headcheese. Either way you ought to cure them first.

Pickled Pigs' Feet

Cure pigs' feet the same way you'd cure ham, for about 2 weeks. Simmer gently so the skin won't deform, pulling the feet out of shape. Cook until you can just barely poke a finger through the skin. Remove as many of the bones as will easily separate, cool, and pack in cold 5% vinegar mixed with half its volume of water. Add ½ tsp. allspice and a bay leaf for every quart of this solution, if you like. You can process this in a pressure canner, but we never did, it keeps in jars for months in just a refrigerator. Check the price of pickled pigs' feet ("pickled trotters") at the market and you'll wonder why raw pigs' feet are so cheap.

The remaining chunk of shoulder is called the *long cut shoulder.* It can be cured like a ham, as is, but gets tougher than ham when cured. Better, I think, to either bone out the whole thing and make sausage of it, or separate it into *butt* and *picnic* portions. The picnic is the leg half. The butt is the part up by the back, and is also called the "Boston butt."

Before separating these, we remove the *clear plate,* which is the layer of fat covering the butt. To do this, slice through the skin and fat 1 inch to the leg side of the shoulder blade, which lies exposed where you sawed the shoulder from the loin. Cut perpendicular to the said saw cut. You'll observe that the fat covering the butt has a certain visible cleavage. Just follow this right up the butt until it's off. The clear plate makes excellent lard and wonderful fried pork skins.

You may cure the picnic and butt into ham, use it for roasts, or make sausage of it. For the latter use it is superior to almost all other known meats for mixing with fatty cuts. Besides, it has a certain amount of gristle, which makes it less attractive for other uses anyway.

Loins

Cut a loin from the carcass and lay it skin side down. Look at both ends of the loin and you'll see a large round area of meat right up in the "T" of the backbone, covered by the skin and a double layer of fat. This is the *loin eye.* At the ham end there is a tapering muscle on the inside of the backbone called the *tenderloin.* To separate the back from the belly, first scribe a guide line from the belly edge of the tenderloin all the way to the belly side of the shoulder blade at the other end. Now you can cut with the knife, starting at the ham end, until you reach the ribs. Saw through these, still following your scribed guide line. It's best to leave all the loin eye meat in the back, but get as much as you can of the red meat next to it into the belly.

You may carefully cut the tenderloin free of the back, or leave it in place. It is the premium cut on any meat carcass. If you leave it in place, it will be part of the pork chops, or a very good pork roast. There isn't much of it, so we normally don't separate it.

The fat layer under the skin on the back has a visible cleavage point. With care, you can use this as a guide to sever it from the rest of the back. This long rectangle is called the *fatback,* and can be cured for adding to pork and beans or used for making lard, cracklings and fried pork skins. We always called the cured fatback "Uncle Ned."

There are several ways to handle the remaining back, or *chine.* The loin eye may be carefully cut free of the backbone and cut into boneless chops, or cured into *Canadian Bacon,* the backbone then being cut into barbequeing size chunks. Or it can be sliced into chops and/or roasts with the backbone left in. To do the latter, start at the rib end, cutting straight across the loin meat, using the rib bone as a guide for your knife. The first chop will be somewhat tapering. When you get to the backbone, lay the partially finished chop out of the way, lay a heavy knife in the appointed place, and smack the back of the knife blade with a wooden or rawhide mallet right through to your cutting block. Sawing is really fiddly unless you've got a meat cutting bandsaw. If you're that well outfitted, you probably don't even need instructions to butcher a hog. Then follow the other side of the rib. Chop through the backbone as before. Then repeat using the next rib as a guide. You'll run out of ribs before you run out of loin. You may continue now that you've got the hang of it, without benefit of the ribs, or cut the remainder into a roast or two. I like wrapping chops by twos for the freezer; suit yourself.

Bacon

The belly still has the spare ribs attached to it. There is a natural separation point between the spare rib meat and the bacon, and more and more I find myself just following it to remove the ribs. Most folks slide the knife right along the rib bones to leave as much meat on the bacon as possible. There are cartilage pins at the ends of each rib, terminating at the breast bone. Get all this off the bacon too. The spare ribs are usually divided into twos, but I like to cut each one loose, because it makes it much less messy come eating time.

The bacon slab itself is trimmed of all little fribs of meat and skin to make it straight and cleanly. Cut off the strip at the belly side to get the teats off. These aren't much of a problem unless the hog is a sow which has had pigs. Then they contain unsightly glands called "kernels." Cut 'em off in a strip, remove the skin, and grind 'em into the sausage, or if you're squeamish, feed 'em to dogs, cats, chickens or pigs.

In middle Europe, the long slab of bacon is split lengthwise into two slabs. I haven't found this useful except when butchering hogs over about 300 lbs. But cutting in half the other way to shorten the slab helps fit the bacon into the curing vessel. More of this anon.

Work the other loin before you start on the hams.

Ham

Part of the backbone remains on the ham. This can be trimmed out with a sharp boning knife, keeping the knife flat against the ham meat, leaving as little meat on the bone as possible. If you plan to cure the ham, keep in mind that thin fribs and corners of the meat will dry to inedibility, so they're best cut off and used in the sausage. Extra fat may be put into the lard, though its presence makes the lard a bit softer.

Saw off the foot just below the button of the hock.

Cutting the ham into ham steaks is difficult, but made much easier if you first bone the ham. Do this without stabbing into the big hunks of meat on either side of the bones. Before you do this though, remember where the thigh bone is. After boning, cut across where it used to be, so the meat fibers are as short as possible in the steaks to prevent toughness.

There is a Slovakian method we have used for curing boneless ham in sizes small enough to be useful. It involves first cutting off that portion of meat parallel to, and on the loin side of the aitch bone. This big chunk of meat is called the *house bacon,* and is usually cured. We usually cut it into small pieces for grinding into sausage and use the clear meat portions for recipes calling for strips of high quality meat, such as stir fry. Stand the remaining ham on the cut just made, with the shank facing left and the kneecap facing right. Cut beside the kneecap, touching the thigh bone all the way down to the cutting block. This big chunk is called the *nut ham.* It too can be cured, or cut into steaks by slicing across the grain of the meat. The pointy end is used like the house bacon. Saw off the hock, and carefully cut out the remaining bones. This is the biggest portion of ham meat of all, and is called the *scarf ham.* It cures wonderfully. It also has the best steaks.

The Head

Cooking a whole pig head used to be the traditional way to celebrate the New Year, hence the "Boar's Head Carol." I have never done this.

To prepare the head, remove the eyes and eyebrows by cutting all around the eye socket with a sharp paring knife. Cut off the ears close to the head. These can be cured and put into headcheese, or pickled in the same fashion as the feet. Same goes for the snout, which may be pared off the skull at this time, or done last. There is a lot of meat in the jaw sockets and in the back of the head. To remove the lower jaw, insert a sharp knife all the way up in the jaw sockets and bone around the jaw bone up in there. You can't cut it entirely free, but you can get most of it. Now take both jaws in hand, and pull them open until they're 180 degrees apart. It takes quite some strength. Cut 'em loose. Cut out the tongue. Cut off the remaining meat, and grind it into the sausage. Many folks split the skull with a saw or an axe (the saw leaves no splinters) to remove the brain. If this is added to the headcheese it is known as *souse* or *hash*. I'm not fond of either, but my hens are fond of pig brains.

Headcheese

Bone and skin the head, ears, tail, feet, and any other odd bits you feel like using. The mix is not really that important. For each pound of skin though, you'll need a pound of fairly lean meat. You can use heart, tongue and just plain meat to make up the lean. Cut the skin and meat ingredients into chunks no bigger than your hand. Mix every ten pounds of lean and skin meats well with 6 tablespoons of salt and a pinch of saltpeter stirred into ¼ cup of cold water. Let it cure in a cool place for 2 days. Simmer slowly with just enough water to cover for 1½ hours, or until the skin can be pierced with a finger. Drain and cut into chunks no bigger than an inch square with a knife. If you like, you may finely grind 10% of the skins. Mix in about 15-20% of the meat's weight of cooking stock. Season with enough salt to taste. Premix the following ingredients and add enough to taste: 4 oz. ground pepper, 1 oz. ground caraway seed, 1 oz. ground marjoram, 1 oz. grated onions. Stir well, and stuff by hand into the prepared pig stomach, or into casings made by sewing clean muslin into long 5 inch tubes. Sew the stomach (or casings) tightly shut, and simmer slowly for 1½ hours. Cool with a weighted board on top of the stomach to flatten it slightly. Some folks just cook it again and pour it into molds. Sliced with onions in vinegar and served with bread and mustard, this is scandalous good eating.

Sausage

There's no helping it now, you need a meat grinder. *Sausage* is a generic term for ground pork, with or without further seasoning or curing. We grind all our pork with a ratio of one third fat cuts to two-thirds lean cuts. All generally has some fat and some lean. If too lean, we found our sausage is dry and falls apart in cooking. Season it however you like, but the usual ingredients for common American breakfast sausage are: 10 lbs. pork, one third fat; 6 level Tablespoons salt; 1 level Tablespoon each rubbed sage, ground black pepper; 2 level teaspoons ground red pepper. Mix spices in a cup of ice water, then mix into the meat. This can be stuffed into the casings using a large funnel. Bunch the casings on the funnel spout and press the meat through with fingers, taking care to avoid too much air getting in. Or you can just fry it in patties. This sausage won't keep, except in a freezer.

We don't usually season our ground pork at all when we put it in a freezer in 1 lb. packs. That way we can use it as we like in all sorts of recipes.

Lard, Fried Pork Skins, and Cracklings

To make lard, one simply places very little water in the bottom of a large pot, dumps the pieces of fat in, and heats it slowly to render out the lard. We use a canning pot filled half way with pork fat. Don't fill it full because the sizzling fat may make messy (and highly combustible) spatters all over. Add 2 cups of water, cook at medium (not quite frying) heat, stirring occasionally to prevent sticking at the bottom. Never use smelly or questionable fat, for a little of this will ruin the whole pot. But the chunks with skin on them can be put in whole. It is far less laborious to cut the softened skin and/or fat while hot in the lard than to cut it up cold on a board first. Besides, if the skin is cooked without severing it from the fat, it will form puffy, crunchy fried pork skins. If you cook the skin solo, it will harden enough to break a tooth. In time the heat can be increased, because more liquefied lard appears. Cut the big chunks smaller. Leave no pieces larger than about 2 inches square. Continue gently heating the whole mass, removing the skins as they fry up crunchy. Drain them on a rack. Other fat will float and sizzle, turning light golden and then sinking to the bottom. Keep stirring them now and then to avoid sticking. These are cracklings and are used in gravies and biscuit recipes. When the lard is completely rendered, remove from the heat. You can tell it is done when most of the cracklings have sunk. Don't cook it too long or it will darken. We dip it into dry, sterile mason jars and screw the seals on tight. It keeps like this for years. When you dip out the cracklings, it helps to first place them in a sturdy strainer, and mash them to press the extra lard out.

Lard can also be used to preserve all types of meat and even fish. Pour it over cooked meats which have been weighted down in jars. When the lard hardens, remove the weight, and fill the cavity with more hot lard. Screw the seals tightly, and store in a cool, dark place. This keeps a very long time too. Almost any cooked meat can thus be very conveniently kept without expensive refrigeration.

Ignorant know-it-alls sneer at lard, I know; but chefs at world-famous hotels use it to make spectacular pie crust, for more digestible deep frying, and of course the finest tasting popcorn.

Smoked Ham, Bacon & Sausage

There is a lot of poor quality smoked meat on the market. The bacon pops in the pan--properly cured bacon does not. Store bought ham has very light pink color and will not keep. You must pay for all the water they added, too. Some commercially made dry cured ham is too salty. Even after soaking the salt out it retains something of a bitter flavor. In learning to cure pork predictably well, I ruined a lot of meat, let me tell you. All the literature gives recipes but no theory, so when the conditions, i.e., temperature and humidity are not perfectly controlled, the meat turns out wrong.

The first rule in curing is to use fresh meat. This is why I suggested finishing work with the shoulders before removing the loins. You don't want to give germs an opportunity to enter before you put the salt to 'em. The second rule is to cure only portions of meat which have been completely chilled. Warm meat doesn't take in salt very well. The third rule is never let meat freeze during the curing process. Frozen meat spoils before it cures.

The main curing ingredient is salt. It works its way into the meat cells by osmosis. The water contained within the cells is attracted to the salt. Some of the salt dissolves in that water, and deeper adjacent cells absorb some of its salt. Complete curing is thus a factor of thickness, not of weight. I was steered wrong by so many "expert" curing books! They all specify curing time as a factor of weight. It simply doesn't work that way, and I wound up getting things over salted and under cured all the time until I wised up.

Two things preserve meat: 1. Drying, and 2. Complete penetration of the cure mix. Safety requires both. Salt draws moisture from meat at the same time it helps kill germs. Another ingredient is useful: Saltpeter. This is the same stuff used in black gunpowder, blasting powder and some nitrogen fertilizers. It is used in exceedingly small amounts to block the development of *clostridium spp.* germs. These germs could cause botulism poisoning. And yet I know several folks who have used nothing but salt to cure hundreds of hams, with no mishaps whatever.

The percentage of salt in the dried product is critical. Less than 2.5% by weight of salt just won't keep. More than about 3.3% is so salty it is practically inedible. This tight a leeway can be met most exactly with *brining*.

Brine Curing

Raw meat submerged in brine will absorb salt. The salinity of the brine will drop as the meat's rises. Soon enough they'll be balanced. If the meat is then removed from the brine it will lose about 20% of its weight by the time it's dry. To put it another way, it will become about 25% saltier.

In other words, the meat must have 2½% salt content wet to dry to a 3⅛% final content. One way to do this is to place meat into a container, with 2½% of the container's capacity (by weight) of salt. Fill the rest of the way with water, remove the meat, dissolve salt in the water and replace the meat. Let it stand long enough for the salt content to reach uniformity in the container. As you can see, the smaller the container, in relation to the size of the meat, the more concentrated the brine will have to be. If too large a container is used, the brine will be so weak as to inhibit osmosis. One time I tried this and the meat never did really cure right. Whenever the meat was packed fairly tight, the cure was predictably perfect every time. Keep the meat submerged by weighting it down with a boiled plate, smooth rock, or large chunk of glass.

Packing the meat tight has its problems. Some portions won't start curing quickly because they are pressed against each other, excluding the brine. So the whole thing must be taken out of the container and put back rearranged. This is called *overhauling*. Most authorities say to overhaul after a week, and again at 2 weeks. I prefer to overhaul after 3 days and again in 2 weeks. Leave meat in the cure for 9 days per inch thickness of the thickest pieces. It can be left considerably longer if it isn't convenient to take it out.

In warm weather the brine may become *ropy*. This means it gets viscous and slimy, with a moldy scum developing on top. This can be amended by carefully pouring off the water (don't pull meat up through the scum) boiling it, skimming off the resulting scum, cooling and returning it to the meat. You can't just mix up a new batch, because by this time you don't know what the salt concentration is without a very accurate salinometer. Check on brine every day to guard against spoilage.

The less concentrated the brine, the longer it will have to stand, up to twice as long. The more loosely your meat fits into the container, which is to say the less meat and more brine you use, the weaker the brine will have to be to avoid oversalting. So match the vessel(s) to the volume as much as you can. When correctly sized, the container will be no more than one-third full of brine.

As to ingredients besides salt, they are added as a ratio of the salt content. For every 4 pounds of salt there should be about 1 pound of sugar or molasses to counteract the harshness of the saltpeter. Similarly, an ounce of saltpeter will be needed for every 4 pounds of salt.

The meat curing wizards of Slovakia combine brine and dry curing like this. Before immersing the whole thing in brine, they dissolve the saltpeter in twice its volume of water, which isn't much. This is then used to completely moisten the dry salt/sugar mix. The damp mix is rubbed all over the meat, which is then left to stand for at least 8 hours in the brine vessel. Then the remaining solids are dissolved in water, forming the brine. The water drawn out by the salt can be left in the vessel. Curing is hastened by this method. Don't pack thin cuts with thick, otherwise the thin will be oversalted and the thick may spoil for want of salt. Or else you'll have to leave it all in until the salt is balanced in the thickest cuts.

When you judge the curing complete, haul out each cut, rinse and maybe even brush to remove any clinging scum. But rinse at last. Hams can be hung up by punching a hole through at the shank and tying with clean twine. Flat and/or boneless cuts need a knitting-needle-sized wooden skewer (store bought bamboo skewers are fine) pushed all the way cross from the thin edge an inch from the top. The twine hole is then punched just under it. This method keeps bacons flat and the string from pulling through or deforming boneless cuts.

Two days after they stop dripping the meat will be dry enough to smoke. My first smokehouse did very well, and looked like a privy perched on a small rise. I dug an 8 inch deep trench down the slope from the middle of the dirt floor of the building to a pit about 6 feet away. I covered the trench with planks and then the dirt taken in digging. The pit had the fire in it, and I covered this with a big scrap of sheet metal, which was sealed with dirt at its edges to slow the fire. The important thing is that the smoke be cold, not hot. I used corncobs, because it's what I had. My favorite woods now are hickory and oak, in that order. Don't use pines, walnut or sassafras, because they deposit bad tasting resins. Most hardwoods are fine. Despite its very sweet smoke, sycamore is somewhat bitter. Eight hours to 2 days' heavy smoke is plenty.

Dry Curing

Most people dry cure with good success. There are two ways of doing it. One simpler method is very dependable. Take the cuts of meat, rub the above described salt/sugar/saltpeter mix all over, making sure to get some down along any bones. Wrap the cuts tightly in paper. Make sure the paper holds cure mix against all the meat. Place the wrapped meat into cloth bags you've made to fit. Twist the cloth bags shut, tying the twisted end to absolutely exclude bugs. Tie twine just below the knot in the cloth and hang this up where the saline drip won't harm anything. When it has completely stopped dripping, it's cured. Some folks just leave it like that forever. It's better to give it about a week per inch of thickness to hang, if you're not too sure. Then take it out, wash off the excess salt mix, and smoke it. Rewrap it, hanging in cool darkness until needed. It is best if it can hang at least a few months to mellow. But the same goes for brined and dried meats.

The other way to "dry" cure is to have a water tight box or container in which the meat can be packed very tightly indeed. A mix of 5 pounds salt, 3 pounds sugar and 3 ounces saltpeter per hundred pounds of meat, moistened as described above, is rubbed all over the meat, which is then packed and pounded tight in the vessel. Salt mix is sprinkled on top and the whole thing covered against dust and varmints. After a week, it will be covered with "pickle;" overhaul it and get the undissolved salt equally distributed. Don't remove the "pickle" which is highly saline water drawn from the meat. Again, one week per inch of the biggest cuts in the vessel is to be the curing time. When finished, rinse, dry and smoke as with brine cured stuff.

Brine Curing Ingredients Table

Size Vessel	Salt	Sugar	Saltpeter
½ Gallon	3½ T.	2½ t.	⅛ t.
1 Gallon	7 T.	1 T.+2 t.	heaping ¼ t.
2 Gallons	¾ Cup+2 T.	3 T.+1 t.	heaping ½ t.
3 Gallons	1¼ cup	¼ cup+2 t.	1 t.
5 Gallons	2 cups+2 T.	½ cup	1½ t.
7 Gallons	3 cups	¾ cup	2¼ t.

Farmer's Dried Sausage

Grind up lean shoulder meat and an equal amount of trimmings (50% fat). Mix well. For every 5 lbs. of this, add ¼ cup salt, 1 T. sugar, ½-1 T. pepper, up to 1 T. minced garlic (or 1 t. garlic powder) and heaping ¼ t. saltpeter dissolved in a little water. Stuff into home made 3½ inch diameter muslin casings up to about 20 inches long. Pack it tighter as it shrinks. In 60 days it's ready to eat. It is similar to hard salami from the delicatessen.

22
MUTTON

...Thou shalt kill of thy herd and of thy flock, which the LORD hath given thee, as I have commanded thee, and thou shalt eat in thy gates whatsoever thy soul lusteth after. Deuteronomy 12:21

There is so much bad lamb in the supermarket meat case that few Americans even bother buying it. They have tried it a time or two, but it had a taste like a cross between beef and a bar of soap. That taste is characteristic of the Merino and related breeds. The rule of thumb on choosing a breed with tasty meat is that the more slender the wool fibers, the "woolier" the taste of the meat. Conversely, the more stout the wool fibers, the tastier will be the meat. Cotswold has superb flavored meat, but because it matures later than small breeds its meat has a more pronounced texture. The smallest meat breeds have very acceptable meat flavor, and very fine texture, but their wool isn't especially valuable. And wool is after all, what makes a sheep uniquely useful among farm animals.

Well, they are also uniquely useful because a small family can butcher a sheep and expect to use it up before it spoils--without artificial preservation methods--even in summer. Good mutton makes a very adequate and healthful substitute for beef.

Speaking of beef, 3 things keep it from being practical to butcher on a small holding. First, it's so large that special, potentially expensive equipment is needed to maneuver it. Second, there are so many special cuts that it takes much more skill to learn to cut it up. Third, there is too much meat for a small family to store practically. The easiest way to handle beef is to take it to a butcher. The best way to learn to slaughter it is to watch the butcher do it. Even then it's best to sell or share part of it with others.

Slaughtering Sheep

The best thing about sheep is they require so few tools to prepare. Ancient shepherds did very well with nothing more than a sharp knife and a rope, but I would add to this a saw and a bench or platform on which to bleed the sheep so blood won't get on the wool and make a mess. The work goes easier if the sheep has been given only water, no feed, for one day before.

Set the sheep on its rump as in shearing. It is easiest for you if you bring its left hind leg forward and tie it to its two front legs. At least one leg should remain free for a thorough bleed. The animal is laid on its left side with its head hanging over the edge of the killing platform.

Some folks shoot the sheep between and a little above the eyes with a .22 cal. rifle. If the sheep jerks just as you pull the trigger, this won't be very humane at all. Kinder, I think to just pluck up your mettle and do it right.

The razor sharp, thin-bladed knife is stuck into the side of the neck a little below the right ear and just back of the jawbone, edge toward the sheep's throat. Sever the throat by a strong upward and outward motion of the knife. Immediately, clamp the neck to the killing platform with one hand, and

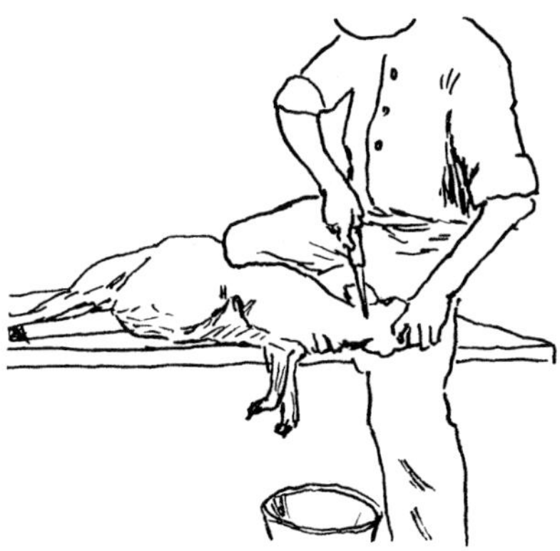

with quick pressure on the sheep's nose, break its neck over the edge of the killing platform. It takes a lot longer to read this paragraph than to perform it. Any thrashing the sheep does is completely involuntary, because it has no time to feel a thing before it loses consciousness. It is all nerve reaction, and serves the purpose of pumping remaining blood from the carcass. This is the only reliable way to kill a sheep without cruelty.

Legging Out

After bleeding, the sheep is *legged out.* This means the legs are skinned. Hold one of its front feet between you knees and pinch a bit of the loose skin of the front leg near the body. Cut into this fold of skin well away from the meat. Keeping the blade from cutting into the meat, carefully pare off a strip of skin with the knife all the way along the front of the sheep's leg to the foot joint, and cut the foot off there. Do the same on the other leg.

Now join the two leg slits by cutting across the breast with the knife. The knife edge faces away from the meat, so only skin is cut.

Using the knife as little as possible, loosen the skin from both sides of the breast, the legs, and both sides of the neck. After loosening, cut right up the goozle to expose the gullet (the food pipe). The outer, fleshy part of this is bent over, using the back of the blade, and tied shut so no stomach contents or gas will escape.

The hind legs are peel out like the front legs, only on their back side, from anus to hoof. They are saved for last because the leg of mutton is a valuable cut, and by leaving the pelt on longer they are much easier to keep clean. Don't skin them as completely as the front legs, just loosen the skin from the sides and back of the leg. Remove the feet at the knuckle joint of the hoof, never closer to the leg--otherwise the tendons will pull out when the carcass is hung up.

Fisting

The belly is loosened by *fisting*. One holds the skin with one hand, and wiggles the fist between the hide and carcass with the other. The thumb of the hand doing the fisting rests on the index finger, and probes ahead as the fist moves in. A young fat animal fists much easier than an old skinny one.

There is something you must remember while fisting. The skin is attached to the *fell,* a membrane separating meat from hide. Try not to rupture the fell while fisting off the hide, because it helps protect the carcass from spoilage, like a plastic wrapper. Fist the hide loose from breast to belly, then from cod (penis) or udder to belly, so the skin is loose all the way up the belly. Then do the sides of the belly, starting at the center of the side, and working towards the loosened belly, and towards the back. This way helps avoid tearing the thin muscles along the flank.

Hanging

The two tendons between hock and ankle on the backside of each hind leg are loosened by cutting between them and the bone. Slip a gambrel or even just a cord through the opening thus made, and hang up the carcass. Once hang-

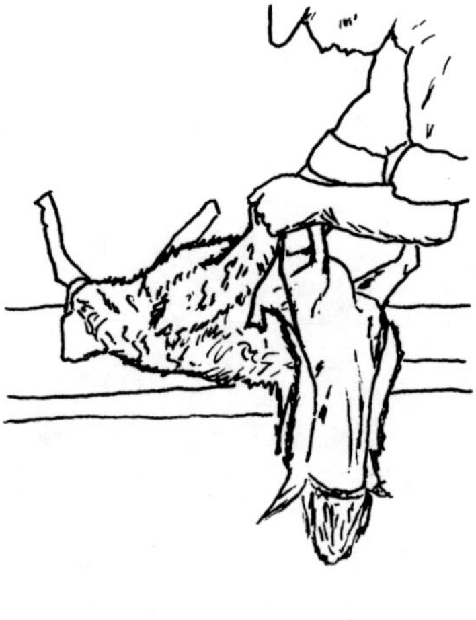

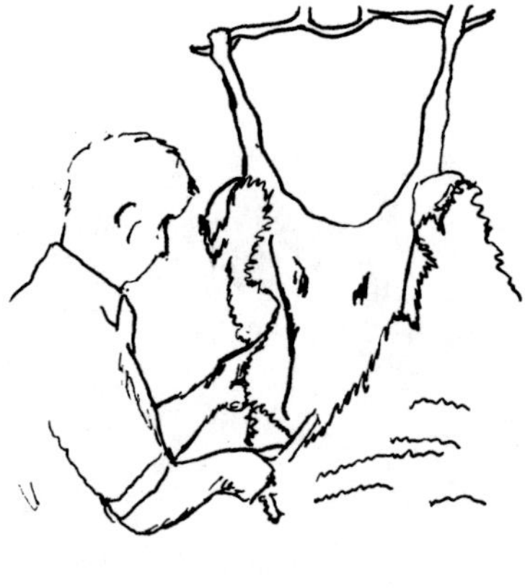

ing, slit the loosened belly skin from the breast to the anus. Take care not to peel muscles off with the skin at any point as you fist right around the sides to the back

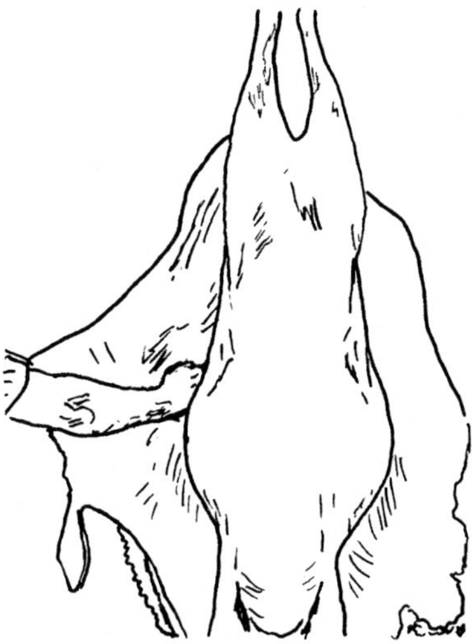

and all the way down over the shoulder. Now fist upwards toward the tail, which will have to be released using the knife. Finally, remove the hide from the neck, with the head, at the atlas joint. To clean and bleach the carcass slosh it with water as hot as your hands can stand. You may wipe it dry with clean cloths wrung out in hot water.

Gutting

The first step is to cut right around the anal vent, taking care not to puncture it. Pull it out a few inches and tie a string tight around it to keep you-know-what from spilling onto the meat.

Open the abdominal wall from cod or udder to breastbone, being careful not to cut into the guts. You can do this by keeping two fingers inside the carcass, to shield the knife tip as you cut. Allow the paunch and gut to roll out and hang. Unless the lamb is full of feed it shouldn't break loose. Reach in and locate the loosened rectum; pull and work it down past the kidneys. Take out the bladder; beware not to spill its contents on the meat. Roll the paunch slightly out, gripping it with one hand where it joins the intestines. Work the other hand into the body cavity, above and behind the liver. Gently tear the liver free where it is attached near the kidney. While still gripping the paunch, work your way under it along the diaphragm, pulling and pushing the organs up and out of the carcass. You may save the intestines, cleaning and salting them in the same way described for hogs on page 135. Use them for hot dogs and small sausages.

Split the breast bone with a saw and/or knife. Now remove the *pluck* (lungs and heart) by cutting the white part of the diaphragm near the ribs, cut the pluck loose on both sides of the brisket, sever the large blood vessels along the backbone, and pull the pluck down and outward. Slosh the carcass with cold water. It can be wrapped in cheese cloth and hung in a cellar, spring house or any dark, clean, cool place free of vermin and disagreeable odors.

Butchering Mutton

Split the carcass in half with a saw, just like with a hog. Cut the carcass into wholesale cuts with a saw, starting with the breast and shank. Pull out the kidney fat, and use it for tallow. Then proceed to cut shoulder, rack, loin, and leg, as needed. The lamb will keep in mountain or dry desert climates, in the shade, if exposed during the cold of night, and covered and insulated with its cloth during the day. Otherwise, slaughter it in the cool months, and save it frozen, salted, or dried. I like it hung for at least 2 weeks before cutting, and frozen into the smallest size packs I'm likely to use; I can always use more than one pack for guests.

Remember to always serve mutton or lamb on a heated plate.

LAMB CHART

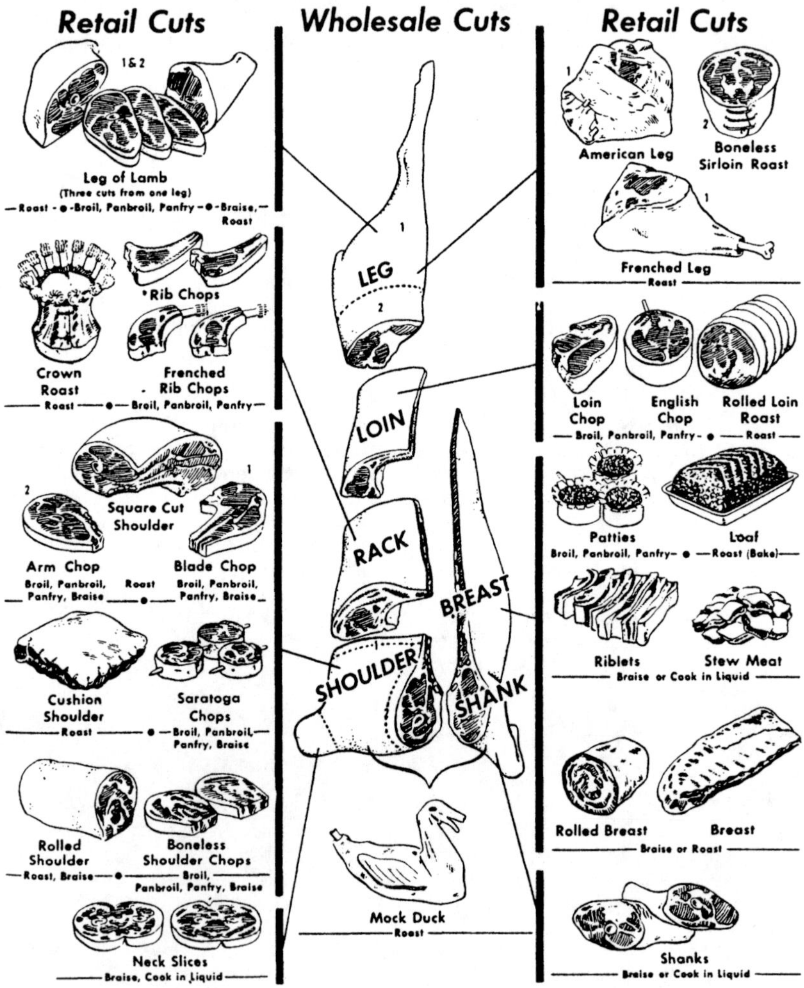

23
PASTURE

He causeth the grass to grow for the cattle, and herb for the service of man: that he may bring forth food out of the earth. Psalm 104:14

Grass is a common generic term for just about any herbage fed to livestock. Botanists don't include clovers, alfalfa or sedge plants among the grasses, but do include sugar cane, maize, wheat, rye, oats, barley, etc. But us humble rustics have ever called all plants used mainly for livestock forage "grass." In fact, the clovers, alfalfa, etc., are *legumes,* or *pulses,* having the exceedingly useful ability to derive their nitrogen needs from the air. More on this later.

A pasture is a grassy field where livestock graze. A meadow is a grassy field where hay is cut. If a crop of hay is taken, and livestock are then turned in to graze the *aftermath,* which is the word for the regrowth of grass, the field is still called a meadow.

Certain grasses are better suited to grazing than others because they form a dense sod or tenacious clumps, aren't seriously hoof-damaged, are long lived, are palatable, and seem to proliferate naturally once started. The most important of these in America are bluegrass, redtop, bromegrass, orchardgrass and Bermuda grass. Each has its own best area of adaptation.

Adaptability is the single most important consideration in choosing any crop, animal or vegetable, and pasture grasses are no exception. Bermuda grass is a Southern grass, not much good north of the Mason-Dixon Line, or in Appalachia. Orchardgrass finds its best adaptation roughly along the Mason-Dixon line east of the Mississippi. Bluegrass does well in this area too, and on up a bit farther north. Redtop is the most widely adapted grass in North America, finding suitable climate and soil conditions almost everywhere. Bromegrass is adapted to all areas of relatively low humidity where high temperatures don't prevail for the majority of the year. Many other grasses are grown, but I'm not familiar enough with them to discuss them. Besides, the few mentioned here comprise probably 90% or more of the intentionally propagated pasture grasses in America.

Legumes adapted to permanent pasture in North America include clovers, trefoil, alfalfa (lucerne) and lespedeza. Of these, alfalfa and certain clovers (crimson, sweet) are difficult to establish on untillable ground. There is at least one clover for every region; trefoil is mostly adapted north of the Ohio River and Appalachia; lespedeza south of the Mason-Dixon Line.

The Way of Pasture

One of the most important skills in grazing livestock is to have pasture available throughout the growing season. Some grasses and legumes are very early growing in spring, and persist late into the fall. Most begin to really lag during July and August because the soil is driest and the sun is hottest. Many husbandmen wind up feeding hay in these months if they have insufficient pasture saved up. Some devote a bit of their land to temporary pasture, like millets or Sudan grass, both of which grow in the drought. Where I live, a mix of orchardgrass and lespedeza, planted thickly to make a close sod, will usually persist right through the dry months, albeit with slower growth than when there's plenty of rain.

Most pasture is kept as pasture precisely because it cannot be tilled. Hills, rocky, stumpy or poorly drained ground are most often pasture. This doesn't mean they can't be made productive.

Grass grows at a rate proportionate to its size. That is, it takes just as long for it to grow from 2 inches to 4 inches tall as it does to grow from 4 to 8 inches, or from 8 to 16. This is why it's best not to graze it closer than 4 to 6 inches. The most important things to remember are to select persistent species adapted to the site, and to manure the ground.

The latter is the most often neglected because most of us have been led by ignorant school teachers into believing the dung of grazing animals gradually improves and fertilizes grassland. Rotating livestock is even suggested for more balanced replenishment.

But we must remember grazing animals accumulate the most nutritive elements in their bodies, voiding those portions least soluble and least fertile on the field. If an animal dies and rots on the field, it will put back just what it took out, no more. The land improves not a whit. In removing livestock from a field, we also rapidly remove its richness. Then only grasses adaptable to poor land can grow, which are invariably unpalatable and not very nutritious. Often the land simply erodes to total barrenness.

Despite the land's poverty, grazing can be improved at a steady rate by buying in hay (and thus fertility), manuring with compost, growing very deep rooted legumes, and/or the use of artificial chemicals. The latter are officially repudiated by organic growers, even if they manure with lime, hardly organic. Anyway, "Lime spread as the sole manure; Fathers rich and sons poor." This old saying illustrates the fact that lime only makes extant nutrients more soluble, so plants can extract them faster, or they leach away to the subsoil.

Lime is the cheapest manure, and a very good one in its place, like many other inorganic substances. Calcium is essential to bone formation in grazing animals, so lime has to be available to the pasture plants. Ground limestone, slaked lime, humus (compost) wood ashes and bonemeal, in order of cost, all supply lime to the soil. In some localities slag and marl can be had cheaply. Not here.

Why is compost so good? Because it is normally made with cheap forms of cellulose, such as sawdust. Trees are very deep rooted, drawing water from great depths. In regions of high rainfall, where the lime leaches away most rapidly, only trees can reach it. That is why the eastern U.S. is naturally forested. In areas of low rainfall, very little lime leaches, and so grassland prevails. If deep rooted (deciduous) trees are allowed to populate highly acid soil, by the time mature forest returns the soil will be found very nearly neutral again, neither acid nor alkaline—perfect for most plants. Not only that, but there will be a great deal of humus accumulated, from decaying leaves, wood, bark, bugs and other wildlife.

In the wild, very little lime is lost.

Humus can be obtained from the forest and spread on pasture, but this is laborious, if not heartless. The trees need it too, as do the fauna of the woods. But they don't need all of it; after all, they'll eventually repopulate even abandoned farmland where practically no humus remains.

But humus isn't all. Dense sod sometimes becomes rootbound, a condition which can be remedied by harrowing your pastures (and then sowing better adapted grass/clover seeds too, if you like) with a spring tooth harrow until it looks like it'd never recover.

Artificials

The quickest way to see if you can improve you pasture with store-bought additives is to get some urea, some superphosphate and some potash from your local fertilizer merchant. Make a big letter N on the ground with the urea, a letter P with the superphosphate, and a letter K with the potash. Wait a few weeks and observe the grass reaction. Usually its the N, or nitrogen. Applying purely nitrogenous material to the ground will suffice for a year or two, but then the other elements will be used up by the more rank growth. Lack of K, or potash inhibits legumes. If you encourage legumes, they'll make enough N for themsleves and the grass. Insoluble phosphorus (P), abundant in most soils, is released by the solvent action of decaying organic matter.

Pasture can be made the most profitable part of almost any smallholding. That is, it can be made to return the most from a given investment of capital or labor. This is true of even untillable ground. Never neglect your pastures.

Specific Forages

Brome grass *(bromus inermis)* and orchard grass *(dactylis glomerata)* are the two most productive

permanent grasses in the U.S. Brome is probably the better of the two, but needs a considerable amount of lime to survive. It is palatable even until it goes to seed, a rarity among grasses. Ryegrass, meadow fescue, wheatgrass, and timothy all share this characteristic but are either far less fruitful, or cannot stand much grazing. Brome comes on early, stays late, and can tolerate pretty dry soils if the weather doesn't stay hot for too many months of the year. It spreads by runners, so increases in productivity from year to year. It cannot stand poor drainage. In every way bromegrass shares the same requirements as alfalfa, and it is often grown in combination with it. The alfalfa supplies nitrogen to the brome, and the brome helps choke out weedy plants.

Orchardgrass is considerably more tolerant of acid soils, and can stand just as dry conditions as brome. It produces just as much animal fodder, coming on as early, staying as late, and being as drought tolerant, but gets very unpalatable by the time it is in full flower. And unfortunately, it grows very strongly and flowers early. Its first cutting for hay, by far the biggest one, must be prompt, or it won't be suitable for sheep or dairy cattle. Beef cattle and horses will eat it if there's nothing else. The problem is that when it is ready to cut, the weather is still likely to be rainy. You can graze it early, then let it grow to hay so it's ready to cut in mid-July, and then graze the aftermath, or just keep mowing it to prevent it going to seed. If you do this, you'll waste most of its best growth. Fully two-thirds of its productivity is in the months of May and June. If correctly managed, orchardgrass is the most productive grass in the states just South of the Mason-Dixon line. It makes a very good mixture with lespedeza. The first cut of hay will contain mostly orchardgrass, but the aftermath will be well balanced with lespedeza. Orchardgrass doesn't spread by runners, and it grows in definite bunches unless sown so thickly the bunches grow in the appearance of tight sod. Alternatively, you could sow an acid and drought tolerant runner-type grass like redtop amid the orchardgrass.

ORCHARDGRASS

Redtop *(agrostis alba)* is very much used in New England, because of the worn out, acid soils there. It can stand cold, heat, dry soil, poor drainage, and seems to even thrive best where lime is lacking. It makes a pretty good hay plant, and it spreads by runners, forming a coarse sod. If a lot of bluegrass is present, livestock will somewhat neglect grazing the redtop until it gets totally unpalatable. Sadly, the bluegrass won't

RED TOP.

tolerate the dry months as well, and then you have nothing but the inedible redtop. One way to stop this process is to mow the pasture before the redtop flowers, to keep it young and tender. Redtop is ready for haying just as it flowers, which is late enough to miss most of the rainy spring weather. Its best companion is white clover. Its seeds are the smallest of the common grasses, having about 10 times as many per pound as orchardgrass. So, though it is expensive, it takes far less of it, and it will spread itself by runners in time.

Bluegrass is a short, very nutritious, persistent and very tasty grass. It's no good for hay, because it doesn't grow tall enough. If the ground is very fertile, as it always becomes where dairy cattle are kept, it will eventually take over the field. Its chief fault is it really suffers from dry heat in July and August. It sometimes looks as

Kentucky bluegrass

if it has completely given up. Never fear, with the fall rains, it will come back, but not enough to make up for its lost productivity before winter stops its growth. It makes its best partnership with white or Ladino clover, both of which unfortunately have the same summertime problems. I don't think this grass is as productive as its high reputation seems to suggest. But it is a beautiful sod, and fills in gaps among bunch grasses. There are two common American bluegrasses: Kentucky bluegrass *(poa pratensis)* and Canada bluegrass *(poa compressa)*. The latter is

CANADA BLUEGRASS

more tolerant of both poor drainage and dry ground, and can take more acidity. But it's not quite as palatable.

Bermuda grass *(cynodon dactylon)* is the main pasture in the most Southerly states. It is good for both pasture and hay, binds sandy soils well, can take acidity, and is weed-proof whereever the ground doesn't freeze solid. Burr clover *(Medicago arabica)* is the legume companion of choice, and some kinds of lespedeza also will grow with it. The legumes greatly

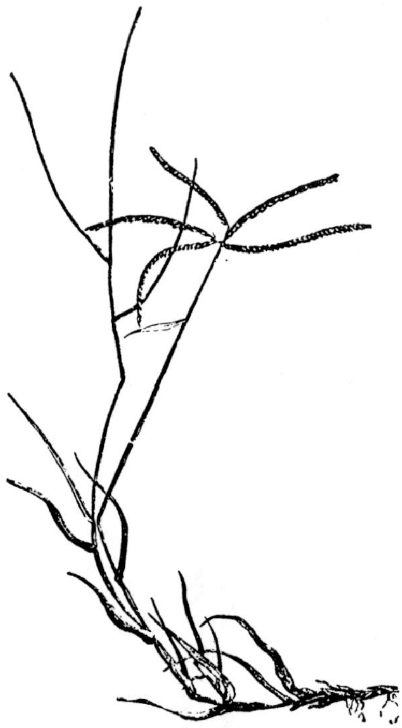

BERMUDA GRASS.

enhance its production. It is often cut for hay, up to 3 times in a year, but it only totals about 1½ to 2 tons per acre in average circumstances.

As a rule, legumes are a good deal more nutritious than grass, but livestock seem to need grass too. Grass cures into hay far better and quicker. Grass amid the legumes helps the latter cure without spoiling. Lespedeza is the most rapidly curing legume. Trefoils aren't much grown, because they take 3 to 4 years to really

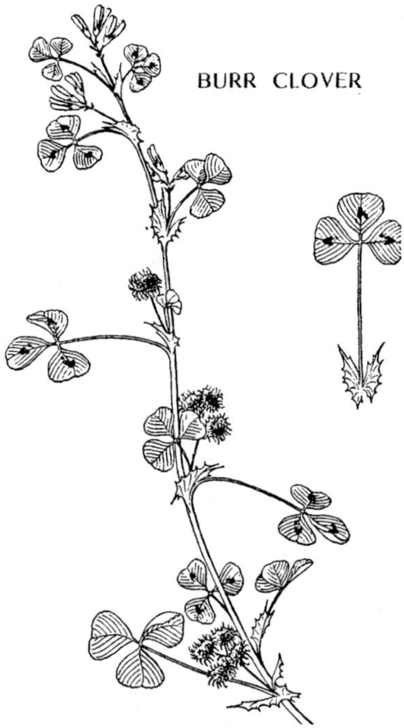

establish themselves. But they are very good forage, being very palatable, productive, acid tolerant, and survive poor drainage conditions. Like Bermuda grass, they are best grown from "plugs" which are planted like bulb flowers. These will take hold and spread on their own.

Whatever grass you grow, don't forget to rotate your farm animals among your pasture paddocks. Calves first on new grass, then cows, then lambs, then ewes, in that order. If you are a monoculturist, rest the paddocks after grazing them right down, both to eliminate parasites and allow the grass to recover.

The artful grazier is quite a skilled person, never letting his grass get too short, knowing how much there will be to graze in summer's heat, or when and how high to cut if it gets ahead of his stock. When asked what was the most profitable type of farming, the ancient Roman statesman and agricultural commentator Cato replied "Good grazing." When asked what was the next best, he said, "Mediocre grazing." And the third best? "Poor grazing." Enrich your pastures, and they'll enrich you.

24
MEADOW

The hay appeareth, and the tender herb sheweth itself, and herbs of the mountains are gathered.

Proverbs 27:24

Hay is made in the meadow. Grass is cut just at the right time, and is dried in the field until it has lost enough moisture to keep from heating and molding in storage. Hay making is a great labor, unless one has all the expensive mechanical implements to do it. These include a tractor, mower/conditioner--which not only cuts the hay, but runs it through rollers to quickly squeeze most of the sap out of it--tedder, which fluffs it up so air can get under it, rake, baler with kicker to throw it onto the wagon, and bale elevator to get it into the barn. Nowadays large round bales are getting more common, and as these weight from 600 to 2000 lbs., a hydraulic lift is also necessary to move them.

All of that is very nice, but for all the expense you could probably instead buy all the hay you'll need for the life of the equipment. If you contract someone who owns all of it to come make your hay, you must remember that everyone's hay is ready at exactly the same time, and the weather is the same for all, too. Your hay is likely to get rained on or just postponed until after it has gone to seed, when it is mainly fit for bedding straw.

One obvious thing to do is buy good hay and improve the fertility of your place (see the previous chapter).

Another option is to carefully manage your holding so as to minimize the need for hay. You do this by conserving your pasture. You may even plant a crop of rye, the most winter hardy of the grass family, so that you'll have grazing available late in the fall, and again very early in spring. That, and some fodder crops like pumpkins or roots, will cut hay consumption dramatically. Another thing is to avoid overstocking. Be honest about your land's carrying capacity, it's probably not very high these days. Better to understock than overstock.

One advantage of making hay by hand and horse methods for a smallholding is that there are many ways to cure it not suitable for people with too much land. But before we get into that, let's look at hay crops.

Hay Plants

Alfalfa *(medicago sativa)* is the king of hay crops. It can be fed to good advantage to all farm animals, including poultry and pigs. It yields higher than any other hay crop, it is the most nutritious of all hay plants, and it survives all

ALFALFA OR LUCERNE.

kinds of difficult weather conditions. Alas, it is a crop which almost has to be planted on tillable ground, because when young is very susceptible to weed competition. It also absolutely must have plenty of lime! So the ground is prepared by first liming, then plowing in the lime. Then disking, then liming again, then harrowing with a spike-tooth harrow.

How much lime? Get a soil test kit at your garden supplier, and keep liming until it's a pH of 8. In the west, with low rainfall, it's probably already about right. In the east, it may take from one to as many as 4 or 5 tons of lime per acre. Even then, it will need about a half-ton per acre each year to supplement the lime leached and taken up by the crop.

It must be planted in late summer, preferably after fallowing all summer, which means just tilling it every week or two to kill every weed that sprouts. Sow 15 to 20 lbs. of a seed variety your seed merchant recommends for your area in August or very early September. Use certified seed. By next spring it'll already have roots 3 feet deep! If you see any weeds by May, harrow the whole crop with a springtine harrow. It's hard to make yourself do this after all that trouble, but be assured you will not hurt the alfalfa, no matter how terrible it looks. Three things will keep your alfalfa crop productive as long as you live: 1. Don't cut it within 30 days of the first fall freeze. It must store nutrients in its roots to survive the winter. Cut it right after the freeze, and the hay won't be too awful. 2. Lime it every year if you live in the humid east. Keep it about pH 7 to 8. 3. Harrow it when you see weeds, and don't bother growing it with a grass mixture, except brome grass. Many people get 6 to 9 tons a year of alfalfa hay in West Virginia. Out west, they get even more. For the biggest harvest, it is cut when it reaches about half blooming stage.

Red clover *(trifolium pratense)* is a good hay crop too. It yields nearly as high as alfalfa in good years, but needs lots of lime too. Not as much as alfalfa, true, but most failures with it are due to acidity. Red clover hay is great for dairy cows and most sheep, but no good for ewes during breeding season, because it affects their hormonal balance, reducing your lamb crop. It gives two heavy cuttings, as opposed to alfalfa's 3 to 5 (even 9 in some areas!) It has deep roots, but not as deep as alfalfa. It dies out in 2 years, being biennial. Sow 12 lbs. alone, or 8 lbs. when sown with timothy. Cut when flowers are in full bloom. It is often sown alone, but can also be used in a mixture with adaptable grasses.

RED CLOVER

Mammoth clover *(trifolium pratense perenne)* is a variety of red clover, but is perennial. It blooms late, and is thus ready to cut at the same time as timothy. For this reason, the two are often kept together in the same meadow, being both long-lived, more tolerant than red clover to acid, and more likely to be ready to cut by the time good haying weather has arrived. Only one cutting per year is ordinarily secured, but it's a big one, 2 to 4 tons per acre.

Alsike clover *(trifolium hybridum)* is much used in place of red clover because it can take both more wet and dry soil conditions, and more acidity. It too makes a good companion with timothy, though on wet land or acid it would be safer to sow it with redtop. Sow 8 lbs. alone, or 5 lbs. with grass. It yields a bit less than red clover, sometimes as much as a third less.

Sweet clover *(melilotus alba)* requires as much lime as alfalfa, is not perennial, and gets quite woody in its second year. Its main claim to fame is it will grow in limestone

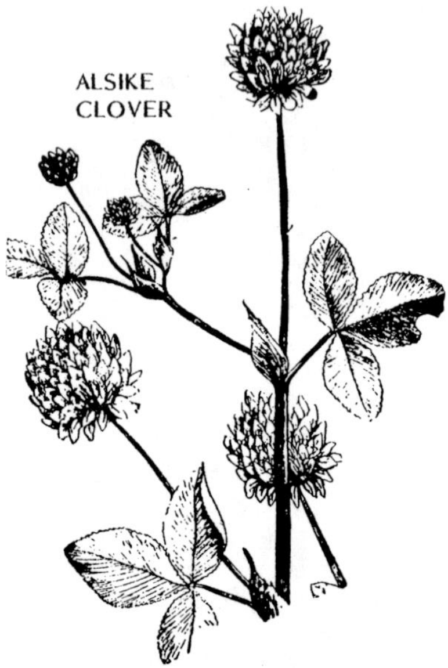

gravel, needing practically no nutrients. I don't bother with this one.

Crimson clover *(trifolium incarnatum)* is a winter annual. That is, it is planted in the fall, grows until the coldest weather prevails, and then resumes very vigorous growth early in spring. Its main use is as a winter green manure crop, being planted from Maryland to the Carolinas. It can't take much colder winters than that, and its limitation in the gulf states is that late summer is often too dry to get a stand. Its hay has little hooks on the stem which can cause horses to get colic, but is good for sheep and cattle. I have never grown it.

Hairy vetch *(vicia villosa)* is winter hardy right up into Canada. I find it grows best if I plant it in the corn field, at the time of the last cultivation. It will sprout and

make spindly growth under the maturing corn, but as soon as the corn is taken off it will continue growth until some time in November. It will resume growth in the spring, and makes hay practically the equal of alfalfa. It can grow on acid soil, and soil practically devoid of other nutrients. Sowing isn't cheap though, needing sixty pounds of seed per acre. Its growth is often so heavy as to make cutting difficult. Cut when in full bloom.

Timothy *(phleum pratense)* retains its palatability right through

CRIMSON CLOVER.

seed stage. You can feed it safely to horses, for it hardly ever molds if let to dry for even only one day. It has always been the standard hay for horses. It yields one very heavy crop. It's not much use for pasture, because it has a *node* at the top of the root, above ground, which are much damaged by the tramping of hoofs, and close-eating teeth. Its seed is very cheap, and very small, so rather little of it is needed. Still, sow about 15 lbs. per acre. Cut it just after it finishes flowering. All that pollen is debilitating to both man and beast, and its quality isn't noticeably diminished. Besides, by then the weather is dry enough to make superb hay.

HAIRY VETCH

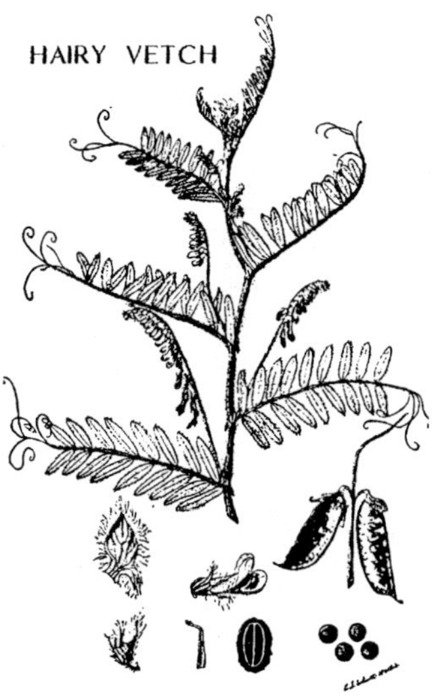

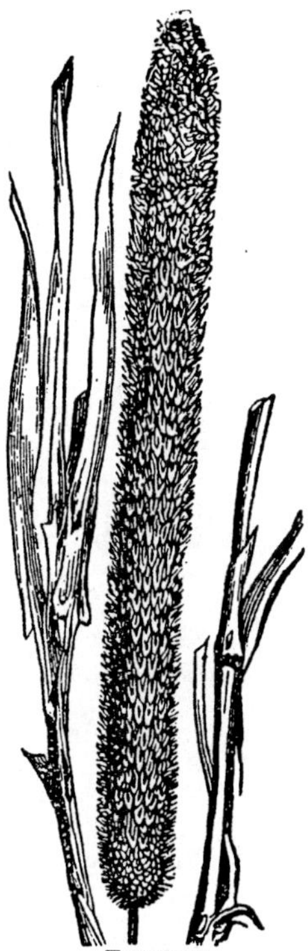

TIMOTHY.

Just plain mixed grass, such as might also be used for grazing,cut on time, makes very good hay, maybe the most desirable for sheep and dairy cattle.

Making Hay

I have cut many acres of grass with nothing more than a scythe. To do this effectively the scythe must be positively razor sharp. Not just sharp; razor sharp. It cannot be maintained so without keeping the last ¼ inch (the very edge) ground or peened very thin, so the scythe stone only needs one or two swipes to reset the edge. I have never found an American blade which worked as well as the Austrian made blades. Even those, unless they are the hand forged ones, are no good. The hand forged ones are difficult to find in hardware stores, and must be gotten through mail order most of the time. They aren't cheap, but if they're trued through peening instead of grinding with a file, they may well last a lifetime. The handle, called the *snath,* is either wooden or aluminum. For ease of swinging, I like the aluminum. It is swung very close to the ground, indeed it glides right on the ground with the edge slightly higher than the spine. This isn't possible with American made blades, because the tang isn't angled properly. The swing isn't like driving a golf ball way down the course, it is a slow, gently swaying motion of the torso. I don't advise taking a cut of more than two inches, nor more than 3 feet wide at a swing until you're familiar with the whole process. Nothing is more unnerving than watching a beginner sink the point right into the ground because he doesn't have the snath adjusted for his own height. The handles must be positioned so your blade practically balances itself. With some time and practice, you can cut up to about 6 inches in one pass of the blade. I hear armchair experts flatly stating one to two acres per day as about the normal amount cut. This is utter nonsense. A half acre is all I can do in good cutting ground with a superbly sharp edge if the hay crop is moderately heavy. I think it's something of an endurance test to do more than ¼ acre per day on a daily basis. Even half that is plenty for most folks! Besides, you need to leave yourself enough time to get yesterday's cut into *a cock* or on a *hay fence.* The former is just a pile of hay about 6 feet in diameter and 7 feet tall, built on a 5 foot high tripod with several cross bars or twine wrapping for the hay to rest on to get the air circulating under it. The latter is a temporary fence consisting of two endposts, well anchored, with 3 evenly spaced wires tightly stretched between them. Intermediate posts are set to help keep the wire from sagging. The whole thing is built right down the center of your meadow. The bottom wire is set two feet from the ground. You pile lots of hay on this until it's about two feet above the wire. The second wire is now strung, and the first re-tensioned. More hay is piled up two feet higher. Now the third wire is strung. Hay is piled all the way up. The length of fence is based on the quantity of hay. You may only need two wires. This method works even in rain, and very little hay is lost. Just leave until your hay is dry. Then undo your wires, pulling them right back through the hay onto a reel. Load the hay on sleds or wagons, or a pickup truck, and take it to the barn.

In better weather you may simply rake the hay into windrows, which are long rows of hay about two feet high when fluffy, a foot high when settled down. This keeps them from getting too brittle from drying too fast in the sun. Brittleness means loss of the precious leaves, which contain more than 75% of the goodness of the hay. This is why hand made hay is so superior to machine made. Also, it's not so compact, and is thus less likely to mold or heat in the barn.

Storing Hay

It takes about twice as much room to store a ton of loose hay in your barn as machine baled. But the work can be so much more pleasant. I prefer to have the hay off the ground, so we get cast off packing pallets from various businesses, and just lay them on the ground, whether we have made loose hay or bought it baled.

You can see hay making is actually quite simple. But it is hard work. Cutting at the right time and in the right weather, and timely storage, is everything. Grass cures much faster and is less prone to heating than legumes. Never cut more than you can timely make into hay. One old secret to haying was to simply cut the hay, rake it up and cart it to the barn floor. As long as no more than 3 inches of new stuff is added daily it'll neither heat nor mold. I have never done that on a field scale, but my experience with just a few square rods has been that it makes the very best deep green hay.

Even seasoned hay makers often put the hay away too wet. It heats, molds, browns, and has even been known to catch fire. There is a reliable method to find out if the hay is dry enough. Even moist hay appears dry in good hay drying weather. Take a representative sample of the hay from the field, a handful big enough that when cut to length it will loosely fill a quart jar. Put a teaspoon of table salt (not coarse) in the jar and shut the lid. The hay sample shouldn't be so tight to prohibit the free circulation of the grains of salt when you shake it. Shake the jar about 100 times to keep the salt and hay moving about. Hold the jar upside down, and shake a little more to get the salt down into the lid. Remove the lid. If the salt has less than 25% moisture content, the salt will still be in small grains, and the hay is safe to store. If too wet, the salt will be gathered into little clumps. If the salt changes in 30 seconds or less, the hay is definitely too wet. In borderline cases repeat the experiment, and let the salt stand a few minutes.

If hay sits in the field any longer than needed to get to the safe moisture level it will start losing quality. If it gets rained on, it loses fast, unless the rain is light. Just ted it about when it clears up, and proceed as before. It'll lose some, but you'll still get a harvest. A heavy rain or two won't hurt hay in a tripoded cock or a hay fence.

25
POTATOES

Moreover the profit of the earth is for all: The king himself *is served by the field.* *Ecclesiastes 5:9*

I was not fond of potatoes until I discovered the ancient method of growing what are called "straw potatoes." In this method, one need not plow the land at all. Nor cultivate. Except for possible bug or fungus control, you do nothing until harvest time. Potatoes grown like this not only taste far better than the standard dirt-grown methods, but they don't "run out" for use as seed. That is, they can serve as next year's seed, year after year, without losing vigor, even in climates not normally capable of growing healthy seed potatoes. In fact, run down potatoes, grown from conventional methods are brought back to full vigor by one year of growing under mulch.

Potatoes not only can form a fairly suitable replacement for grains in our diet, but can be fed to cattle and sheep in place of grain too. It takes from 3½ to 4½ lbs. of potatoes to equal a pound of grain. But then, a field which will grow a ton of corn can grow up to 3 or 4 tons of potatoes. Potatoes can also be fed to hogs and chickens, but they need to be boiled or steamed so the starch cells burst, making the energy available.

Conventional Growing Methods

Potatoes like fertile soil, but prefer it slightly acid. Don't lime the soil very heavily, if at all. Plow it deeply, to mix all that dung you've applied. Speaking of which, a good crop needs about 20 to 40 tons of dung per acre, half that much good compost, or the equivalent in expensive store bought "fertilizers." The latter will do nothing to help make tillage easier. And till you must, to keep the weeds--whose seeds you brought up when you plowed--from stealing all the water and plant food and growing their roots right through your potatoes!

To make tillage more convenient, potatoes may be planted in rows two feet apart. Tractor cultivation requires rows a bit farther apart. I always made the rows by cutting a shallow trench with the wheelhoe, and dropping the seed potatoes about 8 inches apart in the rows. They can then be cultivated with the wheelhoe weekly, whether they need it or not. The secret to growing any cultivated plant is to cultivate weekly whether they need it or not. They need it.

Seed potatoes are just healthy potatoes from last year's crop, or of whatever varieties you prefer. Plant what your neighbors do best with. Experiment on a small scale with exotics before you trust your whole field with 'em. In any event, we've found that *chitting* the potatoes makes much more robust plants. This means simply setting them in the sunshine to turn green a few weeks before actual planting time. Short, thick eye sprouts are much healthier than long, gangly ones. The only exception to this is if you have very few seed potatoes. The eyes can be gently removed and immediately planted, while new ones are allowed to grow on the original seed piece. But they still don't need to be gangly. Don't eat green potatoes, they are quite poisonous, though they have no such negative effect on cattle and sheep.

Larger seed potatoes can be cut up, with 2 eyes on each piece, no more, no less. More will make a bigger plant, with smaller tubers. One eye will make bigger tubers, but less of 'em.

Planting time ought to be chosen so that the nights will be cool 6 to 8 weeks after planting, when new tubers are forming in the soil. If nights are hot, the yield will suffer. We start putting them in the ground about 3 to 6 weeks before the last expected spring frost date. If you're too late, wait and plant a fall crop. Check the variety's maturity schedule and count back from the first expected fall frost. Keep them tilled until the vines start falling over into the rows, making tillage difficult. At this point it's wise to hill up all that loose tilled earth against the plants to protect the tubers from the sun, which will green them.

At some time during the growing season two pests will probably appear. The *Colorado potato beetle* winters over on "rogue" potatoes, those which escaped harvest and volunteered in the spring. Or on the common horse-nettle *(solanum spp.)*. The best way to control these is to pick the earliest ones off by hand and smash them so they won't lay their little orange eggs on the undersides of the potato leaves, which hatch into brown and black speckled, potato eating larvae. Alternatively you can dust with poisons on a weekly basis.

The second pest is the *jassid,* or *leaf-hopper*. This pest is extremely destrucive every year here. It is a tiny green insect only about ⅛ inch long and migrates into the region sometime in June. They come by the trillions. They can creep through common window screen and fill up the electric light fixtures. Yes, they can even give an annoying little bite. That bite on a potato plant causes a problem in the stem's juice flow. They lay eggs in the stems too, and a huge population of green wingless nymphs hatches, hardly visible unless you get close. The potato leaves begin browning at the tips; eventually all the leaves die, then the stems. Some vigorous plants will send up other shoots, which in turn are attacked by the hungry bugs which have now been deprived by the death of their first potato hosts. The only help for it is to dust with poison as soon as the jassids appear, with the intent that there will then be little or no second generation. A field devastated by "hopperburn" appears to have some sort of blight, but the potatoes still survive in the ground, usually. What there is of them. The yield following hopperburn is only a small fraction of what it ought to be.

Digging Spuds

Harvesting is done in many ways. Tractor-drawn potato diggers root up all the soil with a wide shovel point and deposit it on a moving track of close-spaced rods. They bounce along like this losing most of the clinging soil and are dropped behind, on top of the ground. Field hands come along and pick up the tubers. Horse drawn potato plows have a flat plowshare with a fan of iron or wooden rods which somewhat remove the dirt, but nothing like a real potato digger. Then there's the method we use, which is a sort of curved fork called a *croom.* Most folks just call it a "potato hook." It is swung into the ground beside the row, pulled towards you, and the potatoes are flipped out amid clinging dirt. Harvesting like this is hard work, but since store bought potatoes are so inferior to homegrown, it's worth it. I once grew four bushels of potatoes from 80 feet of row! If I could keep that up on an acre basis it would have been 1400 bushels to the acre. I planted the crop late, on July 1. I missed all the Colorado Beetles and leafhoppers, and really, most of the chances for blight.

Potato blight is serious business. It is caused by a fungus call *phytophthora infestans.* It sprouts in muggy, warm weather, and is thus prevalent anywhere these conditons are common in summer. The only way to fight it is to prevent it. One may spray with old fashioned Bordeaux Mix, which is discussed in practically every gardening book.

My own spray for blight is superior to Bordeaux. It is called *Peret's misture,* discovered in 1882. I use ¾ cup of copper sulfate, mixed thoroughly in ½ gallon of warm water. I add to this 1¼ cups of washing soda (sodium carbonate) very gradually. It foams up something fierce in my plastic gallon jug if mixed too fast. Then I lose half of it to overflow. Patience is a real virtue. After the reaction subsides I have an opaque deep green fluid. To this, I mix ¼ cup common sugar. Eventually the green will all settle out, leaving a clear blue liquid. Shake the bottle to remix it before using. To use it, first make a mark on the side of the bottle at the full level of the liquid. Then make other calibrations at each quarter of this. Pour one quarter of this in your hand sprayer, and mix in enough water to make a 1½ gallons. Spray the plants with this as soon as they're up, on both top and undersides of leaves. I have sprayed as much as once a week, but don't think it needs more than 3 sprays, evenly spaced throughout the growing season. It clings much better than Bordeaux after rains. It also stops blight dead in its tracks if you spray early enough after symptoms appear. Symptoms are blotches of purplish black "water soaked" places on leaves. Eventually the whole plant turns black and dies if not intercepted. The tubers may not be affected if the ground isn't disturbed until long after the tops have dried up. Collect blight smitten tops and burn them, and then collect the potatoes.

Straw Potatoes

There are great advantages to growing straw potatoes besides better tasting spuds and less labor. They can be grown on soil too rocky or steep to till, and the residue leaves the ground in perfect condition to grow a crop of *legumes,* which is to say, beans or peas. You see, straw or (especially) wood wastes like sawdust, shavings or bark, when tilled in, rob nitrogen from the soil, killing all the weeds which sprout because you brought up their seeds to the surface while digging up the potatoes. But the leguminous plants fix their own nitrogen from the air through the action of little tubercles on the roots. If the ground has never grown these legumes, it's wise to innoculate the soil, either with commercial innoculants or with a few bushels of soil per acre from other ground which recently grew them. The nitrogen fixed by these crops is what makes them so

—Properly cut seed potatoes. Each piece has two good eyes and is about the size of a hen's egg.

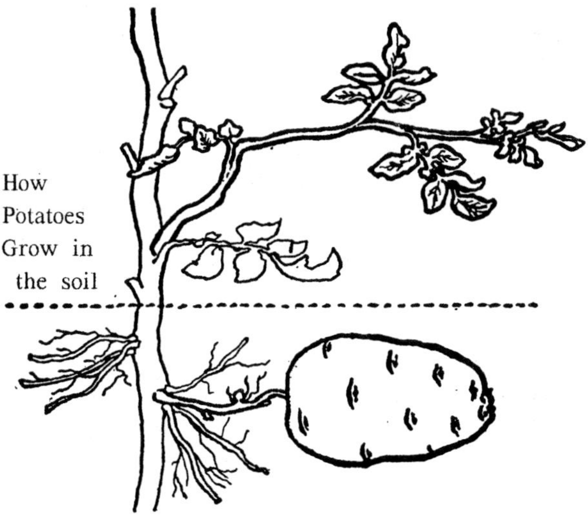

How Potatoes Grow in the soil

The Case for Controlling Insects

high in protein. The main difference between carbohydrate and protein is the presence of nitrogen.

To grow straw potatoes, lay out the rows just like you would have with conventional culture. But just prior to planting, spread a layer of chicken or pig dung right on the surface wherever the mulch will lie. Then lay your seed potatoes right on the ground, spacing them 8 inches to a foot apart in the row, with rows 2 feet apart. Cover with 6 to 8 inches of sawdust, straw, bark, or whatever organic mulch you have lots of. I have a friend who grew splendid potatoes with spoiled silage from a nearby cattle farm.

Harvesting straw potatoes is very easy. Just hook them right out of the mounds of mulch. The last 3 weeks of the growing period form the biggest growth of tubers. The tops then die down, and they can be harvested at any time. Or leave them in the ground until the weather is cooler, so they won't heat too much in the cellar or clamp. When dug, they should be left on the ground for a couple hours to set the skins and toughen the spuds. Then store them either by burying them in trenches covered with 8 inches of straw and 6 inches of soil, or in a cool root cellar. Feed the small ones to livestock and cut up the biggest ones next year for seed potatoes.

26
PULSES

Give us pulse to eat and water to drink.
Daniel 1:12

Pulse crops follow mulched potatoes in the rotation, because all that mulch, especially if it is sawdust or other woody fiber, will bind up all the nitrogen in the soil so that only plants in the family *leguminosae* can survive. They have mutually beneficial relations with certain germs which form tubercles on their roots and fix plenty of nitrogen from the air. So while all the weeds around them turn purple and die, they do very well indeed without all the cultivation normally necessary. So having grown potatoes, we till the mulch well into the soil, and that's the only tillage until next crop! What's more, one need never have plowed the ground, only harrowed it. This way you don't need more than one draft animal, and you get bigger crops with less work than if conventional methods were followed. I am indebted to the pioneering work of F. C. King, a British market gardener of roughly the first half of the Twentieth Century for this discovery.

Soybeans

Soybeans *(glycine max)* are the king of all pulses. They outproduce almost all other beans, and bear the highest quality of protein. Not only that, they have about the widest adaptation. Their only drawback is that for feeding to pigs and hens (the animals which most appreciate

Soybean Roots With Tubercles

this high quality protein) they must be cooked. Soybean oil meal from the mill has been heated in extracting the oil.

SOY BEAN.

Soybeans are somewhat frost hardy, but will not sprout in cold soil. They have the same climatic requirements as corn (maize) which is why they're so widely adapted. Like corn, they are very parochial. What I mean by this is that it takes several years of growing a variety before it really acclimates itself to the microclimate of your holding. It is entirely self pollinating, and its flowers are so tiny as to be almost totally obscured by the rest of the plant. I have planted them as late as July 17, and still gotten a respectable crop, two weeks after the first fall frost. But this was a very short season variety, and they didn't yield as high as if they'd been planted on time.

Many varieties are available, and I would recommend growing the longest season type which your climate permits. Some hybrids have reliably yielded over 100 bushels (6000 lbs.) per acre, but these don't come true to seed if you plant part of the harvest next year. I select seeds by setting aside the plants with the most pods, and with the most beans in the pods. I thresh them separately and plant their seeds.

Plant the seeds when the soil is about 55-60°F at dawn. Then they're less likely to rot before they sprout. Grow them in rows two-thirds as far apart as the variety is tall. Plant about 4 to 6 seeds per foot in the row. I just use the pea plate on my seed drill. I think it is very helpful if the first time you grow them you inoculate with the correct inoculant. Follow the supplier's directions. I shake the seeds in a jar full of water, pour the water out, add some innocculant and shake again. Then I sow immediately. I like to sow them an inch deep if early planted; up to 3 inches deep if planted when the ground is real warm.

If you are just growing them in ground without last year's mulch, you'll need to till the soil once a week. The wheelhoe is the fastest way unless you have a horse. It is best to have someone lead the horse if he's not used to cultivating. The Swiss have special one-horse cultivators which will completely cultivate several rows, but I don't know where to buy them. I have also seen something which looks like a very low-geared pedal car which cultivates two rows at once. Rototillers are o.k, but they take forever, are noisy, stinky, and don't do well in difficult soils, like clay or stony ground. I sold mine years ago.

Harvest when the leaves have turned yellow and fallen off. You can pull up the plants, but if you do this, you'll also be removing all that nitrogen the roots have in them. Then the next crop, small grains, will really lag. Better to cut them with a scythe, bundling them up in sheaves and tying with string. I like to store them out of the rain, on an elevated platform if possible, whose posts are either smooth metal, or are topped with a wide piece of sheet metal to exclude rodents. Mice, rats, and chipmunks love soybeans.

Thresh the beans with either a mechanical threshing box--you can use your own wisdom and ingenuity to devise one of these--or by beating the beans in the straw with an old fashioned flail. The best time to do this is winter when they've dried better, and you don't have much to do anyway. Some folks hold them by the stems in a barrel, and beat them with a piece of plastic water pipe, or just smack the plants back and forth against the barrel walls until the seeds come out. This is very time consuming. You can feed them like hay to your sheep and cow, but feed it in a hay feeder, so you don't lose the seeds.

As for pigs and hens, the beans may be steamed for them, by putting them into a large clean metal barrel, stems down, and put a couple inches of water in the bottom. Light a fire under the barrel, and lightly cover the top with a scrap of sheet metal. Steam for a half hour and remove. Then feed stalks and all, not too much at a time, until you find out how much they'll take without wasting them. Don't feed as the sole feed, but as a supplement to grain, otherwise it's too rich. The black varieties are difficult for hens to scratch, so they'd need to be threshed out before steaming.

Bean Curd

Between potatoes and bean curd, or *tofu,* one has nearly as perfect a diet as one might have without animal products. It would need supplementing with vitamin B_{12}, a by-product of the antibiotic industry, but most vegetarians can afford this. Even if you're not a vegetarian it's handy to know how to make bean curd, to use in oriental food recipes.

Soak 1 cup of soybeans over night; drain and wash. Using twice as much water as soaked beans, mash them up to a fine pulp. Some folks do this in a blender, a half-cup of soaked beans at a time. Strain all this through a sieve, retaining both pulp and liquid. Mash it all again, and strain again. Bring the mix to a full boil, in a double boiler--it sticks easily. Add the juice of two lemons, or 4 tablespoons of vinegar, stirring only enough to mix it well. Allow the mix to stand in the heat until it has coagulated. This bean curd can be used in any recipe calling for tofu or soy cheese. Keep it cold, it spoils very easily.

Peanuts

In areas too hot to grow soybeans, peanuts *(arachis hypogaea)* are just about as high in quality, but contain more energy and a little less protein. They have the added advantage that they have an established market if you should have to sell your leftovers.

The original name for these in the U.S. was "goobers" the anglicized version of *nguba*, the African word. They were taken early from South America by Spaniards to Africa. The peanuts, also called "pindars," were a cheap food for slaves bought from wealthy Africans for use in sugar and tobacco plantations. When New England followed their lead, rum now being used to buy criminals and outcasts from the African authorities, the infamous "golden triangle" was formed, and Africans were sold to Southern states, along with the *nguba*, or goober. So at least something good came out of the dreaded institution of slavery.

The best peanuts are grown in sandy soil, especially if it contains

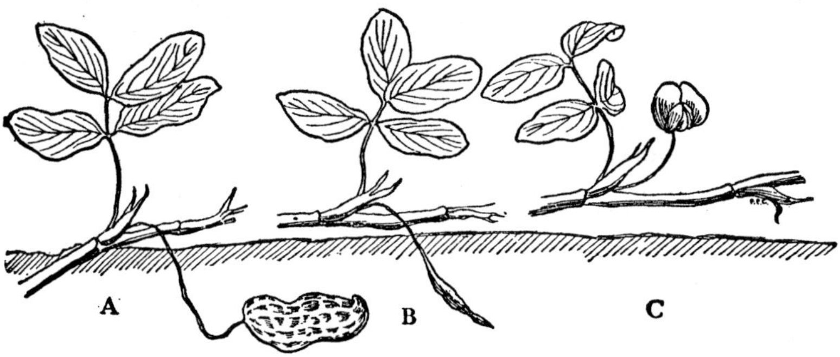

Three stages in development of the peanut: *C*, the blossom; *B*, the "peg" just entering the soil; *A*, the peanut pod.

The peanut plant, Virginia running variety.

plenty of calcium. But peanuts prefer acid soil, so instead of lime, gypsum (land plaster) is the best source. They commonly produce 60 to 80 bushels of pods, weighing about 25 to 30 lbs. per bushel. The heavier peanuts are of the Spanish type, which grow farther north, are smaller and have more oil. The so-called "Virginia" peanut is bigger, lighter per bushel, and has less oil. There are runner and bunch varieties. Grow what your neighbors grow, and you ought not go far wrong.

The blossoms of this plant hang down, forming pegs which pierce the soil; the pods form and ripen underground. They are an exhausting crop because the roots and all are pulled to get the nuts. If you use fertilizer, something like a 2-10-6 (N-P-K) formulation at 300 to 500 lbs./acre is about right. Too much dung makes big tops and not much fruit. If cultivated rather than grown on dug-in mulch, they need to be tilled every week, shallowly, and it's best to not grow them at all on weedy ground.

If pods are planted (instead of shelled beans) they take longer to come up, so mark the rows with gypsum so you can till on time. Large types are sown in rows 36 inches apart, smaller ones about 28 inches. They are spaced 6 to 12 inches in the row, 2 to 3 inches deep, depending on size. The bigger, the deeper. Two bushels of pods per acre are needed or ½ bushel of shelled seeds. The latter are very fragile; don't harm the skin in any way. Plant them after all danger of frost.

They are hilled up just at about the time the pegs start to enter the soil, the same as dirt grown potatoes.

Harvest peanuts with a croom, or potato hook. It helps if a "hoop plow" is first pulled under them

Method of shocking peanut crop, over a stake.

by a horse. This is nothing but an arched blade about 15 inches wide, and 5 inches from front to back, with plow handles affixed, which slides under the plants like a scoop shovel, cutting the tap root. Peanuts are cured by stacking them around a seven foot pole stuck in the ground. At its base, cross pieces are nailed to help keep the vines off the ground so they can air-cure. The peanuts are picked from the dried vines during dry weather and stored in a wet-proof, rat-proof crib. Those used for seed ought not be allowed to freeze. Leftover vines make good livestock fodder.

Peas and Horse Beans

Field peas *(pisum arvense)* and horse beans *(vicia faba)* are grown in regions too cold for either corn growing or soybean growing. Such areas exist mainly in the Pacific Northwest, northern Minnesota and the higher elevations of the Rocky Mountain valleys and New England. Neither crop stands temperatures over 80°F very well. I tried growing horse beans many years before I came to my senses and gave it up. Here in the mountains, the summer weather is cooler than many other regions, but not cool enough for horse beans. Then too, peas and beans can take only relatively little cold. No horse beans survived the winter here, though some almost did, with temperatures dropping to -10°F in a less severe year. Peas are dead before this, withstanding about 10°F. Get hardy varieties.

Field peas spread all over the ground, so are grown commonly with oats. Culture is the same as unmixed oats. They also yield much more this way, up to double. Two bushels of each, mixed, per acre is about right. *Maslin* is the name of a crop formed by more than one seed sown in a single field. Yields are commonly about 40 bushels on good ground, but double that isn't too uncommon with perfect conditions. Seeds are usually kept for 2 years to thwart the pea weevil, which can get pretty troublesome.

Horse beans ripen well only in cool humid regions, thus the Pacific Northwest is its only U.S. home. It can yield enormously, but as human food it sometimes causes allergies among folks of Mediterranean descent. It is best to hold it for a year before feeding to livestock, to mature. Its chief advantage is it needs no grinding for sheep or swine. All other animals need it cracked. In the northwestern states the crop is best planted in October, wintered over, and a crop harvest late the following summer. It smothers out all its competition with up to 5 foot stalks and abundant leaves. Here, it is attacked by fungus and bugs before it ever makes the crop. And then, we have to plant it around March 15 to get it to sprout in spring time. Plant in rows 1½ feet apart, 3 to 6 inches apart in the rows. Harvest like soybeans. Peas and Horsebeans have high quality protein, which needs no cooking. Peas are fed to swine and sheep right in the sheaf. They both contain about 26% protein in the seeds.

Field Beans

Common field beans are what are used for making baked beans. Their protein content is low (as low as 14%) and the quality of the protein is also lower. They are sown in rows the same as soybeans in every way. Harvest is the same, too. You can get seeds by just getting the kind of beans you like at the market. Great Northern has the tenderest skins for making baked beans. Field beans have a great pest, the Mexican bean beetle, which must be dealt with. Dusting with poison is about the only effective control.

Field beans can be fed to cattle and sheep, but the protein quality is too low, even cooked, to compete with soybeans or peas for hens & hogs. Don't feed them to horses or they'll get windy colic.

27
SMALL GRAINS

Doth the plowman plow all day to sow? doth he open and break the clods of his ground? When he hath made plain the face thereof, doth he not cast abroad the fitches, and scatter the cummin, and cast in the principal wheat and the appointed barley and the rie in their place? Isaiah 28:25-26

Follow the pulse crop with one of small grains. Small grain means those members of the grass family whose kernels are smaller than those of maize (corn). Most specifically it refers to wheat, barley, oats and rye. There are quite a few others, but these are the most important small grains of modern America.

The reason to follow the pulses with these grains is because the ground will be filled with nitrogen, and with all the rich elements finally broken down from that large quantity of mulch applied two years before. Not only that, there will be comparatively few weeds in the field. And here's the rub.

If the ground is tilled, precisely the perfect environment is created for weeds to grow. Weeds are God's provision against erosion. They are the earth's "green carpet" as it were. All that humus now formed in the soil where the mulch has been decomposing very much hinders erosion, so the weeds don't need to sprout, and more importantly, without disturbances, very few weeds sprout at all.

A relatively new technology, practically the only widely useful one introduced by the Japanese mystic M. Fukuoka, is that of sowing seeds without tillage by enclosing them in little balls of clay. One simply digs up a few gallons of clay from a creek or other source in idle hours. Then, it is spread in the sun to dry, protected from rain and dew. Once dry, it is ground in an ordinary grain grinder and stored until spring time or whenever the seeds will be sown. Two ways to make the clay seed balls were introduced. One is to shake a tray of the fine clay powder together with the seeds while misting with a spray bottle. The other is to just mix the whole mass, including the seeds, with water until it's plastic, then force it through a ¼ inch mesh screen to form the pellets around small seeds, like carrots or onions, or a ½ inch screen for kernels of small grains. The balls thus formed are then taken to the field within a day or two, and scattered or dropped in rows as the case may be. I have never done this, but many people have been successful at it, and it would prevent stirring the soil and causing weeds to sprout. It takes 2 bushels of wheat seed thus prepared to sow an acre.

Another way is to broadcast the seed, and harrow lightly with a spike tooth harrow. This is especially useful in the fall-sown small grains, like wheat or rye. In some areas, even wheat will winter kill, and it must be spring sown. The plus side of this is that spring wheat varieties are the highest in protein (gluten) and make the finest bread and related pastries. The down side is there may be more weeds, which wheat positively detests.

Wheat

Before growing wheat it would be wise to assess how much you'll really need. A bushel of wheat weighs 60 lbs. and makes about 120 lbs. of whole-wheat bread. Thus 4 bushels would allow about

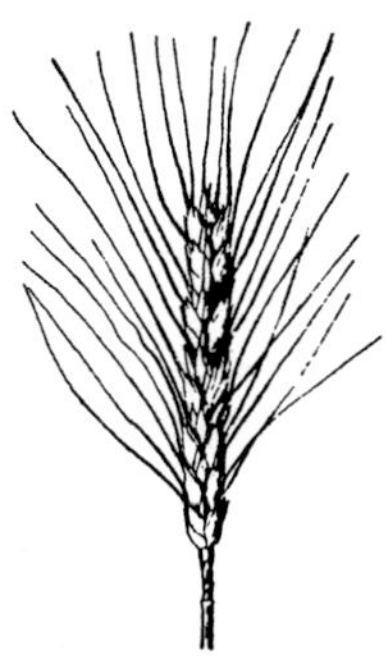

one 1-lb. loaf of bread daily for the whole year, with some for cookies, cakes and pies. Husbandmen always see to it that they allow for surplus, in case there's a bad year. It does keep, or can be fed to livestock, or sold. In most areas wheat normally yields about 60 bushels an acre. So a fifteenth of an acre of wheat is all one might need to sow, assuming no adverse conditions. This would be about 11 square rods, or say, 2900 square feet.

But back to wheat. Here, winter wheat is sown. Some folks sow in late summer so they can graze it to encourage *tillering,* which means the growth of multiple stems. Each main stem has a head or ear of wheat at the top. The more stems, the more wheat berries, so the theory goes. But this regrowth ordinarily consists of stems much weaker than the first ones. And in areas of hessian fly, later sowings avoid this pest. A good rule is to sow about 6 to 8 weeks before the ground freezes in your vicinity.

If you sow spring wheat sow as early as the weather will permit. The seedbed doesn't have to be perfectly fine, and two factors favor such early sowing of spring wheat: 1. Long cold weather growth periods encourage a better root development, hence better yields. 2. Early sowing makes earlier ripening, avoiding *rust* caused by the dry heat of late summer.

Grow only those varieties which have proved themselves in your area. New varieties which catch your interest are best grown several at a time, side by side on a few square yards, testing them against the tried and true. Use clean seed.

Wheat winter kills in poorly drained fields because constant freeze/thaw cycles heave the ground, breaking the roots. The roots also break off in areas like the Great Plains, where winters are very cold and windy, but little precipitation occurs. The soil gets freeze-dried and cracks, breaking the wheat roots.

If you don't eschew chemical fertilizers, 100 lbs. of a 3-8-5 mix applied at seeding time may increase the crop. I never bothered.

Unless you have a wheat drill (and why should you?) you'll need to broadcast the seed of either spring or winter wheat evenly at the rate of about 4 bushels per acre, or half that in dry areas. Then harrow it in with a spike tooth harrow so the birds won't eat it. They'll eat some of it, because not all of it gets buried. As long as it's not more than 2 or 3 inches deep, most of it will come up.

After sowing there is nothing more to do until the summer harvest.

It will be ready to mow with scythe or sickle bar mower some time in late June to Mid July where I live; earlier South, later up north. The way to tell is that the bottom half of the stalk has yellowed and the upper stems are still slightly green. In fact, it can be riper than this if you don't mind a lesser quality of flour. But for next year's seed, save the best looking part of the stand for last, letting it stand until it's as white as possible without the tops shattering when you cut.

Cut only as much in a day as you have time to gather into sheaves. If your scythe can be fitted with a very light frame of wooden fingers, to knock the straw over into a row as it's cut, it will save much labor. A scythe thus outfitted is called a *cradle*. The sheaves are bundles of small grain with all the heads pointed in the same direction, and tied with a little piece of twine. Horse drawn binder machines did this just before the big combine was invented. But it takes a real field of wheat to be able to even turn around with this equipment.

The sheaves may be gathered into shocks to further cure in the field until threshing. Stand 4 sheaves together first, heads up, inserting a handful of the stems from each sheaf into the string of its neighbor. Now lean 3 sheaves on each side of this, forming a round shock, and inserting a few stems into a neighboring sheaf as before. Make a cap for it all out of two sheaves. The first one's head is parted in two, and each half bent over in opposite directions, like a mushroom top. The second cap sheaf is laid in the split thus formed, with its head lying on the split one's string. In

FIRST CAP.

SECOND CAP.

fact, it's a good idea to insert some stems through the split one's string to bind them from blowing down. Shocks thus made will stand up to all but the worst winds, and turn rain off pretty well. Take pride in good shocks, for hardly anyone knows how to make them any more. Don't leave it in the field more than 3 weeks.

In the U.S. there is rarely a

need to make the big old grain stacks like sometimes are seen in wet climates like Britain. And even if there were, the small bit of wheat grown for one family need not use so much labor. Here it is usually dry enough in late summer to just thresh it into a barrel. I first saw this done on an old plantation in tidewater Virginia. One of the sharecroppers just took a little paddle and paddled away at a bundle of wheat in the barrel until

THE SHOCK FINISHED.

it looked pretty well spent, and then tossed the rest into a straw *mow* (rhymes with cow). The straw mow is just a portion of the hay storage area in a barn used for storing bedding straw.

Winnowing comes next. This means pouring the threshed grain so that it must fall through a blast of wind. The wheat, being heavy, falls through the blast unaffected while the papery chaff blows to the side. A windy day in a doorway was the old fashioned way, but another one was to have some helpers wield large winnowing fans made of wicker. An electric fan will work too. It's best if you can collect the chaff, for it can be mixed with grain and fed to horses. It adds bulk, improving digestion. In small quantities, all small grains can be stored in burlap sacks hung in the dry away from rodents and birds.

There are a number of good small mills on the market, but the one I have seen the best results from is an electric one, devised by laboratory scientists for grinding medicinal plants into fine powder. A good friend always used to supply us with very fine ground wheat flour from her mill. It has its limitations, in that it cannot be set coarse for cracking the skins off of oats to make oatmeal, but there are hand mills for that. My little plate mill, which I got at a junk shop, doesn't grind the flour fine enough for perfect bread.

Barley

Barley is needed, in much smaller quantities, for making good bread. Barley is often sown as a winter crop in my vicinity, and I would have never thought it would

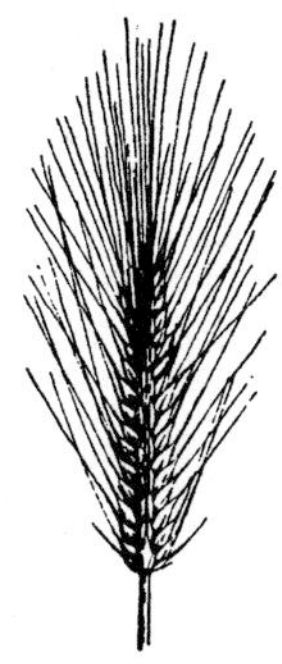

stand up here through the winter. But most barley is spring sown, when the land has dried up some in April or May. If the weather is just right, as it sometimes is, it can be sown as early as March. What it needs is dry ground which can be pulverized by repeated harrowing. It can also be sown in clay balls as outlined above. Sow about 3 bushels/acre. It grows very fast, and will be ready to cut in July. Very little is needed, unless you are an avid home brewer, but I am not. Hull-less types, or at least awnless types are best suited to the non-mechanized husbandman. Harvest when its ears are bending over, that is, quite ripe. When the red streaks running lengthwise in the ripening grain disappear and the straw takes on a yellow hue, it's ready to harvest.

The reason it is used in bread making is because if it is *malted,* which is to say, sprouted, and then *kilned,* which is to say, heated just barely enough to kill its growth, it will be full of *diastase.* This is an enzyme which converts the starch in wheat into sugar in a form which yeast absolutely loves. All commercial bread flours have some malted barley flour in them. Try adding a bit of beer or at least malt extract and see how wonderfully your bread improves!

To sprout your barley, soak it for 3 or 4 days in a wash tub. Let the water drain out of it, but keep it moist, turning it several times a day. It wants to be about 65°-85°F. Sprinkle it to keep it moist. Under the skin of the grain the shoot will appear, say a week, and within a week and a half will be two-third as long as the grain itself. Now dry it at around 120°F. Don't ever let it get over 140° or you'll kill the enzymes. Test it for bone dryness by biting it. It should crack like it's bone dry and taste a little like a malted milkshake. You really don't need to malt more than a couple quarts at a time. Store it in dry jars.

Flour and Yeast

Grind your flour very fine. Grind some malted barley very fine too. One pound of flour is needed for every two pounds of bread. Add a handful of the malt flour for every pound of wheat flour. We like to make up at least 3 lbs. of flour at a time.

I've never made yeast, so I don't know how it's done really. One recipe calls for mixing 1 part rye flour and 2 parts barley with just enough water to make pretty solid dough, about like biscuit dough. This is cut into very thin patties and left outdoors in summer to dry in the sun. Yeast is supposed to infect the patties. When yeast is needed, a pattie is crumbled into warm sugar water, and let stand for a day or two, until it's getting foamy. This is used like dissolved store bought yeast.

We always bought our yeast at a good bulk natural food store. We get it by the pound instead of the gram, like at the supermarket.

Bread

Eva makes all our bread now, and superior bread it is. It is much easier to make than the recipes in all the cook books. Bread takes so little time and effort to make I wonder at the popularity of all the electricity-dependent bread making machines which only make one very small loaf at a time. With 20 minutes' effort, we have half a dozen big loaves for friends and family, even when the electricity fails. We have a wood cook stove.

Here's Eva's recipe: "You'll need a large stoneware or wooden bowl. They retain the warmth better for raising the dough. The only ingredients are the good flour you just ground, live yeast, salt, sugar or honey, corn meal or coarsely cracked wheat. One cup of water makes one loaf. I usually make 6 to 7 loaves at a time, but this recipe will make 3 loaves.

"Put 3 cups of warm water in the large bowl. Add 3 handfuls--about 3 tablespoons--of salt; 3 of dry yeast, and 3 of sugar. (Next time you'll probably want to adjust "handfuls" according to your taste.) Mix well. Add flour gradually, mixing it first with a wooden spoon, then using your hands. Keep adding flour until the dough is firm. The amount will depend on the fineness or coarseness of the flour, so I won't give quantities; you'll figure it out. A bit of flour takes the sticky dough off your hands; don't wash it off using water.

"Knead the dough by punching it vigorously--either in the bowl or on a floured board--until it stops sticking to your hands, about 10 to 15 minutes. The longer you knead it the moister your bread will be. Return the dough to the bowl if you worked it on a board.

"Cover the bowl with a towel and leave it in a warm place to rise. When doubled in size, punch it down. Divide it into 3 parts. Form loaves any shape you desire. Lay them on a cookie sheet which has been well strewn with cornmeal or cracked wheat. Let it rise half way, then bake in a medium oven (±300°F) until done, about 30 to 35 minutes.

"Test for doneness by thumping on the loaf. If it sounds hollow, it's done. If you don't think you can recognize a hollow sound when you hear it, sacrifice one loaf by cutting it in half. If the center looks raw or doughy, put it back together and return all to the oven (one hopes you kept the oven door closed so it wouldn't cool off) and bake 10 or 15 minutes more. It may not look as pretty, but will taste good. If the bread crust is too hard, brush some milk on the top. Adding malt or malt extract into the mix increases bread's elasticity. So does beer. Don't be afraid to experiment, adding other kinds of flours, flavorings or spices."

The secrets to good bread are good hard flour, plenty of yeast, salt and barley malt. Water in which potatoes have boiled is somewhat of a substitute for the malt, but not fully as effective. When growing wheat, see that you grow hard wheat, not soft wheat. The latter is the kind most often grown east of the Mississippi, but is really only good for biscuits and such, which are more crumbly than elastic.

Oats

Oats are grown just like barley, except they are a little bit later to mature. Indeed, they are cut when the stalks have a bit of green in them, otherwise they shatter too easily. I like it harvested as animal feed while the grain is in the

"milk" stage, which means it will pop into a milky substance if you squeeze it between your fingers. It is cut and left lying in loose gavels in the sun for at least a few hours before binding. The sheaves are tied low enough to leave the head part pretty loose. As soon as they are dry enough not to heat, they're removed to cover. They make good feed right in the sheaf. A sheaf or two at feeding is enough for most horses, and the straw with the oats is good feed indeed.

For human food it is harvested closer to ripeness, though still a little green, or there will be much shattering. It is dried in the stook 3 weeks, then threshed. To make oatmeal, dry the grain very dry in artificial heat. Next, grind very coarsely to crack off the skins. Winnow. Then grind it more finely if desired. It is not like the rolled oatmeal of the shops.

Rye

Rye isn't much grown, except for a straw crop--it has the longest, coarsest straw--the best for bedding. It is the most winter hardy small grain, and is sown just like winter wheat, except not

as thickly. From 1 to 3 bushels are sown, the former if conditions are better, and coarser straw is preferred. The latter is better for grazing, because it is softer and more palatable. Anyway, don't wait too late to graze it because it will be too rough to eat. It is ready to cut perhaps a week ahead of wheat.

Buckwheat

Buckwheat isn't really a grain at all, but is treated like one. Its chief virtue is it can be grown on the least fertile ground and still secure quite a crop. One bushel per acre is sown broadcast 70 days before the first fall frost and harrowed lightly. It comes up in 3 days, and soon takes over the field.

Its white flowers are quite attractive to bees, and the honey is very favored in some regions. It continues to blossom and set seed until frost, but if sown too early will lose seeds as fast as it sets more. The maximum crop is about 50 bushels per acre; minimum about 10. Cut just before the killing frost. The crop cures slowly, so it is usually cured in loose gavels for a few days. Then set it up on end in tied sheaves, only two or three large ones together. It can't be stacked or stored without great danger of heating. It threshes very easily, commonly done when still green. The grain is used mostly for griddle cakes, but animals like it pretty well. Hang it to dry in burlap sacks away from rodents.

Grind buckwheat very coarsely first, and winnow to remove the inedible hulls. Then grind finely.

28
MAIZE

The earth bringeth forth fruit of herself; first the blade, then the ear, after that the full corn in the ear.
Mark 4:28

Ordinarily simply called *corn,* maize is the most important crop in North America. It yields better than any small grain, has very many uses for man and beast, helps clean the ground of weeds because it is a row crop which must be tilled, is easier to harvest than any other grain, and can be grown utterly by hand if need be.

There are 6 principal types of corn, and at one time there were over 500 named varieties. That has all changed with the discovery that if 2 productive varieties are crossed, the progeny will be more fruitful than either of the two parents. Also, the natural preference of corn to yield well only where it was bred can be overcome through cross breeding. It is easier for a farmer to dispense with careful preservation of his own locally adapted strain and just buy hybrids.

In an effort to help husbandmen establish independence from corporate seed suppliers, I will describe as much of corn growing as I have room for in this work.

Types of Corn

The six types of corn all have their uses.

The most important type is *dent* corn. It is so called because upon drying, the kernels form a dimple

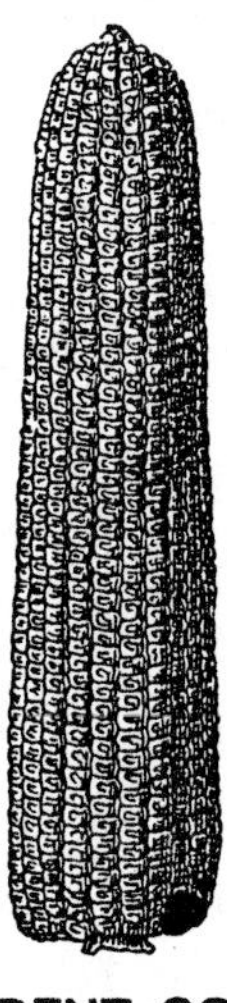

DENT CORN

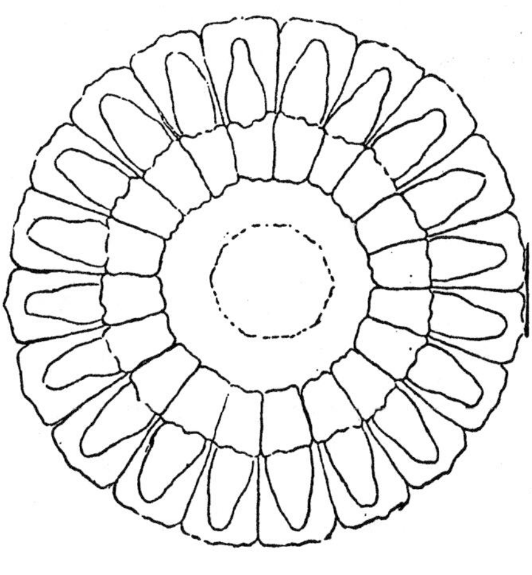

ACTUAL SIZE COB

or dent in the outer surface. This type of corn yields the highest everywhere it is adapted, once locally acclimated. It is grown principally for the starchy grain, but many enjoy it as a vegetable when picked immature and roasted.

Sweet or *sugar* corn is considered a vegetable by most of us. It is one of the easiest vegetables to grow, being grown exactly like field corn. It too can be selected and localized.

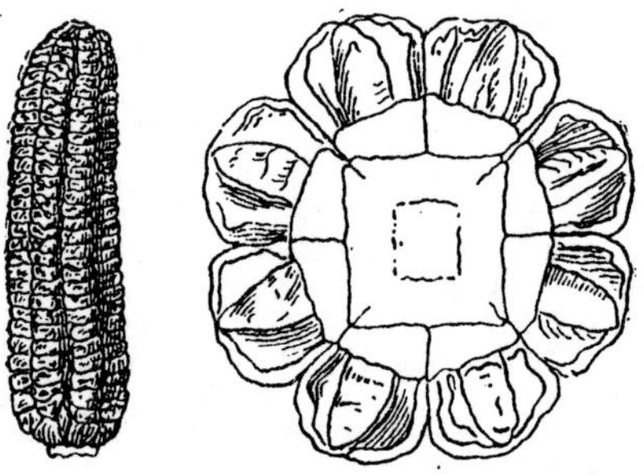

SWEET CORN ACTUAL SIZE COB

Popcorn is a smaller yielding type than either of the first two. But when heated, the moisture trapped in its very hard kernel turns to steam and it pops. A similar phenomenon can be induced

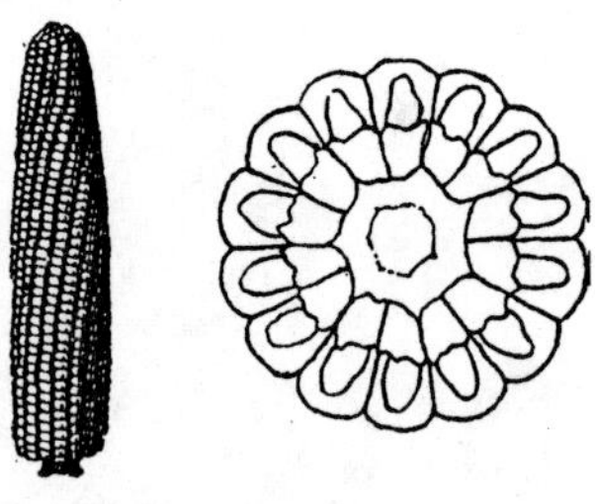

POPCORN ACTUAL SIZE COB

in field (dent) corn by steaming under pressure in a pressure cooker until it is about 240°F. Then the cover is removed, and the pressure drop causes the moisture accumulated in the corn to turn suddenly to steam, puffing the corn. This is how puffed breakfast cereals are made. The puffed kernels are then dried in a slow oven for storage.

Flour corn looks like dent corn, except there are no dents in it. The kernel is much softer than other corns, because the interior is practically all soft starch. This type was widely cultivated by the people living in America when the Europeans first came here. It can be pounded into flour with wooden implements. It is now quite rare.

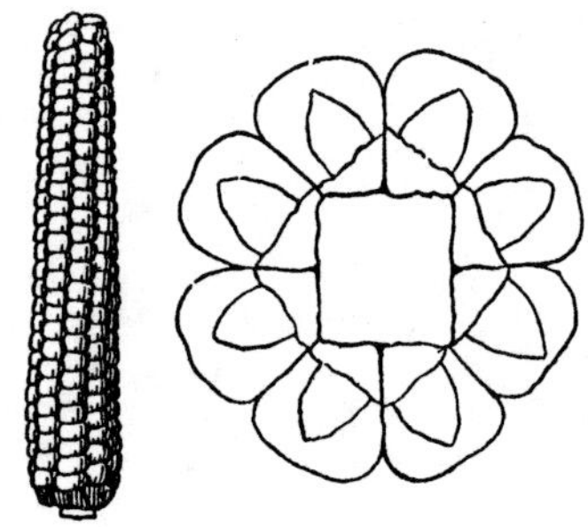

FLOUR CORN ACTUAL SIZE COB

Flint corn is quite the opposite of flour corn, being as hard as flint. Well, almost. Livestock sometimes have difficulty eating it unless it's soaked. It is almost as hard as popcorn. It has long narrow ears, sometimes as long as 18 inches. The oldest people I spoke with when I first came to the Allegheny highlands remembered it well. It was formerly the only type grown in cool climates, because it will mature in short seasons, at high elevations, and where cloudy conditions prevail. Such is not the case with any other type, as a rule. Today, special dent varieties have been developed for

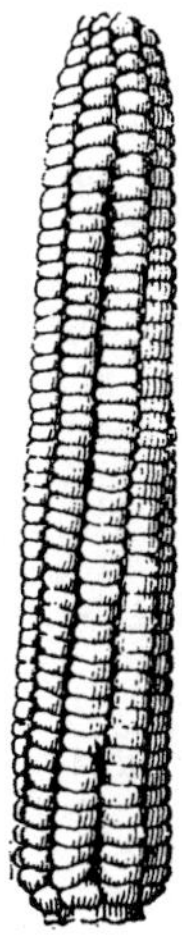

FLINT CORN

ACTUAL SIZE COB

short, cool seasons, and are supplanting the lower yielding flint types.

Pod corn is so called because each kernel on the ear has its own little husk. Its main use is for horses, because ordinary corn is so pasty in the gut as to cause colic. The extra bulk from the husk helps prevent the problem. It's still probably better to feed oats, though pod corn, fed whole, may be a better choice where labor is concerned. Pod corn is fed in the ear to horses.

The Way of Corn

Corn is a coarse feeder, meaning it thrives on land freshly manured, which would burn other crops and cause the small grains to *lodge,* which means they fall over and the grain rots on the ground. Land can hardly be made too rich for corn;

POD CORN

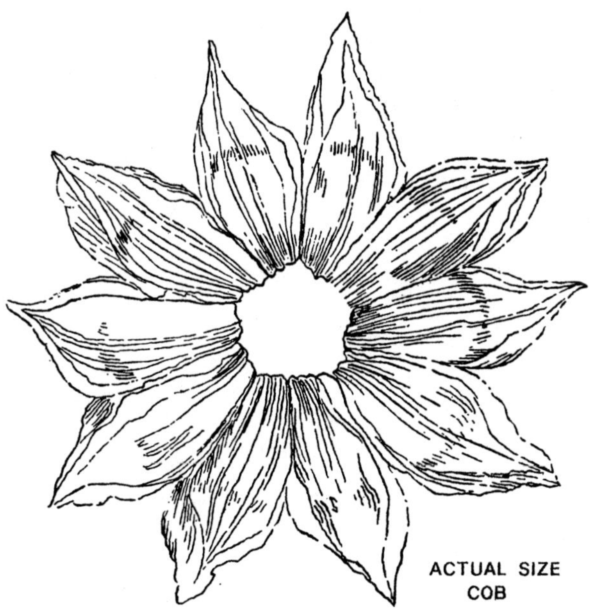

ACTUAL SIZE COB

the richer the ground, the better the yield.

There are varieties of field corn which will ripen seed in as few as 55 days, or as much as 300 days. Corn is very sensitive to frost. But I have planted corn which got frosted several times before its leaves widened for the principle growth period, and made some of the best yields in the community that year. Still, it's wise to choose a variety whose growing season is short enough to avoid spring and fall frosts. Most corn ripens between 120 to 150 days. A variety's growing season is prolonged by about 20 days per 1000 feet elevation. Choose varieties which use up the whole season.

Rainfall is very important to corn. Less than 12 inches of rain during June, July and August means the crop will be short. Weekly cultivation mitigates this to some extent, because loose surface soil checks evaporation losses. Eighteen to 20 tons of water must pass through the corn plants to produce one bushel of corn.

Corn is planted either in rows or in *hills*. The latter are not really elevated mounds of soil, but groups of seeds planted so they form rows both ways. These are also called *checkrows,* or just *checks*. The main purpose for this is to allow cultivation by horse, tractor or wheelhoe in both directions, eliminating hand hoeing. Very small or irregularly shaped fields make hill culture less advantageous. On steeper land, row culture allows the husbandman to follow the contour of the land. Otherwise, rows and checks yield the same, as long as the population of plants is the same, and both are kept free of weeds. One man can plant 5 acres of prepared ground in one day with a jab planter.

Corn varieties for the Gulf States, such as Eureka, grow so huge that only 4000 plants fill an acre. In northern areas, up to 26,000 are needed. In our region of the mountains, 12,000 to 15,000 are planted unless the corn is used for silage. Corn silage is made by chopping up the entire plant, stalks and all, storing it in an airtight silo until needed. Thus stored, the whole plant can be used to best advantage, keeps for years, and is independent of weather conditions. But it requires considerable machinery. A better way is to use corn to save land and hay. We do this by growing some especially for the dry spell in July and August, cutting and feeding it to the livestock.

If you decide not to grow small grains at all, corn may follow the pulse crop in the rotation. It loves the organic matter of the decayed mulch, but it loves the nitrogen of the decaying pulse roots too. If the potatoes and the pulses were grown entirely by tillage the corn benefits by large quantities of dung, best added in the fall after removing the pulse crop. The longer that dung can infiltrate every pore of the soil, the better the corn can use it. Indeed, it makes an even better crop to place poultry dung in narrow bands alongside each row of corn, about 2 or 3 inches from the seeds themselves.

Growing Corn

Soil preparation begins early. It's time to plant as soon as the sugar maple leaves are the size of a squirrel's ear. This was an old Indian sign, and it really works, even if the weather seems too cold then.

Start with good seed. I plant 1 inch deep, rows 30 inches apart; 14 inches apart in the row. This is 15,000 plants/acre. Take good care of the crop, cultivating once a week until the plants are so tall you can't conveniently till any more. It helps to hand hoe at least once between the plants in the row.

The corn will grow slowly at first, but the leaves soon begin to get their characteristic waves at the edges, and the plants grow very fast to shade the ground. Keep tilling it weekly until it's at least hip high. At the last tilling you may plant a winter cover crop under it like hairy vetch or white clover.

The tassels come, and the crop pollinates by wind. Other folks' corn may affect the purity of your crop, but don't worry too much, the dominant pollen will be yours, especially in the center of the field. Speaking of pollen, it comes from the tassel at the top of the stalk. Go through the field just when it starts tasseling. You will find an occasional truly inferior stalk, which ought not be allow to fertilize anything. Cut its tassel off. Look for stalks with no ears at all forming; cut their tassels off too. This is very important to good seed next year, never neglect it.

Corn Seed Lore

As the season wears on, go through the field, selecting and marking those plants in each row which exhibit the greatest vigor. They represent those lineages which possibly suit your local conditions best. Select the longest ears. Look particularly for ears with deep kernels, especially ears filled at the tip, with no cob visible. Extra large diameter ears, particularly those with large, coarse, pithy cobs produce seed which matures late. Seek out also those which are uniform both in color and size of kernel; inconsistency in either are signs the genes may be "running out" due to a poor individual lineage. Ears whose tips point downward are better, because they are protected from rain, and even snow, should harvest be delayed. Ears with mold or discoloring lack vitality, their seedlings won't grow as well. The less taper, the better, a perfect cylinder is the ideal. Too large a shank makes hand harvesting difficult; too small makes ears fall off. Multiple ear stalks, if each ear is respectable, are preferred. If only one ear is exceptional, select it anyway; prolificacy is in its heritage.

It only takes 12 to 15 ears to plant an entire acre! If you aspire to a 100 bushel (5600 lb.) yield per acre, each ear selected for seed represents from 350 to 500 lbs. Get several times what you need. When the first frost threatens, go and save those ears so they never freeze. Frost greatly diminishes seed quality if the corn hasn't finished drying.

Hang the ears on strings, or impale the butt of the cob on nails driven through slats of wood which can be hung up away from rodents.

Some time during the late winter or early spring, test all the ears at once, by germinating in a seed tester. This is nothing but a 5 foot piece of 9-inch wide cloth with numbered squares for each ear's test kernels to lie in, drawn with a waterproof marker. A good way is to make a line lengthwise down the center, and mark this off in 3 inch segments on both sides of it. Take kernels from opposite sides of each ear; two from the butt, two from the center, and two from the tip. Wet the cloth. Place all six kernels in one square, and number the ear so you can't lose it later. Do the same with all the ears. Fold the edges in, then fold

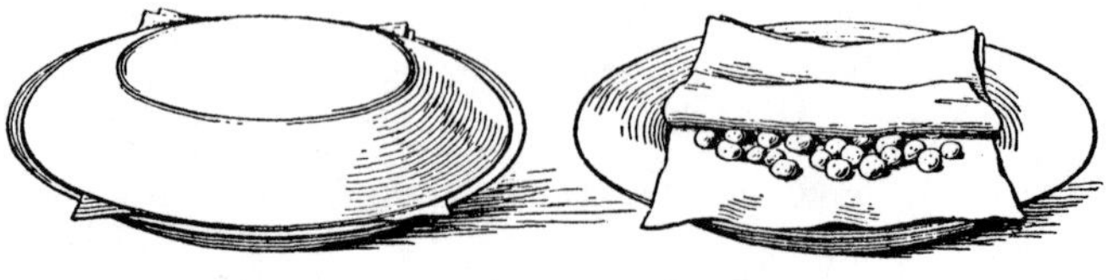

A SEED GERMINATOR

Consisting of two soup plates, some sand, and a piece of cloth

in half longways and roll up the cloth in such a way as not to jumble or lose the seeds lying within. Tie it shut with twine. Stand it in shallow tepid water for a few hours. Then put it in a closed box or under an inverted bowl so it doesn't dry out. Make as many of these rolls as you need to test all ears. Inspect in 6 days. An ear should be rejected if even one kernel fails to sprout. If one sixth of an ear's seeds don't sprout, you've lost at least 56 lbs. of your corn crop. Also reject ears whose kernels sprout weakly. Weak plants bear little or no corn. One barren ear will account for 350 to 500 lbs. lost yield.

The superior ears ought to number more than you'd need for two years' planting, especially since the tips and butts ought to be discarded if you're using a seed drill, jab planter or any other measuring device to set kernels. End kernels are not uniform enough to insure the right number of seeds in each row or hill. Two seeds in a hill where there should have been 3 means a loss of a third of the crop. Four would mean a loss of about a sixth, because there will most likely be one barren stalk and at least one with only a stunted ear--a *nubbin*. Even after such careful selection, you must pick out the surprisingly large number of deformed, irregular, broken, diseased or otherwise imperfect kernels. Prudent husbandmen like to have at least an extra year's worth of seed corn in reserve in case of disaster.

Harvest and Storage

Corn ears contain almost two thirds of the feed value of the corn plant, while the stalks contain the rest. Cut with the ears, it is called *corn fodder*. Ears removed, it is called *corn stover*. Many erroneously call fodder stover, and vice versa. Idle horses and mules and even oxen can winter mainly on good green stover. To get it green requires cutting the corn at the ground with a "corn knife." This is an L-shaped tool, with a handle like a chair leg, and a short knife blade attached at one end. The sharp edge of the blade is on the inside of the "L," facing upward as you grab the end of the wooden handle.

Before cutting any stalks, bend 4 standing cornstalks from two adjacent rows, or 4 adjacent hills, so their ends come together at the top. Tie them to form a frame against which to lean the cut stalks. Cut all the stalks around it, grabbing a stalk, and lopping it off with your sharp corn knife. When you have 30 to 60 stalks leaned together

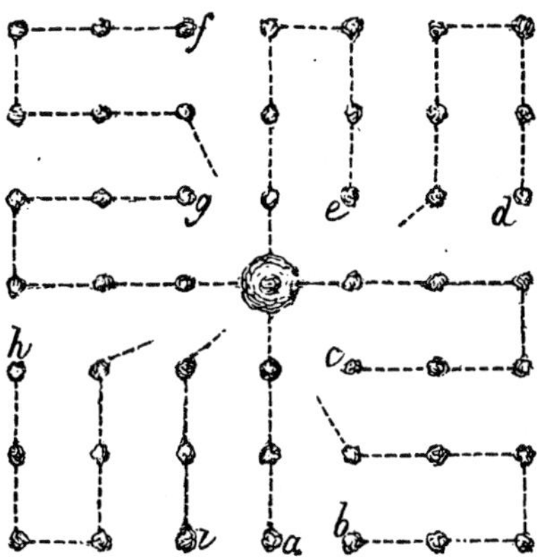

CUTTING AND SHOCKING CORN.

looking like a midget teepee, tie the top together as tight as you can. Some husbandmen devise special tools for cranking it really tight. These shocks will keep for months, and can be fed to horses after removing the ears. They won't eat the lower half of the stalks, unless shredded.

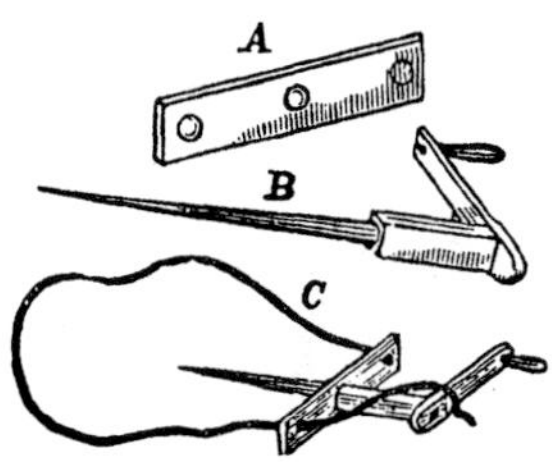

CORN-SHOCK BINDER.

In Southern areas, the custom of "pulling fodder" was formerly quite common. Wisely employed, it is a good way to use the crop. If frost doesn't come to the corn, and the lower portion of the stem has begun turning yellow, the top is cut

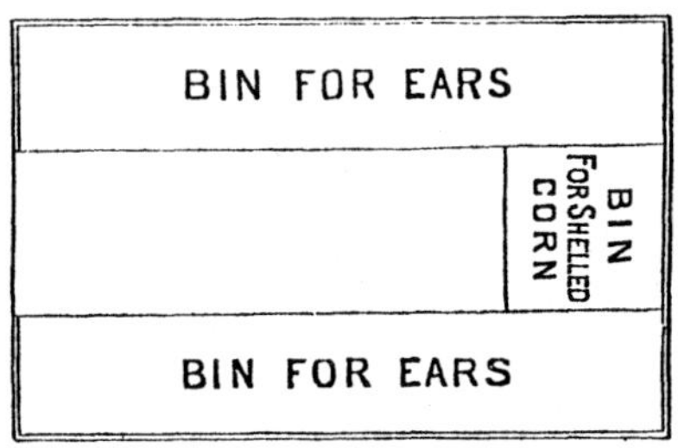

TRADITIONAL CORN CRIB

Raised 3 ft. on piers, with sheet metal vermin guards, sides sloped to encourage upward ventilation thru slatted walls.

off just above the maturing ear, and the leaves on the stalk below the ear are taken too. These are cured in shocks or chopped into excellent and palatable stover. The corn is taken later when it's dry enough to crib. If done too early, pulling fodder results in a lower grain harvest.

The last way is to walk down the rows, after the frost has turned the stalks white, and snap and shuck the ears, tossing them into an accompanying wagon. This method wastes all the food value in the stalks, which lie about on the ground making homes for corn borers and other nasties. Corn stalks ought to be used at least for bedding hogs kept in concrete sties. They trample and dung it and it makes superb compost.

Corn cribs are made in very many ways. The only one I want to describe in writing is made of snow fence. That's the fence made of vertical wooden slats, usually stained red, strung together by line wires at top and bottom. Make a 10-foot diameter circle of the snow fence atop a wooden platform at least 2½ feet off the ground. The platform posts should be topped with a wide scrap of tin, or be wrapped with sheet metal so rats can't climb up. Fill the circle of fence with ear corn. Then stand an 8 foot circle of fence on top of this, and fill it with ear corn. Make an impromptu shock of stover stalks on the top with lots of butts hanging over the edges to shed most of the water and snow off the corn. Tie it with wire so it won't break; snow rots string. You can remove corn from the bottom ring of wire, and it will replenish itself as it empties the top ring. Assuming each snow fence is 3 feet tall, this crib will hold 175 seventy-pound bushels of corn. If the corn is still quite wet, make a 1 foot diameter hole in the center

SNOW FENCE CORN CRIB

of the platform and make a 3 foot tall cone of snow fence to fit over it to admit air circulation. If you make a circular corn crib 5 feet or less in diameter, no such cone is needed even with still-moist ear corn.

A FIELD OF SHOCKED CORN.

29
ROOT CROPS

...*A time to plant, and a time to pluck up* that which is *planted. Ecclesiastes 3:2*

Roots form the final crop in a rotation, because they must be tilled almost throughout the summer, helping destroy all weeds. Also, no crop enjoys old fertility (rotted manures, etc.) as much as the root crops.

Carrots

Carrots stand supreme among root crops for human food because of their almost universal palatability. Furthermore, they are one of the very few crops which will yield remarkably well in near-barren soil. The largest yield I have heard about was over 1000 bushels/acre on very poor stony land in Westchester County, New York in 1819.

Selection of carrot seeds is very important. I found the best way is to plant at least six varieties at once, side by side to test them. More is better. Treat them all the same, and choose the one you like best on the bases of robust growth, ease of culture, and perhaps most importantly sweetness without bitter aftertastes.

Prepare the soil well, harrowing it weekly from the time it dries out enough to work until later than most folks plant. Indeed, carrots planted in very early June often catch up to those planted a month or more earlier. The early plantings take up to a month to come up, and in the mean time all the weeds are getting bigger. You can't see the tiny carrot seedlings for the weeds, and if you pull the weeds by hand you'll uproot the carrots. Besides, pre-planting tillage by horse is much quicker and more thorough than row cultivation by hand.

Even late plantings come up slowly enough for weeds to get the advantage, so when I plant the carrots using a seed drill (which I always use with carrots) I always come back immediately and carefully lime the row with slaked lime or wood ashes. I have tested carrots limed thus and they are certainly sweeter than unlimed ones. The lime line enables me to see where to hoe. The best distance for rows is 12 to 16 inches apart, though it takes a steady horse indeed to cultivate rows narrower than 24 inches. Young carrots are so delicate that I don't trust anything but hand implements.

As soon as the rag leaf is quite evident it should be thinned, the most laborious part of growing any root crop. Perfect thinning is the secret of a large crop. You may thin with a hoe first, if you like, so the clumps of carrots stand from 4 to 6 inches apart. For unusually large varieties, 8 inches won't be too far apart. If thinning is delayed, there's risk of breaking off at the crown, forcing you to repeat all that work when it resprouts. During June and July carrots must absolutely be at least wheelhoed once a week, if not more often. When weeds become visible without you stooping down, they have already done immense damage. One reason sandy or loose soil is best for carrots is because unlike stiff clay it can be tilled 1 day after rain quits. And if you can, you should do so, even if it means tilling 2 or 3 times a week. Let the weeds go when young, and you cannot possibly undo the havoc they have wrought unseen, and your crop will be largely diminished. Many gardeners have never had what I would call impressive carrot crop, yet take pride in a paltry 200 to 300 bushels per acre. A really good yield starts at 600 bushels.

The carrots should be taken up before the ground starts freezing severely. Once they're harvested they need protection from even slight frosts or they'll be much more likely to decay in storage. Storage in a root cellar is most convenient, but they are very commonly stored by *clamping* which is the trench storage method described in the chapter on potatoes. Carrots may be best kept for human consumption by topping, then planting in sand boxes in a cellar.

Fodder Beets

I think we can dispense with sugar beets, because though their nutrient level is twice that of mangel beets, they yield only half as much and their roots are very much more difficult to harvest. Mangels are very large beets whose roots grow mostly out of the ground. I have grown them several ways and can vouch for the fact that to get a large crop they must be treated just right. They aren't really a human food crop, but because they are fully capable of yielding over 50 tons, or 2000 bushels per acre--especially on a well tended plot of less than an acre--they are suitable for growing for your house-cow to eat in winter. Mangels aren't worth bothering with if you can grow, harvest and store corn silage. Since this would require a huge machinery investment, I think we can rule it out.

Start mangel seeds like this: Burn a brush pile in a place where you can conveniently exclude livestock. Plant the seeds on the burnt ground in rows a foot apart, seeds roughly 1 to 2 inches apart in the row. Each beet seed is actually a capsule with up to 6 seeds in it. An average of two plants will come up per capsule. There is little to worry about if the ground is well burnt, because the weed seeds have been killed, the soil loosened and sweetened by the effects of the leaching ashes. You will scarcely have to pull a single weed.

In the mean time, prepare your field by spreading 20 to 40 tons of dung to the acre. If you can't do this, sow hairy vetch at the last cultivation of last year's corn, and let it grow until June 1 where your mangels will be. Disk it to chop it well, then plow it into the ground about 6 to 8 inches deep. You may need to hire a team or tractor, because one big horse is doing all it can to plow 5 or 6 inches deep in already loose ground. It shouldn't cost you much. Then dung it and harrow that in. If you have no ideological objections to "bag fertilizers" you may really boost the crop with up to 600 lbs. of 4-8-10 fertilizer per acre a week after transplanting.

I know the word transplanting has a chilling effect. Take my word please, that it is not a quarter of the effort of planting the mangels where they are to grow and then thinning and weeding. And transplanted mangels always are more productive. Mangels and rutabagas show great improvement in root development from transplanting.

When the roots in the seedbed are anywhere from finger size to an inch thick, yank 'em up and clip the tap root near the swelling. Now cut off all but 3 inches of the top. This way it won't draw water so heavily from the damaged root. The mangels will wilt pitifully, and some will even appear to dry up. A very few will die. If they are dunked when pruned in a mix of compost and cow dung, with a little chicken dung too, and water to make a semi-fluid paste, they'll be more likely to succeed, even when set out in pretty dry weather.

We set the roots by cutting a 3 to 4 inch deep furrow with the wheelhoe shovel in the early hours of the day. Without delay, pull up some mangels from the seedbed, trim, and set them upright at 1 foot intervals in the trench, then repacking the dirt around them. Quit before it gets too hot. Do the same the next day. If cloudy or drizzly, so much the better, but it doesn't matter much. Save the sunny hours for tillage; it helps kill uprooted weeds and crumble soil clods. Make your mangel rows at least 2 feet apart, 2½ is better.

For 10 days your mangels will appear to have stopped growing. Then the leaves will begin to spring forth. Encourage as much leaf growth as possible by applying 700 lbs. of chicken manure between the rows in mid-July. The sooner they shade the ground the sooner you can slack off on your weekly wheelhoeing. Besides, the root doesn't swell until the days begin appreciably shortening in late summer. The plants with the biggest leaves will swell into the biggest roots. The long varieties are taller and easier to pull, but the globe varieties hold the record for the highest yield, something well over 60 tons per acre!

MANGEL BEET

Harvest the beets before the ground freezes or they won't keep. Wring off the leaves by hand, don't bother trimming rootlets--there aren't many. Handle very carefully, so as not to damage the skin or bruise the roots. Lay them down, never toss them. Carrots and rutabagas can be thrown about, but not mangels. Leave them lying in the sun all day to dry, but don't let them get frosted.

I have never been successful keeping mangels through the winter in field clamps. This is because it is so much work digging a trench big enough to stow them all, and so much trouble to take them out in deep snow and frozen earth. In cellars they'll keep for up to six months. If you save some good ones until after the ground stops freezing, you can plant them out and may get seed from them. I never was successful with this, either the beets died for some reason, or varmints ate them.

Rutabagas

Rutabagas or "root-o'-beggars" as we always called 'em, are grown exactly like mangels, but are much more frost hardy, grow faster, usually make just as big a crop (20 to 30 tons, average) and are commonly consumed by people as well as animals. They keep growing until freezing weather stops their growth, but cannot stand the heat like mangels. If you feed them to cows, the milk will taste just like rutabagas. They are the best root for sheep.

Kohlrabi

These and mangels don't give the milk of your cow a cabbage-like taste. Kohlrabis look like a cross between a turnip and a kale plant. The bulb grows above the ground. They are grown just like rutabagas, but the size one grows them to for cow feeding is too big to be salable. Large tender varieties, the best we've tried was Superschmeltz, really don't grow a whole lot bigger than the cheap common types, like White Vienna, when both are heavily fed compost and dung. A few White Vienna survived the winter here one year, but most are killed by the constant freezing and thawing. If we banked them up with mulch, I'm sure we'd have had plenty of seed stalks from them for the next crop. They are planted in April or May in rows 18 to 24 inches apart, and thinned to 8 to 12 inches apart in the row. Keep them cultivated as clean as carrots. They are easier than carrots because they grow faster and sturdier. They can even take a couple of freezes before you need to harvest them. The leaves and bulb are of equal feeding value. They get sweeter after frost, like other roots. Cows like kohlrabi better than any other green feed, including fresh, green sweet corn stalks.

KOHL RABI.

Kale

This is not a root crop, but it is in the *brassica* family like kohlrabi, cabbage, rutabagas and turnips. These crops should never follow one another, because they get a build-up of *club-root* causing organisms in the soil. Kale yields pretty well and is even more winter hardy than kohlrabi, but mostly winter kills here for the same reasons. It can be made to bear seed by mulching as with kohlrabi. It is planted and grown exactly like kohlrabi, but it can be cut twice for feeding to sheep or hogs as a green supplement. If fed to cows, it taints the flavor of the milk even worse than rutabagas. Its best feature is that it is easily grown in rather infertile soil, if well tilled, and doesn't mind acid nearly as much as any other brassica. I have watched in amazement as hogs pass up even milk for frosty kale leaves.

30
GARDENING

Build ye houses, and dwell in them; *and plant gardens, and eat the fruit of them.* *Jeremiah 29:5*

The rules for success at gardening are simple: 1. Select varieties suited to your conditions. 2. Never fail to cultivate at least weekly, whether the garden looks as if it needs it or not. It needs it. No crop of any kind needs cultivation deeper than 2 inches. 3. Strive to enrich the garden soil more than it could possibly need it. 4. Plant on time. 5. Be prompt in remedying problems.

The only exception to any of these rules is the unlikely event you can get enough rich mulch for free each year to entirely cover the ground 8 to 10 inches thick. When I was a teenager I grew a garden each year with grass clippings--I lived in a town. The neighbors all wanted great lawns, so they fertilized heavily, and the grass grew thickly. It also grew rapidly. They had to cut it at least once a week, sometimes twice. Young growth contains the most richness, so when they set those grass clippings out for the garbageman, I came skulking in the evenings in a borrowed truck, and brought plenty home. I grew all sorts of plants by simply pulling the mulch to one side and setting seeds or started plants in place. If weeds tried to come through, I plunked a foot of clippings on 'em and settled the matter. I never cultivated anything. The decomposing grass clippings, having absorbed all that fertilizer they'd been given, made my barren back yard into a garden spot rank with growth. But in the country, it is exceedingly difficult to obtain such mulch, so I had to learn conventional gardening methods.

One time I got a load of free hay to try mulching with, but it was a disaster. The abundant grass seeds in the hay provided several years of extra weeding before the vegetable matter finally decomposed.

Selecting Varieties

Not only should you select vegetables which are "easy keepers," which means they are not difficult to grow in your locality, but you should find the ones which produce the most appealing fruits. For example, in Maryland the best tasting non-hybrid beefsteak tomato we grew was Marglobe. In West Virginia the best tasting was Belgian Giant. Don't be arbitrary, or you'll only deprive yourself of having the best. For years Kentucky Wonder surpassed other beans we tried, but here in this little valley in West Virginia we found McCaslan and (especially) Romano better. It costs very little to try several promising looking varieties of every vegetable each year, until you hit the combination of hardiness, ease of growing, good flavor, and ease of preparation. Once you have the one you really like, save seeds from it each year, and it will get better and better adapted to your site's peculiar conditions.

Tomatoes

Tomatoes, after potatoes, are our most important vegetable. I have already discussed potatoes in their own chapter. They are grown in the garden the same as they'd be grown in the field.

Tomatoes are a very close relative of potatoes, but their culture is almost entirely different. They have too long a season for us to get timely harvests if we plant directly in the garden from seed, so we start the plants indoors. A few plants will generate plenty of tomatoes.

Eight weeks before the date of the frost-free period, we plant tomato seeds in sifted compost. We always mark the plant pot by the variety it contains. The pot is filled with compost, 8 to 10 seeds laid on top as far away from each other as the pot allows, and then another ¼ inch of compost sprinkled on top. The pots can be store bought kinds, home made, or just paper egg cartons. After the seeds are sown, the pot is set in a shallow tray with water, and the whole is enveloped in clear plastic. Years ago, I did very well putting the seeds into an inverted glass aquarium. I'm sure a large glass bowl with a pane of glass on top would also do fine, use your imagination. The seed trays are to be kept warm with bottom heat. Between 80° to 90°F is best.

In a few days the seeds will sprout. If they're in a tall, closed, clear container, they'll stay moist through returning condesation. If not, you must uncover them now, and water them from time to time to avoid drying out. Keep them bottom heated. Very bright direct sunlight is best for them, because in weak light, expecially without heat, the plants stretch for the light, getting long and spindly. Then they fall over, dehydrate and die. So keep the bottom heat, and if possible, give them full, direct sun all day long. Never let them dry up. Some type of very dilute plant food in the water, with very dilute copper compounds, will protect against *damping-off* fungus.

As soon as the second set of leaves forms, the little clump of plants must be transplanted, each one to its own pot. Up end the pot and tap the clump gently from it. The planting medium will easily crumble. Taking the plant by one of the 2 bottom leaves, gently separate it from the compost. Don't hold it by the stem or roots. Now put it into its own pot. The best way is to have about ¼ to ½ inch of weed free sifted compost in the bottom of the new pot, and holding the plant so its roots just touch the bottom, fill more compost in around it using your other hand. Get a little of its stem buried too. Firm it all in gently. Repeat with them all, labeling the pots with the variety name. Water them.

Now the plants may be set in a cold frame. This is nothing more than a 1 foot deep box with a pane of glass over the glass set in the sunlight. In direct sun this can get so hot it cooks the plants. To prevent it, we lay a window screen over it to provide partial shade. Even then, we monitor it just in case it needs to be left a little open for ventilation. Bring the plants indoors in the evening at first, until they're well along.

When the danger of frost is over, tomato plants may be set in the field. We dig a hole a foot deep and put a little well rotted cow or horse dung in the bottom. Cover that with a couple inches of soil, then set in the plant. Fill the rest of the hole with dirt. It's good to bury the stem too, leaving the majority of leaves above ground.

Now sprinkle wood ashes or slack lime lightly around the plant and till it shallowly in. This helps prevent *blossom end rot* of the fruits. Water them.

We once had a plant break off, and we planted it anyway, so that most of it was covered. Only a small portion of leaves were not buried. This plant sprouted roots, and bore tomatoes.

As soon as possible after setting in the garden, we punch a hole in the soil 6 inches away from each tomato plant, at least a foot deep and an inch diameter, using a home made *pritch*. This is an iron rod about 5 feet or so long and an inch thick with a point forged on one end. Into the hole a 6 or 7 foot wooden stake is pounded. I cut the stakes from hardwood saplings around the place. I pound it in using a fencepost driver described in the chapter on fences.

As the tomato plant grows, it is tied carefully with strips of rags to the stake. This is done by wrapping the rag one loop all the way around the stake, then gently tie the main branches of the tomato close to the stake with the ends of the rag. Keep the ground well cultivated, or better, mulch it heavily. If you live in a blight area, spray at least once every two weeks with Peret's mixture, as described in the chapter on potatoes, but dilute it to one half the potato strength. Peret's mixture also cures blight if you spray it soon enough. Completely soak the plant, topside and underside; stem and leaves.

We like to grow at least three types of tomato. We always grow a paste tomato. We like *San Marzano* the best. Cherry tomatoes bear soonest, and we grow *Sweetie*. For our main season beefsteaks we like Slovak *Vineyard* tomatoes, *Pruden's Purple* and *Belgian Giant*.

Save seeds from your tomatoes. The best way is to wait until they are overripe and mushy, then scoop out the seed with a spoon in its enveloping gel, placing it into a cup or jar. Cover with an inch or two of water, and let it stand for 3 to 5 days. Pour water and seeds into a strainer. Rub the seeds around in the strainer with your fingers while running water through to get rid of the gel. Tap the seeds out onto a clean piece of paper or cloth. When dry, gather them and store in a dry container, labeled sealed and kept at low humidity. No seed keeps well in humid conditions.

Sweet Corn

This is grown exactly like field corn. I don't like growing one variety planted on 3 different dates, 2 weeks apart. This is because corn loses its momentum as the days shorten, and late planted crops don't "make" well, even if they're short season varieties. I think it's much better to plant 3 different varieties, early, medium and late. They'll tassel and ripen at different rates. This method also helps thwart raccoons, because the tassels put out so much smell as to help mask the scent of the ripe corn. The 'coons go for everybody else's corn. Even when 'coons were killing my chickens, I never had them destroying my sweet corn.

Select seed for next year the same way as field corn.

Garlic

An easy crop to grow, obtain a bulb of garlic to get started. Each clove of this bulb may be separated and planted. Plant them in autumn or very early spring in rows about 16 inches apart, and 8 inches apart in the row. Then just cultivate weekly. In late summer the garlic can be pulled up and left to dry some in the sun. Then you may braid the top and hang them in your pantry. If you wait until fall to harvest, you'll see the cloves spreading apart and little kernels, like corn kernels among the cloves. If these are left in the ground, you'll have garlic for next year. If not, just save some cloves to plant. Every clove will make a new bulb.

Onions

Grown from "sets," onions sprout fast, grow fast, and die fast. But there is very little variety available when buying sets. Seed-grown onions grow quite slowly, so many folks start them in flats or pots indoors 12 weeks before the last spring frost date. In flats, sow them in weed free sifted compost, in little rows about 3 inches apart, ¼ inch apart in the rows. In pots, sow as many as you can fit into the pot at ½ inch apart. Once sprouted, keep them in cold frames with plenty of light. If a hard freeze threatens, bring them indoors. When grown to 5 inches, clip them back to 4 inches each week, which makes them stocky. Set them in the open some time after the ground stops freezing, and only light frosts are likely to occur. Grown like this, seeds may produce up to double what they'd yield if planted right in the field.

Some onions grow from sets. Just buy them and plant them. Potato onions are one type which produces sets. We have planted these; true to name they multiplied underground. The onion is dug and replanted. You soon feel like the old lady who lived in a shoe....

So called "Egyptian" onions grow very tall and make sets at the top instead of seeds. As a boy in Maryland I dug these wild in fields, but I can't say I liked the taste. I wasn't much of an onion eater as a child.

Harvest when the onions start developing skins and tops are falling over. Pull and dry in the sun (in the shade for white onions) for at least a week. They can be further dried in shallow boxes in a warm dry place, like a kitchen with a wood cookstove going. Remove rotting ones periodically for a few weeks. The remaining dry ones can be stored in net bags or shallow boxes where they are cool and dry but won't freeze.

You can grow onion seeds from seed-grown types if they're not hybrids. Store the best looking bulbs separately, selecting especially those with short necks. Plant them when the ground stops freezing hard in spring about 4 to 5 inches deep, and 6 inches apart in rows 16 inches apart, in only moderately fertile soil, too much nitrogen interferes with seed growing. Support the seed stalks by hilling up about 7 or 8 inches, and/or driving stakes at the row ends and at frequent intervals in the rows and then stretching twine on both sides. When the seed heads turn yellow promptly remove them with 6 to 8 inches of stem attached and store them in a large vessel in an airy place to finish drying. Several seed head cuttings are needed, because the tops don't all ripen at once. They are removed by similar methods as small grains. Then put them in water. The ones which float are poured off and discarded. The rest are dried, then saved in very dry containers.

Cucumbers

We only grow pickling types of cucumbers. We like them fresh and pickled. We have grown many pickling types and keep coming back to National Pickling. It has good taste, shape, productivity, and resistance to the wilt. Cukes cross breed readily so don't save seed from your side-by-side comparison plantings. When you settle on your best variety, grow only that one.

Cucumbers are close relatives to melons and pumpkins, but contrary superstitions notwithstanding, they don't cross with them. Out of all the members of this family of vegetables, cucumbers have the fewest days from planting to bearing.

Plant them about an inch deep in rows 6 to 8 feet apart, and about 2 to 4 inches apart in the row. They can be grown on trellises, but the sloping kind were a bust for us. Vertical is best. During the first 3 weeks after they sprout watch out for the striped and/or spotted cucumber beetles. These suck so much juice from the plant as to kill it, and their bite spreads the germs which cause the wilt. The best remedy is to poison them with store bought dusts until the stems lose their smoothness, getting both hairy and striated.

Till the crop weekly with the wheelhoe, tucking the vines out of the way until they're growing too fast and thick to keep up. Then just let them grow, checking often for fruits. Pick every third day, or they'll get away from you and get over ripe.

The first fruit to come ripe should be left on the vine to enlarge to the size of a small melon, and turn orange. Save its seeds the same way as described for tomatoes.

Pole Beans

I have never tasted any bush beans the equal of pole beans. So many people have convinced me to make "just one more comparison" that I have given up entirely on the bush beans except for dried beans.

Pole beans are grown best on fence trellis. The most labor saving idea I've come up with yet is to grow them on "stock panels." Go the local farm supply dealer and ask for stock panels. They are about 4½ to 5 feet high, 16 feet long, and made of heavy iron bars thickly galvanized. If they're cut to 8 foot lengths they're more manageable.

Set them up by installing stakes on recently well tilled ground, just like described for tomato culture. Then plant the beans about the time the frost free season starts, 1 inch deep and 3 inches apart. If you have more than one row, make rows 4 feet apart. Cultivate at least once a week. I think Romano is the best I've tried, being very early, very heavy bearing, stringless, and buttery smooth tasting. Romano bush beans cannot touch them for quality.

Keep the mexican bean beetles off beans, and don't disturb the vines while the dew is on them. Harvest every 3 days for sustained yields.

To save the seeds, stop picking when you are fed up with beans. Let them dry on the vines, pick, shell, and dry further in wicker trays until they're shiny and hard. Store in low humidity, as with all seeds.

Peas

Long vining types can be grown just like pole beans except it's wise to plant them just as early as the ground stops freezing. They are slow but sure to come up in cold soil. Short growing types are supposed to be stakeless. I think some actually are, but need to have the rows about a foot apart so they all lean on each other to defy the wind. For the ones requiring support the best way is to place strong wooden stakes at short intervals in the rows, and tack a lath on each side, or tight twine. They can move to and fro, but not sideways.

Peppers

Peppers are grown precisely like tomatoes, except they must be started two weeks earlier. They really enjoy more heat, too. To save the seeds, string up the ripe (and I mean ripe, not just ready to eat!) fruits with a thread through their stems and hang them up to dry in a warm place. Pepper seeds are among the shortest lived of all common seeds, so be extra careful to keep them from humidity. Never store any seeds in a damp room.

Lettuce

The name lettuce comes from the Latin name, *lactuca*, from the word for milk. This sweet plant becomes bitter as it matures, producing a milky sap which has been used for centuries as a substitute for opium. Mature plants are chopped and wrung in a bag for the white sap, which is then dried. I'm confident that if such uses became widespread again, it would be outlawed, and we'd have to do without it on our sandwiches and salads, or transplant them.

Lettuce can stand frosts and even a light freeze or two. It will sprout even in quite cold soil, so plant early, just firming the seeds into the surface in rows 16 inches apart. Plant fairly thickly, then thin to about 8 inches apart at about 3 weeks. The thinnings can be used for salads. Cultivate at least once a week. Harvest and plant continuously through the growing season. The butterhead and leaf types are easiest and quickest to grow.

To save seed, let several plants bolt, which means they'll turn bitter, sending up a flower stalk. The last one to bolt is the one best for seed. Stake the stalk for support. When most of the seeds have developed, pull up the plant and hang it in the shade to dry. The seeds can be removed and cleaned in winter.

Leeks

These are grown like onions, only we've done best just firming them into early cold soil using a seed drill set to just drop the seeds on the surface, right in the garden. Plant them in rows 24 inches apart, and thin to 6 inches apart in the rows. You can transplant the thinnings. Cultivate with the wheelhoe at least once a week. Hill them up as they grow, to blanch the stems. Harvest when the variety reaches its full growth. We have only kept them by drying. We cut off the bottom roots, cut in half lengthwise, and dry them on paper in the heated kitchen. Cut up with scissors and store in jars.

To save seeds from them, hill them way up to protect them in winter. Keep them about 1½ feet apart each way, and support the seed stem. Collect the seeds as with seed onions.

Cabbage

Our best success with cabbage came when the seeds were sown about ½ inch deep and three inches apart, in rows which were 2 feet apart. We planted them after the soil was pretty warm, about the same time as corn is planted. Cultivate the soil at least once a week. Early and mid-season varieties make the quickest crops, and *Golden Acre* is my favorite. Cabbages love lime and a whole lot of well composted dung. They must be protected from cabbage butterflies, pretty light yellow insects who lay eggs on the plants, and their well camouflaged green worms make short work of the heads and leaves. Many folks lime the very cabbages themselves about once a week or so to discourage the butterflies, and I have used wood ashes the same way. Get some around the base of the stems, too. Sour unpasteurized milk or buttermilk is said to be effective, but I haven't tried it.

The only way to save seeds from cabbage is to pull up root and all, and save in a cellar near freezing cold, or in dirt in trenches underground, well heaped up with mulch to prevent freezing. Save only the firmest and best heads. They are set back in the garden, buried right up to the head, as soon as the ground stops freezing in spring. When they bolt, support the seed stalk, and protect the seeds from birds. Cabbages cross readily, and two varieties need to be planted several hundred feet away, to be sure they'll come true.

Beets

Culture is the same as for mangel beets in the root crop chapter, but they need neither thinning nor burning if you keep them tilled weekly.

Spinach

Sow spinach ½ inch deep, 1 inch apart, in rows 16 inches apart. Spinach grows very fast, and I think it's best fresh as a salad base. It does like plenty of lime in the soil, and doesn't like acidity. It also likes lots of composted dung, and constant moisture, which is good, because the two tend to go together. It is very cold hardy, and many folks plant it in the fall for an early spring crop. Fall plantings do best in somewhat raised beds for perfect surface drainage.

Spinach is saved for seed similarly to lettuce.

Squash and Pumpkins

In short season areas, these must be planted right on time to insure a crop. We don't eat many of either, but the cow and pigs enjoy pumpkins. They were once planted in cornfields as soon as the corn was up. Usually they are planted in clumps of 4 seeds, thinning to the two strongest vines. In cornfields, they are 16 feet apart both ways; in the open they are planted in rows 5 feet apart, and thinned to a foot apart in the rows. The best crop I ever saw was one in which a huge amount of turkey dung was churned into the soil; alas, it was planted too late to "make." Pumpkins are ripe when the whole fruit has turned orange. They don't keep if even slightly frosted. If growing for livestock, feed them early, or else you'll have to come up with a way to store them at 50°F.

Muskmelons and Watermelons

Melons need protection from the striped and spotted cucumber beetles until their stems are tough and fuzzy, otherwise they'll succumb to *wilt*. This disease is evident when the plants develop their fruits almost to ripeness, then the whole vine dries up and dies. I think they are best started indoors in a heated, weed free compost, the closer to 85°F the soil is, the more likely they'll sprout. Start them about 2 months before they can be set out, which will be when there's no likelihood of frost. Set them out in rows 6 feet apart, 1½ feet apart in the row. The young plants enjoy having a box of scrap lumber set around them to keep the air calm. Some folks even paint the inside white to reflect heat. At night, a pane of glass will hold in the heat better. Cultivate with a wheelhoe at least once a week. We mulch with discarded heavy black rubber mine belting. The blackness of them holds heat, speeding ripening. Plastic is just so trashy as it breaks down. When the days and nights have warmed up well, and the plants are too big for the box, remove it, but continue training the vines out of your way so you can cultivate as long as possible. Watermelons take longer than muskmelons to ripen. You can tell when watermelons are ripe when the vine tendril nearest to the fruit attachment has deadened and the light skin where the fruit rests is turning yellow. Pressure from both ends of the fruit will make a squishing sound if your ear is pressed against the melon. Muskmelons are ripe when they begin taking on a buff yellow color and losing the grayish color; the stem also easily separates from the fruit. Grow several varieties of each for a couple of years to find out which are best suited to your conditions and tastes. Don't save seed until you're down to only one variety of each. Musk- and watermelons don't cross with each other, but they mix with their own kind readily.

Handy Garden Tables

COOL SEASON CROPS

Quick Maturing Cool Season Crops

Spring Salads

Corn salad, *Valerianella locusta*
Garden cress, *Lepidium sativa*
Leaf lettuce, *Lactuca sativa*

Spring Greens

Mustard, *Brassica spp.*
Spinach, *Spinacia oleracea*

Short Season Root Crops

Kohlrabi, *Brassica caulorapa*
Radish, *Raphanus sativus*
Rutabaga, *Brassica campestris*
Turnip, *Brassica rapa*

Peas, *Pisum Sativum*

Cool Season Crops Often Transplanted

Spring Crops Maturing Before Summer Heat

Early cabbage, *Brassica oleracea, capitata*
Early cauliflower, *B. oleracea, botrytis*
Cos lettuce, *Lactuca sativa, longifolia*
Head lettuce, *L. sativa, capitata*

Crops Whose Main Growth is in Autumn

Broccoli, *Brassica oleracea, botrytis*
Brussels sprouts, *B. oleracea, gemmifera*
Late cabbage, *B. oleracea, capitata & bullata*
Late cauliflower, *B. oleracea, botrytis*
Celeriac, *Apium graveolens, rapaceum*
Celery, *Apium graveolens*

Cool Season Crops Which Endure Summer Heat

Root Crops Which Cannot Endure Winter Freezing

Beet, *Beta vulgaris*
Carrot, *Daucus carota*

Freeze-Hardy/Heat Tolerant Root Crops

Artichoke, Jerusalem, *Helianthus tuberosus*
Horseradish, *Armoracia rusticana*
Parsnip, *Pastinaca sativa*
Salsify, *Tragopogon porrifolius*

Greens Enduring Summer Heat

Chard, *Beta vulgaris, cicla*
Collards, *Brassica oleracea, acephala*
Dandelion, *Taraxacum officinale*
Kale, *B. oleracea, acephala*

Salad Plants Enduring Heat

Creasy greens (Upland cress), *Barbarea vulgaris*
Endive, *Cichorium endiva*
Parsley, *Petroselinum hortense*

The Onion Group

Chives, *Allium schœnoprasum*
Garlic, *Allium sativum*
Leek, *Allium porrum*
Onion, *Allium cepa*
Shallot, *Allium ascalonicum*

Potatoes, *Solanum tuberosum*

Perennial Crops

Globe Artichoke, *Cynara Scolymus*
Asparagus, *Asparagus officinalis*
Rhubarb, *Rheum Rhaponticum*
Sea-kale, *Crambe maritima*

WARM SEASON CROPS

Warm Season Crops Not Usually Transplanted

Beans

Green & Wax beans, *Phaseolus vulgaris (nanus)*
Lima Beans, *P. lunatus (macrocarpus)*
Dry Beans, *P. vulgaris (nanus)*

Corn

Popcorn, *Zea mays, everta*
Sweet corn, *Zea mays, saccharata*

Okra (Gumbo) *Hibiscus esculentus*

Vine Crops

Citron, or preseving melon, *Citrullus vulgaris*
Cucumber, *Cucumis sativus*
Gherkin, *Cucumis Anguria*
Muskmelon, *Cucumis melo*
Pumpkin, *Cucurbita pepo & C. maxima*
Squash, *C. maxima & C. pepo, condensa & C. moschata*
Watermelon, *Citrallus vulgaris*

Transplanted Warm Season Crops

Eggplant, *Solanum melongena*
Sweet & Hot Peppers, *Capsicum annuum*
Sweet potatoes, *Ipomaea batatas*
Tomatoes, *Lycopersicum esculentum*

VEGETABLE PLANTING TABLE

CROP	Seed Per 100 ft.	Seed Depth (In.)	Days to Sprout	Inches Between Rows	Space in Rows	Days to Harvest	Target Crop/ 100 Ft.
Artichoke, Globe	*	*	*	48	48 in.	300	50+ lbs
Artichoke, Jer.	6 lbs*	*	*	24-36	18-24	65-100	4-5 bu
Asparagus	30*	6*	*	30	20 in.	2 yrs.	300 lbs
Beans, bush	2 lbs	1½	5-8	30-36	3-4	50-75	5 bu
Beans, Lima	1 lb	1	5-8	30-48	8	68-90	2 bu
Beans, pole	1 lb	1½	5-8	42-48	6-8	60-80	7 bu
Beet	2 oz	½	7-10	16-24	1-3	60-70	2½ bu
Broccoli	¼ oz	½	5-10	24-36	18-24	65-100	60+ heads
Brussels Sprouts	¼ oz	¼	6-10	24-36	18-24	90-120	16 gal
Cabbage	¼ oz	¼	6-9	24-36	16-22	65-100	90-150 lbs
Cabbage, Chinese	¼ oz	¼	6-9	24	12	75-85	85 heads
Carrot	¼-½ oz	¼-½	10-20	16	4	70-90	2 bu
Cauliflower	¼ oz	¼	5-10	24-30	20-24	75-90	75 heads
Celeriac	¼ oz	⅛	15-20	24	4	90-120	200 bulbs
Celery	¼ oz	⅛	15-20	24-40	4-6	125	150 stalks
Chard	1 oz	½	7-10	30	10-12	50-60	10 bu
Collards	¼ oz	¼	6-9	24-36	24	65-75	50 plants
Corn, Sweet	½-1 lb	1	5-10	30	9	55-110	140 ears
Creasy greens	⅛ oz	¼	15-20	16	10	50-60	120 plants
Cucumber	½ oz	¾	7-10	48-60	4-12	60-70	2-6 bu
Dandelion	⅛	¼	7-10	16	8	48-60	150 tops
Dill	½ oz	½	7-10	36	15	70	70 bunches
Endive	¼ oz	¼	5-9	16	8-10	42-50	120 plants
Eggplant	¼ oz	½	12-15	30	18-24	80-120	200 fruits
Garlic(cloves)	5 lbs	*	*	16	4-6	150-300	200-300
Horseradish	*	*	5-10	18	18	180-365	2½ bu
Kale	½ oz	½	6-9	24	12-24	55-65	3 bu
Kohlrabi	½ oz	½	6-9	24	8	55-100	2-10 bu
Leek	½ oz	½	7-10	16	6	80-150	200 stems
Lettuce	½ oz	¼	6-8	24	8-12	45-70	4 bu
Lettuce (Cos)	½ oz	¼	6-8	24	8-12	80-95	100 heads
Muskmelon	½ oz	1	7-14	48-60	48	80-100	60 melons
Mustard	½ oz	¼	5-10	24	9	40-50	3 bu
Okra	1 oz	1	8-14	24-36	12-18	50-60	800 pods
Onion	½ oz	½	7-10	16	3-4	100-140	2 bu

Onion (sets)	2 lbs	2-3	5-10	16	3-4	70-100	2 bu
Parsley	½ oz	¼	14-21	16	4-6	50	200 plants
Parsnips	½ oz	½	14-21	16	4-6	200	2 bu
Peas	1½ lbs	1	7-20	16-36	1-2	60-80	1 bu
Pepper	¼ oz	**	10-20	30	24	115-130	2 bu
Poppy	⅛ oz	¼	3-7	16	6	180	3 qt***
Potatoes	6 lbs	4-5	8-21	24-36	12	60-120	2-3 bu
Pumpkin	5 oz	1	7-12	60-192	12-60	120	50 fruits
Radish	1 oz	½	3-6	16	1-2	21-60	600-1200
Rhubarb	*	*	*	48	36	2 yrs	350 stalks
Rutabaga	½ oz	½	5-10	24	6	80-100	200 roots
Salsify	1 oz	½	10-20	16	2	120	500 roots
Sea Kale	*	*	*	36-48	24-36	2-3yrs	300 lbs
Shallot(sets)	4 lbs	*	*	16	6	80-100	25-40 lbs
Spinach	½ oz	¾	7-12	18	3-5	40-50	2 bu
Spinach (N.Z.)	1 oz	1	5-10	36-48	18-24	50-60	2 bu
Squash (bush)	2 oz	1	7-10	48	42	55-65	200 fruits
Squash (vining)	4 oz	1	7-10	70-90	60-90	90-110	80 fruits
Sweet Potato	100 slips	4-6	**	36-48	12	120-200	2 bu
Tomato	⅛ oz	**	5-10	40-60	36-48	110-130	10 bu
Turnip	½ oz	½	5-10	16	2-4	50-80	2 bu

*Started most commonly from root cuttings
**Started mainly indoors, setting out started plants
***Seed yield varies, select for local conditions

NUMBER OF YEARS SEEDS WILL KEEP

(Properly Stored at Low Humidity)

Asparagus	3	Leek	2
Beans	3	Lettuce	5
Beans, Lima	3	Lettuce (Cos)	5
Beets	4-6	Muskmelon	5
Broccoli	4	Mustard	4
Brussels Sprouts	4	Okra	5
Cabbage	4	Onion	2
Cabbage, Chinese	4	Parsley	1
Carrot	3	Parsnips	1
Cauliflower	4	Peas	3
Celeriac	6	Pepper	2
Celery	4	Poppy	4
Chard	4	Pumpkin	4
Collards	4	Radish	4
Corn (sweet)	3	Rhubarb	3
Cress	5	Rutabaga	4
Cucumber	7	Salsify	1
Dandelion	4	Spinach	3
Dill	4	Spinach, N.Z.	3
Eggplant	3	Squash	4
Endive	7	Tomato	8
Kale	4	Turnip	4
Kohlrabi	3	Watermelon	4

31
FRUITS AND NUTS

I planted me vineyards: I made me gardens and orchards, and I planted trees in them of all kind of *fruits. Ecclesiastes 2:4,5*

I know several folks who lost their city employment--two of them after years of dedicated service--to make room for the boss's kids. This will ever be the prime shortcoming of working for others.

This lack of job security despite faithful service is akin to the good tenant who plants fruit and (especially) nut trees on the landlord's land. In the first place, no one goes to the trouble and expense of planting trees without hope of personal reward in the fruit. Second, there is a yearly investment in most tree crops, in labor and possibly fertilizing costs. That great British author of superb books on husbandry, John Seymour, had his first holding pulled out from under him by the landlord to make way for a pheasant-shooting preserve. His years of bringing the place up to fertility made it quite attractive to wildlife, as it always does. The title holders enjoyed a bit of a hunt a few times a year, and didn't want to see signs of human occupancy when they came for a weekend of recreation. And after all, whose place was it?

So the trees he planted only bore him a year or two's worth of fruit. Trees are enormously profitable, 5 acres of citrus or nut groves being fully as remunerative as 1000 acres of prime wheat land. Where I live, citrus and certain kinds of nuts are out of the question. Still, some fruits and nuts can be grown just about anywhere to the husbandman's complete satisfaction.

The Way of Food Trees

As you stroll along a forest path you may come to a tree which was broken in two by wind, or cut for timber. If the tree was healthy when cut, there will be "suckers"

all around the stump, growing exceedingly fast. They grow so fast because a root network capable of supporting a much bigger tree is now driving the growth of a new tree. It is said that a 50 year old tree cut down will be as large as the original in 12 years. Doing this is called *coppicing*. A woodlot kept for this purpose is called a *copse*. But for tree crops, such drastic cutting isn't common.

When I was a boy I cut down a pear tree in my parents' yard because it was *fire-blighted,* but far from dead. I burned the top to kill the disease. My mother was horrified until I showed her the branches I hadn't burnt yet, and explained that the tree was going to

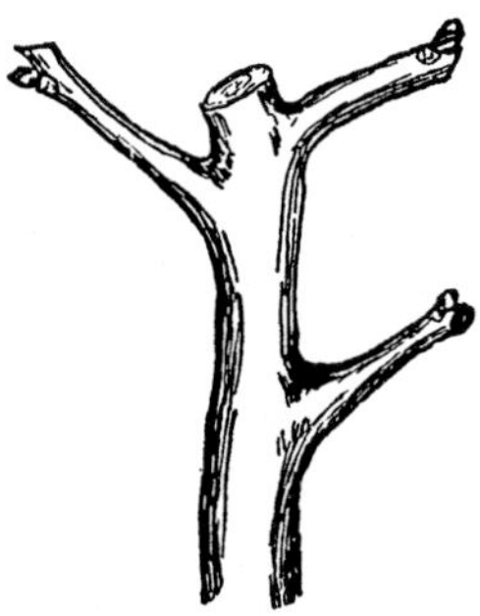

Pruned to direct growth. The growth will be in the direction taken by the topmost bud left when the branch is cut off, as this bud grows most rapidly. This being true, in what direction will each limb in the cut above grow ?

die, and I either expedited that, or prevented it. Sure enough, within a few weeks the tree pushed up some really vibrant suckers. Clipping all but the best looking one, and training that one, within two years we had fruit from that tree for the first time in 4 years. It never blighted again as long as we lived there.

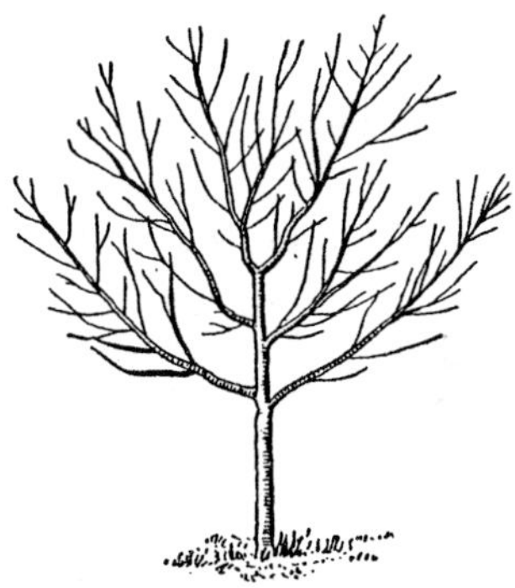

Properly thinned

The purpose of pruning trees is threefold: 1. To open up the branch structure, allowing the roots to really pump the sap to the parts remaining. 2. To prevent unruly

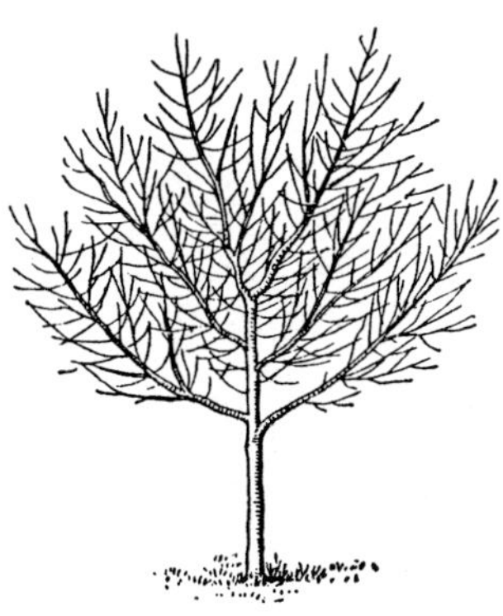

Unthinned

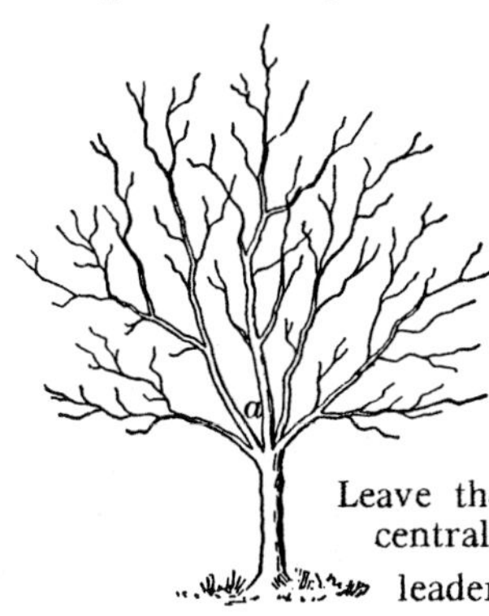

Leave the central leader on pears

growth, which makes branches rub on one another and get too heavy to bear even their own weight, inviting disease and pestilence. 3. To obtain "wood" for propagating more of the same variety of tree--grown from seed they won't always bear fruit like the parent tree's.

Unpruned & Unruly

The most important things to remember when planting a nursery-grown orchard tree are to provide an acceptable hole in an adequate site, and to defend the tree until it establishes itself at its new home.

When setting a tree, the best way is to retain as many of the tiny fibrous roots as possible, and keep them in contact with soil. The bigger roots don't actually feed the plant any more, they just anchor it. The wee little feeder roots should spread as much as they'd like, so don't fold them back on themselves when setting the plant in a hole. Do what you can to insure every little bitty rootlet is in firm contact with soil. Well rotted manure lining the hole is an encouragement to the plant to form new feeders immediately. I've even heard of folks who do the pet-killing at humane societies eagerly snapping up the corpses to plant with orchard trees, and fine orchards they have, too. Decomposing animal carcasses, if not buried to shallow (drawing varmints) provide all kinds of nutriment to trees.

A further precaution, and the one hardest to commit to, is to snip off most of the top of the just-planted tree. The root has been damaged regardless of how gentle it

Sloping line shows where to cut tree

was moved--unless it was grown in a pot! The less top there is, the more vigorous the balance between it and the root, and the better it will grow.

Beside a proper burial, it needs protection. The best way is to completely mulch around the tree, 8 inches thick isn't too much, and a circle 6 to 10 feet in diameter isn't overdoing it either. Grass clippings, sawdust, shavings, tree bark; whatever stays in place--just don't, and I mean don't, till it in. Any dung should be first applied, then lay the mulch. If there are weeds or brush around the site, don't cut them before mulching, bend them over flat, and then mulch. Cutting them makes them grow back through the mulch.

As a tree grows, we train it for strength and to be convenient for working. Little buds on the sides of a young whip can grow leaves or branches. The central whip is called the *leader*. Snipping it off when 3 to 5 little side branches have developed will encourage them to take the lead. I think the side branches are best encouraged to grow within a couple feet of the ground. The tree will not have grown very big by the time this is done, but it will determine its adult configuration. Each side branch should point away from the

Lines show where to trim

others and part company with the central stalk as nearly at right angles as possible. A branch forming a definite fork, or Y-shape is an invitation to the tree splitting dreadfully one day when it is heavily laden with fruit.

Wooden slats can be employed as *spreaders* to see to it that branches which grow upward too much are made more horizontal so light can reach all the leaves better. Once these have "set" you can remove the spreaders.

Just as pruning off the tip of the leader encourages side branches, pruning off the tips of the branches encourages new growth from their sides. Remember, every time you cut away some top, the now pre-eminent root system forces new growth. Many believe it makes more abundant fruiting. I don't think this is so, for a drastic pruning will often cause barrenness for a year or two. Not that the plant might not be better for it--especially if an unruly tree must be tamed again.

Some folks say don't prune at all. I have never had good success with that--the fruits seem to have more problems and are smaller, not to mention harder to get to in the tangled snarl of branches.

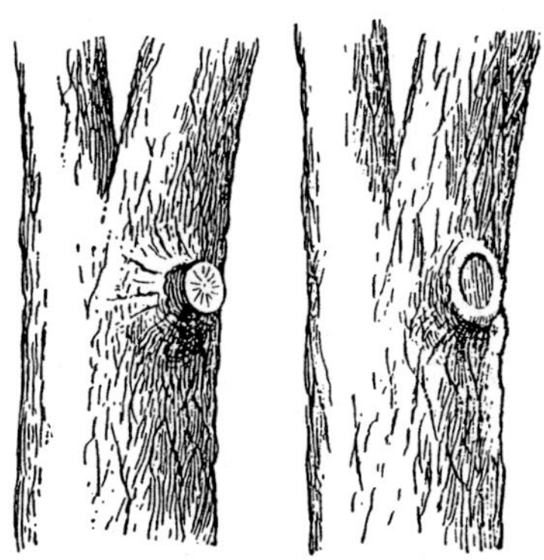

Refuses to heal Heals promptly

Trees may be propagated by growing from seed, in which case you'll have a new variety which may be superior or inferior to the parents, but won't be exactly the same. We have simply thrown peach pits, apple seeds, etc., against the Southern side of a building perhaps covering with an inch or less of soil, and when several plants grew we transplanted them. Peaches are unreliable here because of our late mountain frosts. But seed grown ones are more reliable for some reason.

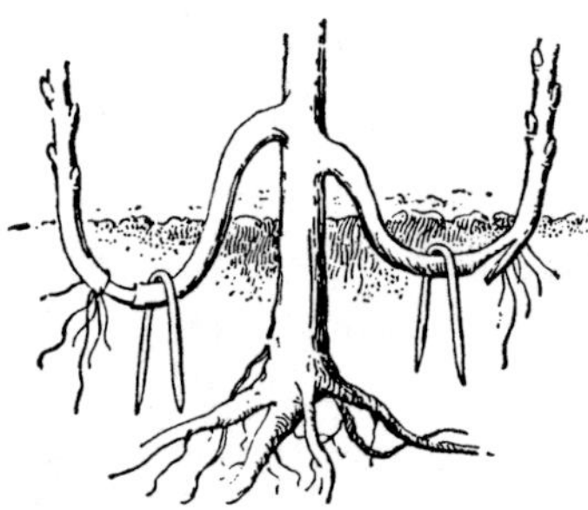

LAYERING

Trees may be most reliably reproduced by *layering*. This is time consuming, but nearly always works, even with really difficult trees. In essence, layering means burying a branch of an established tree until the buried part forms roots. Then the parent branch is severed, and the remaining top is pruned so the roots will dominate. Another method is to mound up soil against the suckers which invariably result from a severe top pruning. Or mounding up soil against suckers growing from a cut stump. If these suckers come from the root of a grafted tree, there's no telling what kind of fruit you'll get, because the top was the named variety. This is called *mound layering*. Yet another way is to scratch away some bark from a young, still-green twig, and pack a gob of moist compost around it in a waterproof wrapping. This is called *air layering*. It's not as reliable as the other two ways, but is easier. In days gone by special flower pots were available with a removable side opening so the layering stem could be enclosed within. I have successfully layered hawthorns for hedges, which are notoriously hard to propagate by other means. I dug wild young ones from the woods in early spring and buried all but the very last buds on the leaders in shallow trenches.

Always keep newly planted trees mulched for the first year or two until they're established enough to compete with the grass. Trees get nearly all their nourishment from the top 6 inches of soil. Their deep roots primarily only drink in water, so they don't die in times of drought. This is why trees don't fruit or make new growth when the top soil is dry, but usually don't die once established. It is also the reason why wood ashes contain about 25% calcium, even when the tree grew in calcium-deficient soil. They got it in the water taken up from great depths.

If you have healthy wild fruit trees, but their fruit is poor quality, you can *bud* or *graft* on wood from "good" tree prunings. To graft, first cut off a young, still green twig about the same thickness as your "good" twig, or *scion*. Be sure to pick a healthy part of the tree. Don't cut straight across, but rather make wedge shapes, so the final union will be strong in the wind. Right under the bark is a soft juicy layer called the *cambium*. For success, the cambium in the scion must precisely join the cambium in the stock twig. This is why they need to be the same size. Fit the parts closely together,

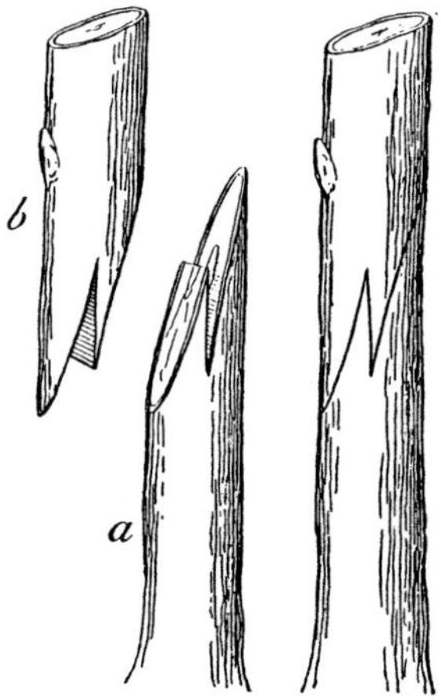

TONGUE GRAFTING

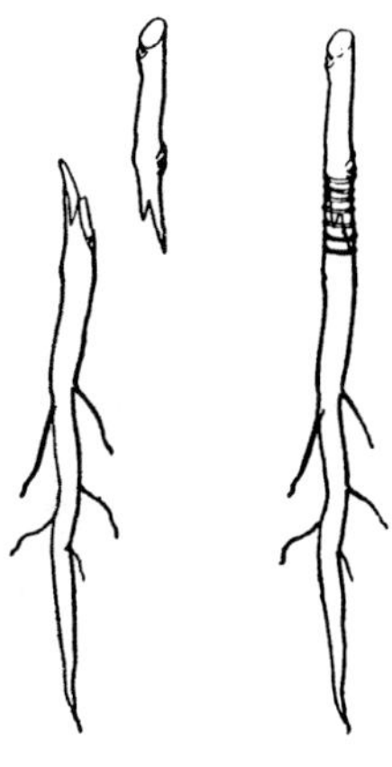

A COMPLETED GRAFT

Showing scion and stock from which it was made

then wrap them tightly with strong cotton string which has been coated with grafting wax. You can make your own grafting wax of equal parts of tallow (or other hard fat), wax (best is beeswax) and raw linseed (or other vegetable) oil. Smear the wax all over the finished joint so it will be air tight. The best time to graft fruit trees is when the scion and stock are both dormant, but the worst winter cold has past. I suggest you try apples first, because they graft easiest.

Budding is similar to grafting except a bud from the side of a one-year twig is used for the scion.

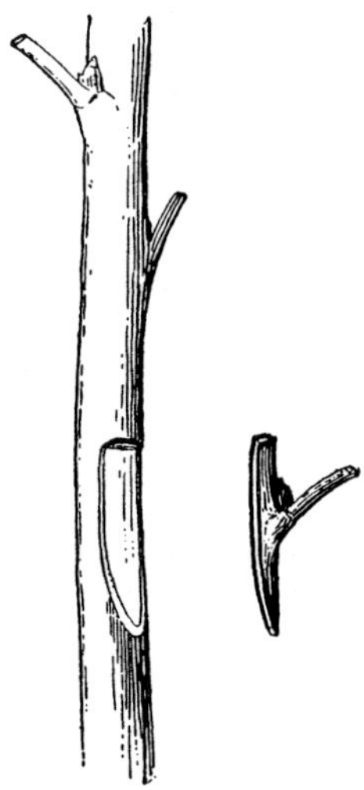

HOW TO CUT A BUD FROM A SCION

It is cut with a bit of bark attached, in the form of a shield. This shield is inserted under a bit of bark which has been slit and carefully peeled back to accept it. The whole thing is then wrapped as with grafting. Buds are best installed low (2 to 3 inches from the ground) on young stocks, but several buds may be placed on older stocks a foot or two higher. As soon as bud and stock have united, remove the wrapping so it won't cut into the bark as it grows. sealed with grafting wax to prevent drying out.

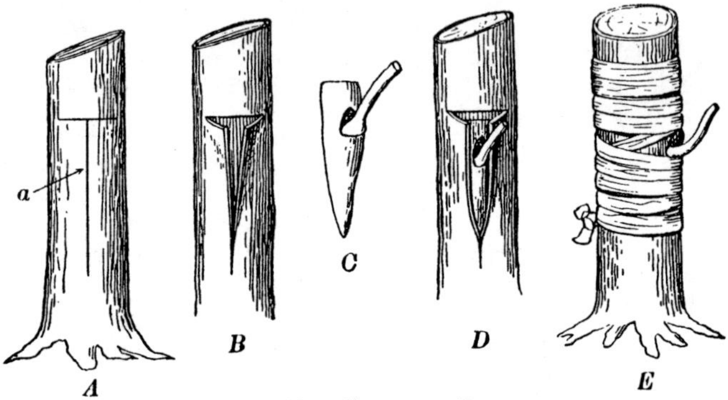

THE STEPS IN BUDDING

The tree is then cut off close to the bud to force nourishment into it. Buds will thus grow fast, becoming the principal leader on the tree if you remove all suckers. In the summer and fall the bark peels back easiest, so that is when most buds are installed. Budding is used mainly for plums, peaches and nut trees, but will work with practically any tree.

A better way to bud is to use a *bud cutter*. This is a handle with a square end to which four razor blades are attached so as to cut a square of bark all at once. Both scion and stock bud patches can be cut like this, since they'll fit exactly, and no bark need be peeled back.

The best time to graft or bud nut trees is when the daytime temperatures will reach around 80-85°F, but the scion wood has been kept dormant in a cold cellar or in a refrigerator. Scions kept long like this ought to have their ends sealed with grafting wax to prevent drying out.

Apples

I mentioned how easy apples are grafted. There are some bugs which attack them, but that is best left to a gardening or orchard book. I will say that there's an old saying with much truth to it: "Beefsteak, brats and barren trees, the more you beat 'em, the better they be!" Apple trees which refuse to bear despite good care, moderate pruning, good weather, etc., can often be stimulated into bearing by bruising the trunk bark all around by thrashing it good with a stout cudgel. Even if it doesn't work, you can vent your anger on it before turning it into firewood. Jesus Christ recommended digging and dunging, which I would have already tried by the time I started waling on it with a club.

Pears

Pears are grown like apples in every way except the pruning is much less severe. If you could avoid pruning altogether they'd be better, but you can't avoid it if you want a sturdy branch structure. You're more likely to need branch spreaders than with apple trees, because pears seem to want to grow their branches straight up. *Fire blight* is a danger too; it looks like the affected portions have been charred with fire. Clip off all affected parts and burn 'em. *Seckel* pears seem immune to it, but though they're good to eat, my favorite types are all susceptible. Keep the tree well mulched, and remember most pears will fall off the tree in autumn before they're truly ripe. They may get bruised if this happens, so after the first few fall, you may pick the rest and store for several weeks until they're ripe. They certainly get better with age! Tree-ripened is best, if you can figure out a way to effect it.

Cherries

There are 2 main kinds of cherries: Pie cherries and bird cherries. The first, the sour species, seems easier to grow in most conditions in the eastern U.S. It is very tart and is most commonly used in pies. The bird cherry, or sweet cherry is a must for Black Forest cake. It is also the kind most eaten fresh as a dessert topping. Its name derives from the fact that the majority of your ripe cherries will be eaten by birds unless you can take some precaution to stop it. Nets to cover the tree are somewhat successful sometimes, but good cherries grow so big it becomes impractical. An everbearing mulberry is probably the only way to lure them away; you'll still lose some. The best thing about cherries is they hardly ever need pruning, which is good because they are susceptible to *silver leaf* disease through the pruning wounds anyway. They don't like being opened up in a bowl-shape like other fruit trees. They are the first tree fruit available in the summer, but untimely frosts often nix your crop.

Peaches and Apricots

Peaches and apricots are way too vigorous to not prune off about 75% of their last-year's growth every year. That figure is no exaggeration, since even having removed three fourths of the tree, its growth will run away with you. And you'll still have to thin the fruits to 8 inches apart to get big, good ones. The main problem is that this tree responds to early warmth, and often blooms just in time to get frosted badly. The blossoming period of most trees is long enough that one or two nights of frost may not materially affect yield. I know of folks who string garish Christmas tree lights on their trees to thwart frost. I well remember the stench of *smudge pots* from my boyhood, which are nothing more than 5-gallon cans of burning oil with a metal lid open an inch or so to slow the combustion. I know a local orchardist who hires a helicopter to fly over his acres of trees fanning the warm air down to the ground; that's a bit much I know, but there's more than one way to skin a cat. To propagate peaches, either bud them on peach or plum seedlings, or just try growing from seeds. Ours always bore a peck of fruit on the third year from seeding. Our peaches always get leaf curl disease unless we spray with fungicide before the buds swell up in spring. Lime sulfur or Bordeaux from a garden supply house will do it. Or Peret's mix as described in the chapter on potatoes, but diluted by half with water. Peaches are great to experiment with because they grow fast, bear young, and are on their last legs by 10 years.

Plums

We never had much luck with plums, so I can't say much about it except that the biggest problem with 'em is the curculio bug. It attacks the fruit long before it's ripe, ruining it. An old man once told me that in difficult regions like ours, plums could be grown by budding a good scion onto a wild plum tree. They can also be budded on peach stock.

Quinces

Quinces are grown by cuttings or layering, but I've never grown them.

Mulberry

Mulberries were once very commonly grown in the South for swine mast. There was a variety called Illinois Everbearer, which was somehow grafted on common wild mulberry stocks. Mulberries are one of the sweetest, most nutritious of berries, and hogs really fatten on them. Small children are fond of getting their sticky substance all over pressed white Sunday clothes, and birds enjoy them enough to leave your cherries alone, because they are attacking the very abundant mulberries at about the same time. Also they are in season with some of the small fruits, like gooseberries and currants, which birds also love.

Persimmon

Persimmon is America's only real ebony tree. Golf club heads used to be made of them. The heart wood is jet black, but there isn't much of it. The fruit is never good until after a frost, but I don't know if the frost is responsible or if the fruit just ripens late. I used to eat these when I was a boy, and they were sweet and chewy when ripe. Unripe they are very astringent. They aren't native in the high elevations of the Allegheny mountains, but where they are native the 'possums always eat plenty of them. "'Simmon" seeds were what identified 'possum scats to all us children growing up.

Paw Paw

Paw paw tastes much like a banana, but I don't know anything about cultivating it. The only ones I ate were always wild in the woods. Still, it might be worth having, being native, and adapted to much of eastern America's climate.

Medlar

I have never grown medlars, which are something like an apple, but you have to wait until they seem "rotten" before they're ready to eat.

Nut Trees

Black walnuts can be gotten from wild trees, and you'll find a great deal of variation in quality. Some trees are self-fruitful, some aren't. Look for the biggest nuts you can find when they start falling in late summer. Black walnuts can be propagated by mound layering, or by budding when the days start getting up to 80°F or more. New, first year buds which sprout from the sides of a stump which has been cut or died back, supposedly can be rooted by cuttings, but I've never tried it. Many folks don't like the strong taste of black walnuts, but there is much which can be done to make them milder. Most importantly is prompt collection. The best way is to remove the green hulls (I just stomp 'em off as I collect 'em) as soon as possible. Then place the still-light colored nuts in a bucket of water and stir vigorously to beat off the remaining hull fragments and wash off the hull juices. Discard the "floaters." Repeat the operation until the nuts look nice and clean, then dry them in a rodent-proof place in layers no deeper than 6 inches, preferably on screens or grids in a warm place out of the sun. Treated like this they're much milder.

English (Persian, Carpathian) Walnuts now have cultivars hardy over most of the continental U.S., but they're still best budded onto Black Walnut stocks, also when the daytime temperatures are over 80°F.

Hickories and pecans are plagued by a bug, the pecan weevil, which pokes its beak into the soft, immature nut, and there lays an egg which produces a worm. When the nut is cracked open, both you and worm are unpleasantly surprised. The only ways to eliminate this pest are to either poison all around the tree each year for three or four years, or gather every single nut under the tree promptly as they fall each year for three years. In the latter case, they must be either all destroyed, or the nuts may be laid on a screen 6 inches from a tight floor, and twice a day the worms found thereunder are swept up and burned until no more emerge. Whatever nuts are left are yours. The good news is that these bugs don't migrate much. Chestnuts are infected by a similar weevil, and are handled the same way.

SOFT FRUITS

The only soft fruits we bother with are red and black raspberries *(Rubus idæus & occidentalis)*, gooseberries *(Ribes grossularia & cynosbati)* red and black currants *(Ribes rubrum & nigrum)*, blueberries *(vaccinium spp.)*, strawberries *(Fragraria virginiana)*, and grapes *(Vitis spp.)*. Blackberries *(Rubus villosus)* are so ubiquitous, there is no need to plant. Domestic blackberries are of much higher quality than wild ones, but we just don't favor them like other fruits.

Soft fruits have one property which makes them indispensable: Rapid maturity. Peaches, the fastest to produce of the tree fruits, may bear a small crop in their third or fourth year. Lepaa or Hinnomaki gooseberries may very well bear fruit the year you plant them out! Certainly strawberries, raspberries and black currants will yield the year after planting. These fruits are all very high in vitamins, some retain plenty of vitamin C even after processing.

Currants and Gooseberries

Black currants probably contain more vitamin C than any other fruit commonly grown in cool climes. I find their fruits very unpleasant fresh. They smell to me like cat urine. They are not sweet. But with honey or syrup added, their slightly fermented juice is transformed into a very delightful and refreshing drink. All currants seem to yield heavily here in the mountains. I understand they aren't successful in areas with hot summers. Get some black currant cuttings about a foot long and simply stick them in the ground with a couple of buds above ground. The ideal time to do this is right after winter pruning. The first year the whip will become established. Black currants bear fruit on year-old wood. So don't prune out new wood; wait until after it has borne fruit. Keep the number of suckers (new shoots) down to about 4 or 5. Trim out the old black wood. I have never seen the yield of black currants negatively affected by bugs or disease or bad weather. I've heard it makes excellent wine. We make jellies and special syrup from it. The best way to treat it--and all soft fruit, actually--is to dung around the plant and then mulch heavily. Don't dig deeply around soft fruits, for the feeder roots are all very near the surface. You probably don't need to dig at all if your mulch is a good 8 inches thick. Plant them about 5 feet apart. The berries get very dark and a little bit softer in the center when ripe. Birds will eat them before they're truly ripe, so don't let that be your clue, and do something about those birds! Tho' we never had problems, black currants do sometimes get diseases. The best way to fight this is to make sure to have plenty of black currant bushes, and plant some from your prunings. If the buds swell up like a Brussels sprout, or the leaves remain wrinkly like a stinging nettle, grub 'em out and burn 'em. Rely on your new plantings.

Red currants are a different creature altogether. Their fruits are mainly suited to making jelly. In fact I think it's fully the equal of cranberry jelly for roast fowl or pork. Even more birds eat red currants than black. Their somewhat astringent flavor takes some getting used to raw. Propagate by layering, which is easy if you simply bend a shoot over to the ground and lay a rock or brick on it. When this roots, sever it from the parent plant, dig up the root and stem, and transplant. The stem is snipped quite short, so only a couple of buds show. Plant them 5 to 6 feet apart. Red currants fruit on old wood, so leave plenty of this when you're pruning. Get rid of spindly shoots and branches too horizontal to resist bending to the ground when laden with fruit. Dung red currants well, and mulch deeply to prevent having to pull so many weeds. Like tree fruits, the plant needs a sort of vase shape, so the center gets sunshine. Keep the birds out or they'll play hob with your crop. They start eating the berries just as they begin to turn ripe, but they certainly destroy plenty of unripe berries too. Ever bearing mulberries probably will occupy them enough to save your currants. Some folks must resort to netting but this doesn't work unless the net is on a frame away from the plants--birds use it as a perch to peck berries through. The red currant turns red, and then gets a deeper red, then they turn soft. Deeper red is perfect, soft is too late.

Gooseberries are used for jam, pie or eating fresh. The *Whitesmith* variety, my favorite, tastes much like a large grape. But it's ripe sooner than grapes, and is much more reliable. Treat gooseberries just like red currants. Even more birds eat gooseberries than red currants, and you may not get a single berry if you don't do something about it. Some gooseberries are red and some white. They attain full size well before ripening, and can only be discerned as ripe through tasting for sweetness.

Maybe I ought to mention the unfair and hypocritical regulations foisted on the public by zealous bureaucrats concerning gooseberries and currants. The fruit bushes are an intermediate host to *white pine blister rust* disease. If the bushes are within 900 feet of white pines, the fruit bushes may contribute to their illness. Thus, if you are forbidden from growing the fruits, those growing white pines are profiting from your investment and taxes on your land, to your hurt. On the other hand, most such states have immense quantities of wild gooseberries and even wild currants growing on state land; no government agents are searching these out and destroying them. They only prohibit yours, because they are more easily found.

Raspberries

Black raspberries or blackcaps, grow wild almost as abundantly as blackberries. Unlike the latter, they have no striated stems, and the stems range from a powdery pastel green in new canes to a beautiful powder covered purple in older ones. Blackcaps fruit on one year old canes. We cut off young growth to about 4 feet tall, and then trim the winter-frosted portions in spring. These *stools* are about 2½ feet tall. Many authorities advise cutting new growth back to 2 feet, so it will grow lots of branches. Then these are cut to about 8 or 10 inches long during early spring. This is supposed to make more fruit. New, long shoots grow and interfere with harvesting, so it's a good idea to set hefty posts at intervals in the row, with a stout wire 6 feet high on both sides, tucking the new growth up and out of the way into it. We never did this, but I think it's a good idea. Blackcaps reproduce naturally when the tips of these long canes hang over to the ground and root. Your garden would soon be over run with them if you let this happen. Still, you ought to start new plantations about every 5 years so virus build-ups can be dealt with by hoisting out the old plants, burning them, and growing say, strawberries in their place for a few years. In any case, dung them well and keep the weeds down with very heavy mulching.

Red raspberries are different from black raspberries. In the first place, they reproduce from canes popping up from the spreading root system. They can be easily propagated by digging up such plants with their root balls and transplanting as deep as you found 'em. Ours are "everbearing" which means they have a crop in summer and another one in fall. Where we live if we let them grow this way, the summer crop's flowers would be frosted off, and the fall crop would come so late as to be wiped out by fall frost. So we just mow off the canes right down to the ground in winter. Then there's no summer crop. But there is a very abundant and long bearing fall crop. They can be trained between two wires like black raspberries. Either kind can be planted in rows about 7 to 8 feet apart, and 3 feet away from each other. The rows will eventually become tight as a hedge.

Blueberries

In America, some kind of blueberry will grow in almost every climate except dry ones with alkaline soil. If you can grow mountain laurel, great laurel (rhododendron) or azaleas, chances are you can grow blueberries. They need acid soil, a pH of about 4.5 is best. You can make soil acid by mixing garden sulfur with it according to the directions on the container. I will deal only with highbush and rabbiteye blueberries, because I don't have experience with any other. The latter is grown in mild coastal and low lying Southern areas; the former is for uplands and northern locations.

Plant blueberries about 5 feet apart in rows about 10 feet apart. Enrich the soil with well rotted dung and sulfur if needed, and set the crowns of the plants at soil level. Water them if the weather is dry; they are bog plants and can't take much dry weather until established. The best way is to mulch them heavily. To propagate further, mound up the plants with constantly moist compost about a foot deep, and when the stems have rooted, cut them loose and transplant. Never hoe blueberries; their roots are just too shallow. Just mulch heavily.

Highbush types need to be pruned annually starting in the third year, cutting out old stems at ground level when they get too twiggy. The plant will send up new stalks as replacements. I recommend cutting the side branches back until they have only three buds apiece.

Rabbiteye blueberries hardly ever need pruning except to remove dead or wimpy branches.

Always plant several varieties, for two reasons: 1. They will yield much better with cross pollination; most won't yield at all without it. 2. You can stretch your harvest season out by using early, midseason and late varieties.

Strawberries

These are just what their name implies--straw mulched. Indeed, I don't ever recall a really good crop which wasn't well mulched. A second thing to remember is the tremendous variation in flavor between varieties. *Surecrop,* a common commercial type, is reliable, but I think it's truly second rate. My favorite tasting types are *Allstar* and *Fairfax*. *Fairfax* is getting hard to find, because *Allstar* is its equal in flavor, but yields much heavier. Even among one variety, there can be better strains.

Strawberries reproduce mainly by sending out runners, which root and become new plants. The runners must be kept in check, or they'll overrun your strawberry bed, and you'll get lots of greens but fewer fruits. Besides, strawberries fizzle out after a couple years bearing, so you need to start a new patch every other year just to insure a good crop. The most important things to remember when planting are to spread the rootlets so as much soil contact is made with as many roots as possible, and to have the crown of the plant right at ground level. They like leaf mould or other humus, but don't like lime. Most types bear during or near June. Everbearers are in my humble opinion a lot of work for little pay. Fresh strawberries are nice, but the main use of this crop is for strawberry jam and frozen toppings. The world will never contain enough of these products to satisfy a dignified epicure.

Don't let the plants fruit the first year--pinch off the blossoms. Otherwise they often pout by withholding fruit the next year. Grow them about a foot apart in rows alternating between 1 and 2 feet distant. Plant plenty of sprigs each year or two; after all, what do they cost you if you get them from runners?

Grapes

There are grape varieties for

nearly every inhabitable climate, and it can be made into wine with little or no adulterants, which is probably why "the fruit of the vine" was designated by Christ as the symbol of His blood--practically everyone could have access to it at all times of year.

Grapes are the easiest fruit to propagate, as long as you have a variety hardy to your environment. Grapes are cut back severely every year, because if not, they'll grow vine only and very little, if any fruit. The right cuttings are selected, cut about a foot long, and planted. Give a twig a quick bend; if it snaps and hangs by the bark, its right; if it bends without breaking, it's too young; if it splinters, it's too old and woody. It is best started in sand, because organic matter could carry disease. Bury it up to the last good bud. If done in warm weather it should be rooted in about a month or so, and it can be transplanted.

Grapes are also propagated by layering. Just take a branch and bend it into the ground, with a rock to hold it in place. It roots, you dig it up and transplant it.

Grape vines are grown on pole trellis in much of Europe. This is just a post set by each vine with cross slats for the vine to twine around.

In America wire trellises are used.

Whatever way the vines are trained, prune them back to only two buds at planting. Train them straight up a post of some sort. The next year, cut them off 2 feet from the ground. Leave 2 buds above and 2 buds below the first cross slat or trellis wire. Also leave one bud near the top. These buds will all form *leaders*. One leader will be allowed to form a central stalk, or trunk. The others will be *lateral leaders* and will be trained on the slats or wires. Each pair of leaders is trained so they run each way down the slat or wire. These leaders are main branches and are known as *fruiting arms*. They are pruned way back each year to only 6 to 10 buds each. All other lateral canes are cut out, but leave 2 stubs near the fruiting arms on which to grow next year's fruit. As you can see, grapes are drastically cut back every year to keep the new wood as near to the central branch, and the root, as possible. Otherwise very little fruit results.

Grapes need above all else, well drained soil. They like sunny warm weather right through the growing season, and varieties must be selected which are adapted to your local conditions. It is also important that they not be in frost pockets. These are points in dips and at the base of hills. Better to grow them on a hill in gravel than in a swale with super rich soil. They hate poor drainage. Despite all the drastic cutting back, a good vine will well outlast you. Some are centuries old and still bearing fruit pretty well. They'll even bear in nearly barren soil, but they like humus. Also, mulching is best because they too have shallow feeder roots which are disturbed by hoeing. Besides, the heat reflected and radiated by good mulch warms the leaves--a must for abundant harvests. Bird damage can be severe when sitting birds' weight breaks off young branches. Also, grapes are prone to frost damage.

I have never grown the muscadine varieties, but they are said to only need renewal of fruiting arms once every 4 to 5 years. They also need fruiting stubs, called *spurs*, each having 4 buds, left along the fruiting arms. It's not a bad idea to spray the grapes

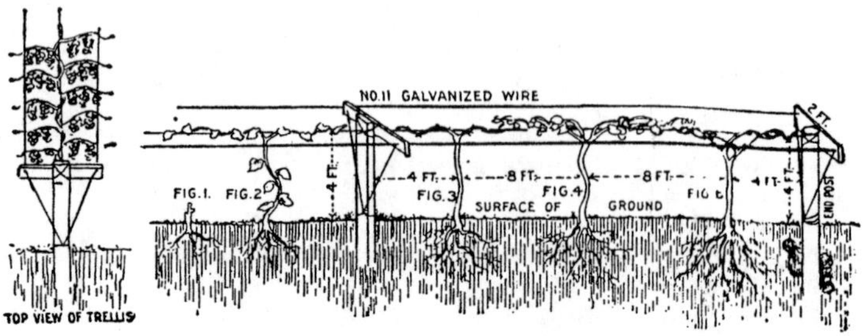

The Munson system of grape culture.

with a solution of Peret's mixture (see chapter on potatoes) diluted by half, every 2 weeks during the season, especially if you grow the European varieties. They are highly susceptible to fungus rots without help.

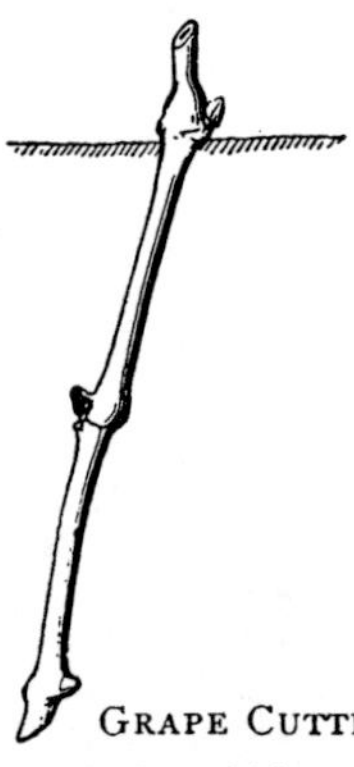

GRAPE CUTTING

Showing depth to which cutting should be planted

Hip Roses

There is really very little to say about rose hips. All roses grow them, but the *rosa canina,* or *dog rose,* is the only common one with really big hips. There are even some selected varieties of these with superbly large hips. Plant them just like setting out an orchard tree. The dog rose is naturalized throughout most of the U.S., so even wild examples can be found fairly commonly. It is self-fruitful, which means it needs no companion plants to pollinate. Plant them along fences for convenience.

After the first frost pick the healthy red hips, discarding the black or shriveled ones. Clean them by breaking off stem and flower remnants. Cut each hip in half; then scoop out the prickly seeds. Cover the now-empty shells with water in a pot and boil until mushy. Press through a strainer, and add an equal weight or less of sugar, to taste. Boil again, stirring to prevent burning, to dissolve and distribute the sugar. Pour hot into sterile jars. You now have rose hip jam. This method is a labor of love, but rose hip jam is quite a rare delicacy, well loved by those Europeans old enough to remember it. Everyone loves it, and you can be sure it'll be well appreciated, even in very tiny jars, as special gifts from the country.

32
HUSBANDING THE HARVEST

And take thou unto thee of all food that is eaten, and thou shalt gather it *to thee; and it shall be for food for thee, and for them. Genesis 6:21*

If we could eat all the food we need for the year in one sitting there'd be no need for this chapter. To completely describe food storage methods would take a book by itself, at least! Having tried plenty of recipes which didn't work, I can say with confidence that I deserve a place in the record books for the most food spoiled by one man. Not all ways of saving the "wanton burden of the prime" as Shakespeare called it, are cost effective or worth the time.

In the first place, certain crops withstand freezing temperatures well below 0°F; there's no way I'd go to the bother of canning, drying or salting them. Kale, Savoy cabbage, Brussels sprouts, spinach, collards, rutabagas, and even kohlrabi will hold up until about the middle of January before repeated freezing and thawing finally makes mush of it all. They'd last right through winter if mounded up with soil, leaf mold, compost, or what-have-you. I have seen carrots remain in the finest possible condition in the ground with the following protection: Stake foot-wide planks on edge on either side of the rows, and fill in with compost or other insulative matter before the soil freezes. Parsnips, salsify and leeks need much less of such protection. Merely hilling them up a bit with soil is good enough. Jerusalem artichokes need no protection at all, in fact they keep in the ground than out, freezing or not. They rapidly dry up too much once dug.

Some plants are exceedingly rapid to sprout and grow, even in cold weather. Radishes come to mind, and watercress.

Still others, wintered over, are the first to "come on" at winter's end. Endive and parsley are examples, but the common dandelion, if cultivated and allowed to grow a big fat root (and kept from going to seed!) will leaf out profusely at the first signs of spring. Indeed, there is quite a market for "dandelion greens" in some localities, because it's the first really green, fresh leaf vegetable. In Europe, a perennial called sea-kale is commonly grown for the same reason. It's very hard to find in American seed catalogs.

All of these vegetables enable us to eat fresh herbs from possibly as early as late February (in milder years) until early January, showing what a short season we need for vegetable storage.

Poultry keepers and fishermen can have fresh meat practically any day of the year.

Still, variety is the spice of life; it's good to have plenty stored up for times of illness or inclement weather.

Root Cellars

I know folks who dig cache holes in the bank of a hill, insert a wooden barrel (never use plastic or metal, because they greatly encourage rot) fill it full of cabbages, potatoes, carrots, or whatever; cover with straw. Then they seal with planks and dirt.

Others dig a 1½ foot deep trench right in the middle of the garden, fill it up with potatoes, carrots, cabbages, turnips, beets, etc., cover with straw or leaves, then planks, and lastly put all the dirt from the hole on top.

I'm sure this is much the best way to store long-keeping vegetables and fruits, though I've never done it. The moisture from the very cool (but not freezing) earth keeps the produce in tip-top shape.

Root cellars are really just much larger versions of this. The easiest way to make a root cellar is to dig a pit under the house or nearby, about the size of a small room. The pit is braced with timbers, logs, or better, masonry. It's often lined with shelves for storing canned goods. Some have bins for potatoes and other roots; hardly necessary unless light can enter somehow. It needs good drainage. Cellars normally have a roof or some kind of shed built above. But they must not be "air-tight", because all produce gives off moisture, and this causes mildew and rot. Very tightly made cellars need a chimney to create a slight draft. If the floor is 6 feet underground, it won't freeze in my region. Root cellars are, as the name implies, for storing roots, but also are good for vegetables like cabbages. Additionally, plants like carrots, beets, etc., can be preserved for replanting in early spring in order to collect seeds, and for storing dormant scions for fruit tree grafting.

Pantry

I'm not too enthusiastic about

—INTERIOR OF SPRING HOUSE, WITH LOW TROUGH.

using a cellar for canned goods. There's too much dampness, and jar lids often rust. A pantry is nothing more than a room, usually with no windows, adjoining the kitchen on a northern corner of the house. This keeps the room coolest in summer. The room ought not freeze, but it shouldn't be heated either. Pantries hold hams & bacons, canned foods, dry beans, braids of onions and garlic, maybe some sauerkraut, a few squashes or pumpkins, herbs, a half-bushel basket each of potatoes, and sweet potatoes or carrots, lots of cheeses, and some grains, both whole and ground. It's also a good place to keep utensils used infrequently, like a meat grinder, scales, grain mill, perhaps a lard press, butter molds, butchering equipment, canning supplies, sausage stuffers, etc.

The Spring House

I have neighbors with a spring near their house, in which glass jars of food and drink are stood to keep cool. Once upon a time folks cooled large quantites of dairy products in fairly small masonry buildings called *spring houses*. There were clay or cement troughs, some shallow, some deep, in which vessels full of perishables were set. If the water temperature is naturally cool, as it almost invariably is in the mountains, food cools quicker here than in any ordinary refrigerator. Even cold water from creeks is o.k. if it's moderately free of pollution. After all, the water never touches the food.

Spring houses also keep food from freezing, and can double as root cellars. Make sure they aren't sited where they'll flood, or make a muddy mess, and that provision is made to carry away the "tail water." The best way is to run it to water troughs, and then away into the creek.

Ice Houses

For those who can't rely on the power company for dependable electricity, an ice house takes on real importance. Ice can be made to cool things down to temperatures

—SECTION OF AN ICE HOUSE FILLED.

below 0°F, by adding salt. When we make ice cream, I have taken the temperature of the ice in the stir-freezer, and have recorded it as low as -5°F! And yet, I'm not sure we could *keep* stuff frozen unless we salted the ice, like making ice cream. I just haven't tried it. Ice used to be the main method of keeping food cool in town and country not so very long ago.

Great expense isn't vital. In Southern Virginia a hole dug in the ground above water level, lined with native clay kept ice acceptably through the following fall. Its only covering was leaves and pine boughs. Ancient Romans used to do this to keep snow. I once heard of a man in Wiconsin building a 4 walled enclosure using green posts fresh from the woods. These supported walls of a single thickness

—CHEAP AND PICTURESQUE ICE HOUSE.

of inch-planks nailed inside. A foot of sawdust formed the floor. The ice was stacked in the middle, surrounded by a foot of sawdust. It was topped with another foot of sawdust, and no windows, doors or roof. The ice kept through summer without much waste. I'm sure it'd keep much better with a roof. Blocks of ice are taken as needed for cooling true "ice box" refrigerators, and other uses.

Four principles for keeping large amounts of ice even in hot summers, are: 1. Keep the ice well drained. This is why sawdust is used on the bottom, and even this is best set on top of coarse sand and/or gravel. 2. Pack the ice as close as possible, to prevent circulation of air within. Only moving air readily conveys heat to the ice. 3. Keep all sides insulated with at least a foot of sawdust, shavings, bark, etc. None of these affects the purity of the ice, and all drain well. 4. Allow space between the top of the sawdust over the ice and the roof for good ventilation. Insure dryness; prevent heat conductivity.

The actual amount of ice needed in a year varies. In older days small commercial farms needed about 25 tons of ice, which could

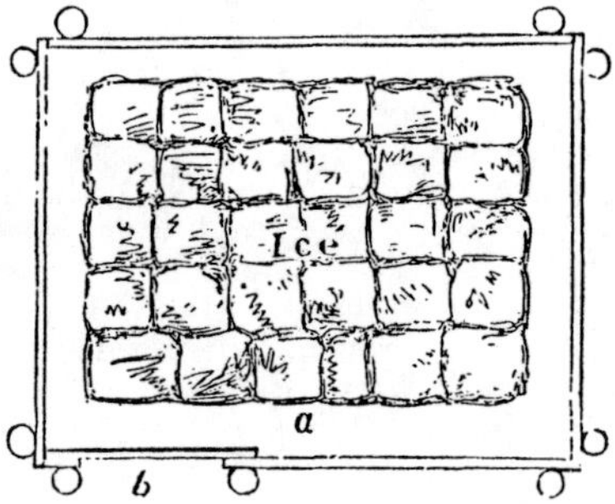

TOP VIEW OF ABOVE ICE HOUSE

be easily stored in a building 12 x 14 x 10 feet tall, including space for sawdust. Many a farm pond was established way back when, for the purpose of insuring an ice supply. Others used boxes of wood or metal pans to set out full of water to freeze. On cold mornings the ice was removed and placed in the ice house. Another way is to shape a shallow basin of sawdust in the house, open the doors and windows to the cold, spray the sawdust with water until it is water tight with surface ice, fill the basin with water, let it freeze overnight, and close it all up on warmer days. Repeat this process, using the first ice as the floor for the next basin, until a huge cube of ice is made. It's tough to get the ice out, I'm sure. It's probably better to just stack 6 inch thick cakes on edge, staggering the joints like with bricklaying. If laid flat it's much harder to get the cakes out later.

Drying Vegetables and Meats

Sun drying vegetables is nearly impossible in our mountain climate. So folks make *leather britches* out of green beans. They run a bit of thread through the upper third of each bean, spacing all the beans a half inch apart on one thread. Strings of beans are then suspended over the wood cookstove in the kitchen. They need to be soaked for hours before they'll cook tender.

Carrots may be cut lengthwise into quarters and similarly threaded on a string to dry overhead in a warm kitchen. Mushrooms are dried the same way. Leeks, celery and celeriac can all be dried on oven racks or on trivets over a warm (not hot) stove, or in a warm open oven. But then, all these might be better kept by mounding them up in the garden. In the Andes mountains, potatoes are set out to freeze, and are then stomped with bare feet to mash them up to dry. The result is somthing like the store-bought instant spuds, but I've never done it. Anyway, potatoes keep long enough in a root cellar to get you through winter and spring, when your next "earlies" are ready to eat.

Sweet corn can be dried by first boiling it well on the cob, then slicing it off the cobs, spreading it out in trays and drying it in a slow oven for a few days. To reconstitute it, pour boiling water over it, and let it stand for 3 to 4 hours. Then heat and serve. I like canned corn better because it's ready to eat as soon as it's opened.

Soup beans are mostly "bush" beans. Let them ripen on the stalk until the pods get papery. Pick them, storing for several weeks in the pod in baskets or gunny sacks until the beans are hard and dry. Then either shell by hand, or put 'em in a gunny sack, tie it, and "wale the farr out of it" with a stick. If you do the latter, you need to winnow the same as wheat.

I have made beef jerky successfully in a slow oven during winter. I simply marinated inch-wide strips which were about ⅜-inch thick, for 6 hours in a solution of ¼ water, ¾ red wine. The strips were pierced and hung near one end on bamboo skewers so none were touching each other. I salted them lightly before drying. The oven door was kept open an inch to allow air circulation. It shrinks a surprising amount. The word "jerky" is from an old American Indian word "charqui," which means dried deer meat. A diet of this would be meager indeed.

Curing and Larding

Almost any meat may be stored in lard. As mentioned in the chapter on charcuterie, meat or fish may be seasoned and cooked, placed into jars, and hot lard poured over it. It's best if a weight of some kind can be used to hold the meat under the lard. When the lard has solidified, the weight is removed, and the cavity sealed by pouring in some more hot lard. Close the jars and keep 'em cool and dark.

Meat curing was discussed in its own chapter. But cured meats may still be attacked by bugs or rodents. I have had *carpet beetles* eat right through tight cloth bags when cured meats were hung in outbuildings away from cats & rats. Other times, *skipper flies* got to them. They look like fruit flies, but their maggots eat cured meats and cheese. They're called skippers because the maggots are able to leap considerable distances. In any case, the results are tragic. Some recommend burying the cured, smoked and dried meats in wheat bran, but I don't think that'd discourage rodents or bugs. A much better plan is to bury in wood ashes, and store in a relatively cool place. Cured meats improve with age up to about 2 years old.

Other Salting

Some recommend salting green beans or sweet corn, but to tell the truth, we put ours up in jars by canning. Even so, let me describe the way: With corn, boil it, husked and in its prime, for 10 to 15 minutes on the cob. Cut the kernels from the cob with a sharp knife. Mix 1 cup of salt for each 4 cups of the cut corn, and place in a clean crock or large, wide-mouthed glass jar. Put a sterile plate on top to hold it all down, and cover with a cloth to keep dust out. If there isn't enough juice to cover it tomorrow mix 1½ tablespoons of salt to each cup of cold water, and submerge it. Keep it in a cool dark place. Some folks say to keep adding corn as you harvest it. Others say not to. I don't think it matters much. To use, take out what you need, soak it for 20 minutes in fresh water, and repeat until it tastes sweet. Heat and serve or eat it cold.

To salt beans, simmer 'em for 10 minutes just like you were going to eat them right away. Slice them 2 inches long. For every 4 pounds of cooked beans, add a pound of salt. Sprinkle a thin layer of salt in the bottom of your jar or crock, then add a layer of beans. Repeat until the jar is full enough. Weight and cover with a cloth. If there's not enough brine tomorrow to cover the beans, do the same as for salted corn.

Canning

In essence, all canning consists of cooking vegetables or fruits in sterile jars long enough to kill germs. The jar lids are rubber lined, and air tight. The boiling hot air trapped under the lid shrinks as the the jar cools, causing a vacuum seal. In the case of acidic fruits or jellies, pure melted paraffin is sometimes poured right on top of the jar's contents. This method still needs lids to keep out varmints.

Pickles are the only canned food we don't cook to seal. My experience is there is enough acid in pickles to keep them from spoiling. Pack the jars with cured pickles (which see). The pickling liquor is heated boiling hot, poured in to cover the pickles, and the lids are tightened to seal. This method may not seal every jar, but we eat them (the ones which don't form a vacuum seal) within a month or so without worry. Even if a yucky scum forms, "kosher dill" pickles are still probably good. I use common sense; just discarding stuff that looks bad. When I boiled the pickles like all the modern books say, they always turned out mushy. I like 'em crunchy.

Tomato juice is made by pressing cooked tomatoes through some kind of screen or fresh tomatoes through a strainer like the *Squeezo* or *Victorio*. The best tomatoes for this are paste tomatoes. We like *San Marzano* (best) and *Roma*. Other kinds have too much water and too little substance; there is no "all purpose" tomato. Juice can be boiled down to get paste, but we rarely do. I like to drink it instead of orange juice in the winter time.

Tomato juice is poured into sterile jars, then boiled, submerged in a covered container, for 20 minutes to seal. Some of the old ladies in my neck of the woods place lots of jars in a wash tub, sitting on a submerged towel, to keep them from overheating against the fire. More rags are placed between the jars to keep them from bumping into one another. A sheet of tin or plywood on top keeps in the heat and steam so the jar lids stay submerged.

We have a pressure canner, but it's so much hassle we don't use it. Let me recommend that you buy a large hot water bath canner, and a canning book by the jar manufacturer. These books tend to be very cheap, and all the recipes have been tested. Fruits like strawberries, blackberries, etc., need the least cooking. Like the sweet old ladies of our region we can green beans and sweet corn in the hot water bath canner by boiling it for 3 hours. The corn is first boiled for 15 minutes on the cob, to set the sugar. Cool the corn on the cob rapidly by plunging it into a bucket of cold water. Then we cut the kernels off, put it in jars with a teaspoon of salt, fill with boiling water and seal by boiling for 3 hours. Whenever jars are removed from the pot we keep them out of the draft, to avoid cracking the glass. Our canning is done over an outdoor fire, rain or shine, and it always works for us because we use common sense about the finished product. That is, we don't eat it if it looks or smells in any way amiss. Folks really nervous about botulism always boil the opened jar's contents for at least 10 minutes before even tasting it. There's no harm in this, but it is a bother sometimes. I'm making no recommendations here, just saying what we do.

A glut of apples, especially the non-keeping summer types, is an opportunity for canning. Peel, core and slice at least as thick as your finger. Heat in a pan with ½ cup of water (to prevent scorching) until you can see plenty of juice while you stir it. Sweeten and/or spice to taste. Then pack into sterile quart jars. Seal by boiling 20 minutes. One quart of this makes one pie, or just serve it alone.

Fermentation

As a boy I tried making wine in the closet, but it was a dismal failure. The only ones which turned out well were made of apples. Still, when my folks found out what I was up to, the experiments came to an end.

Since winemaking is such an involved topic, and I've never been good at it, I think I'd better explain something I have done well; "hard" apple cider.

When apples are pressed, the resulting juice begins fermenting within a couple of days. To keep it from turning sour, a vapor lock must be installed. I always sealed the jug except for a bit of plastic tubing like that used for aquarium pumps. The bottle end of the tube is situated above the liquid contents, while the other end is set into a glass of water. The carbon dioxide given off by the yeast escapes, but no air can come in. At some point, no more fermentation takes place. The liquid is siphoned off, leaving behind the undesirable sediment. It will keep a long time if properly sealed.

An old rural way to make "strong drink" as the Bible calls it (Deut. 14:26) is to allow the "hard cider" to freeze, and then scoop out the ice crystals. Let it freeze again, and scoop out the crystals. In really cold winters, the liquid can get pretty strong, like brandy. It's called *apple jack.*

Finally, if vinegar is what you want, just let your hard cider stand at 70°F (or thereabouts) for a few weeks and it will turn to vinegar. The action takes several weeks but can be speeded up. Just add some of the scum from the top of another batch of vinegar which is nearly finished "working." Cover with cloth to keep fruit flies out.

Sauerkraut is a product of fermentation, too. I haven't made it before because I'm not that fond of it. There are too many recipes to do it justice. Some of them use a lot of salt, perhaps more than necessary for good taste. The amount commonly recommended is 1 tablespoon per pound of cabbage. Don't use the dirty outside leaves. Quarter and shred your cabbage. Place 4 gallons of the shredded cabbage into a crock or other nonmetalic vessel which has been well cleaned by rubbing with vinegar and rinsing, and sprinkled with a tablespoon of salt. Then pound it all down with a *kraut masher* which is just a large club. Repeat until you have as much as you desire in the vessel. Lay clean cabbage leaves on top, and a sterile plate or other "follower" to fit the vessel. Then weight the whole thing with a very clean weight. Cover it with a cloth and let it ferment. It takes from 10 days to 3 weeks, depending on the temperature. If you want to can it when it's finished, bring it just to a boil. Stuff it tightly into sterile jars. Add just enough boiling sauerkraut juice to fill jars to within ½ inch of the top. Tighten the jars and let them seal as for cured pickles.

Kosher Dill Pickles

Start with good pickling cucumbers, like National Pickling. Use only firm, young unblemished fruits. Wash and rub the spines off. Sterilize a crock or glass jar big enough to hold the cucumbers, by dribbling boiling water all over the inside surface. Pour out the hot water. Add the pickles, along with say 1 or 2 flower tops of dill (grown right in your cucumber patch, if you like) and 2 cloves of garlic for every gallon of crock capacity. We place washed grape leaves beneath and on top, to prevent the pickles from losing their crunch. Mix salt and water until a fresh-laid egg will neither float nor sink in it. This will be somewhere around 1 cup of salt to 10 cups of water. Add this to cover the cukes, with a sterile plate on top to hold the pickles under. Cover to keep out the dust. In a day or two, bubbles will start coming from the ferment. Periodically remove any scum without disturbing the pickles. In 3 to 5 weeks the pickles will be good to eat, but we just preserve as explained above under *canning.*

If they're too salty, we add a little water to the brine before canning. They can be just taken from the crock as needed, but they get really soft and unappetizing in a few months.

33 SWEETS

Eat thou honey, because it is *good; and the honeycomb,* which is *sweet to thy taste. Proverbs 24:13*

It would be a dreadful loss to the world if no one could satisfy their cravings for sweets. What would become of grandparents who love to fill their grandchildren full of energizing sweetnin'? What would become of Valentine's Day?

For many centuries people have done without sugar. The word doesn't occur at all in the Bible, though in fairness, it does mention "sweet cane." Notwithstanding, honey is mentioned in all literature in all ages; it was certainly the first sweetener and still is the most healthy.

Honey

Honey is the product of bees, as everyone knows. There is something else everyone knows about bees, and it isn't at all pleasant. In fact, when there are enough bees to make significant amounts of honey, the sight of them makes normal people lose interest completely in that form of sweetening. This is why the initial investment in beekeeping equipment should be, first and foremost, a bee-proof suit, veil and gloves. And these are not all: Smoke calms bees amazingly; a *smoker* is next. The expense of a good suit will set you back a pretty penny, but with care the investment lasts many, many years. I recommend the bee-suit should be a couple of sizes too big. This makes it baggy, and less stuffy on the hot or sunny days when most bee work is conducted. In colder weather it gives room for extra clothing.

The Way of Bees

Bees are communal beasts, with a hive full of buzzing females called *workers*, a few shiftless but eager beaux called *drones,* and one boss lady called the *queen.*

The colony is understood to be one big organism. It starts when a beehive--whether "domestic" or "wild"--gets overcrowded by too many colony members. A new queen is raised up by some of the congested workers, perhaps in protest, who knows? Upon the young queen's emergence the existing queen gathers a few thousand loyal workers and abandons the old homeplace in a quest for "elbow room." Usually she alights on a tree branch, post, or the side of a building. Before her companions join her, they drink as much of the honey from the combs as they possibly can. The first hour or two of such a swarm shows bees buzzing around through the air all over the place in a seeming random and rather unnerving fashion. After they settle down, they form a huge, tight blob, with the queen at the center, from which a few scouts search a new home. When they find one, they vanish from their swarm and quickly take possession

CATCHING A HIGH SWARM

of the new location. While they are still in a tight ball, if they aren't too inaccessible, it is easy to catch them. A feed sack is slipped around them if they're on a tree branch. Then the bag is tied and the branch is cut off. Otherwise they may be gently brushed into a tight cardboard box containing a few empty honeycombs. Take them home, and open the container, placing the honeycombs into the wooden box, straw skep, clay pot,

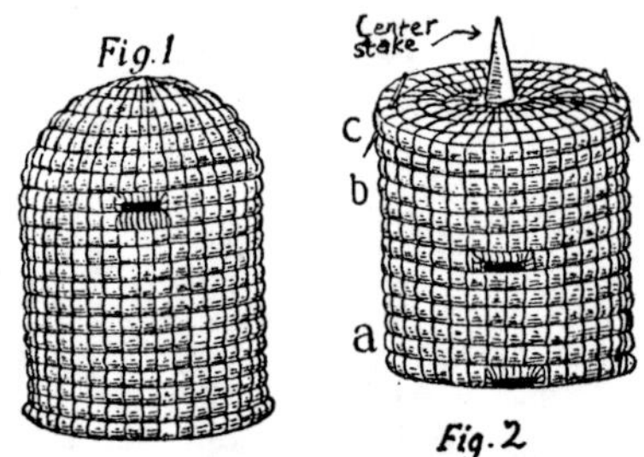

STRAW SKEPS

Fig. 1: 1-story skep; bees be killed to get honey. Fig. 2: Multiple-story skep--a. body; b. super; c. cap.

hollow log section, or standard hive you've got prepared for them. With the drawn comb (the best welcome to a bee seeking a home) they're unlikely to fly off.

Their old hive is left with a few thousand workers, and a new queen. She goes on a mating flight, is impregnated by a drone, and then returns to begin laying thousands of eggs to build up the colony anew. Many variations of this scheme take place. Sometimes the new queen swarms away with a few thousand workers, leaving yet another queen in charge. Once in a while the first queen changes her mind and comes back to battle it out with the new pretender.

From the honey eater's standpoint, the most important thing bees do is make honey. They do this by repeatedly vomiting the nectar they have drunk from myriad flowers into their honeycombs, and then fanning it with their wings until it evaporates into honey. When the honey is finished, they seal over each cell of the honeycomb with wax. They don't make honey from pollen, though they need and use pollen for other purposes.

Left to themselves, the combs are built in every conceivable shape and size. But there are two things which never change. One is the size of the honey cells. The other is the crawl space between combs. The discovery of the latter is what enabled beekeepers to coax the bees into manufacturing several times the honey they once did when kept in clay pots, straw skeps and segments of hollow log. The square wooden box hives one sees in almost every apiary are really extremely accurately made. All the dimensions have been standardized so that rectangular honeycombs of an exact size will sit precisely far enough apart to prevent the wild or *burr* comb from being built. Nowadays, the beekeeper simply slices the wax cappings off the combs with a hot knife, places the comb in a centrifugal *extractor*, and the honey spins out, leaving the comb pretty much intact.

The spun, empty combs are replaced in the hive and the bees clean them up, spic and span. Now they're ready for the next *honey flow,* which is the period of blossoming of high-yielding nectar sources, like white clover, alfalfa, buckwheat, tulip poplar trees, and basswood. There are hundreds of others, but these are the biggies in the Appalachians.

Most beekeepers I know did only a day or two's work a year to get hundreds of pounds of honey. That is, until the advent of the Varroa mite. This nasty little parasite has wiped out over 90% of all the bees in the United States, at this writing. Gone are the cheerfully buzzing "bee trees" of just a few years ago. Gone are the hoards of bees on dandelions, clover and fruit trees. And gone too, are the increased fruit yields the bees once insured. Today the beekeeper is faced with more labor and expense in protecting his bees from this mite. Miticides, or in other words, poisons, must be introduced to the hive immediately after the fall harvest, and periodic inspections of the combs are a wise precaution to save your bees.

Bees are probably the most profitable of all commercial "livestock" operations, in that one may own dozens of hives on a lot no bigger than a two-car garage. Each of these can be manipulated into growing up to 200 lbs. of honey per year, all from other people's trees and flowers! The initial investment in equipment is high though, for practically all of it is stainless steel, and will last longer than you will. Unlike other food producing enterprises, honey keeps until you sell it. It keeps for hundreds of years if protected from varmints and moisture.

Bee Hunting

In the pious hope that there are still wild bees out there somewhere, I will describe how to catch them.

First let me say that the very best time to find them is during the early autumn, when some wildflowers are still blooming well, but most of the leaves have fallen. You can watch the flight of the bees much better because the leaf cover has lessened.

Bee colonies are rarely far from

reliable water supplies, usually well within 100 yards. Locate some blooming flowers, and watch for bees. Wild sunflowers and asters are most reliable in fall. Catch the honeybee as it lands on a flower in a 3 inch square box with a glass top and a shingle for the bottom. Set the trapped bee on a rock or plank and slip a bottle cap of honey, diluted by half with water, under for her to fatten on for a few minutes. While she's sucking it up, put a dab of white paint on her fuzzy back, and take away the glass-topped box. When she's had her fill, she'll rise up in a sort of spiral from the honey, getting her bearings, and then shoot off in the direction of the hive. Time how long she's gone. In the mean time let all willing bees drink from the bottle cap, renewing it, if need be. It takes about 8 minutes for a marked bee to return if the hive is ½ mile away; 13 minutes for a mile. There may be several bee trees around, so the best hunters keep several tints of paint, marking other bees, and following the ones which are closest. It's easy to walk right past the tree. Try capturing the bees repeatedly and timing them from the new spot. Once the tree is located, you could cut it down and rob it, killing the bees. Quite a waste, I think. Or you can mark it, and the following spring hang a wooden box with some drawn, empty honeycombs in it--they always attract swarming bees. Chances are the hive will swarm if it doesn't have a lot of room; chances are also good that your invitation will be accepted. When bees have taken up residence in your box, you can seal the entrances with heavy tape or other contrivance, and take the bees home. They'll be angry at first because they can't find their way. In a couple of days all will be well.

To avoid them swarming, you must give them ample room to make more combs and honey. Extra boxes can be set atop the first, with no entrance except through the bottom box, and they'll fill it with combs and honey. If the passages are metal, and small enough to exclude the queen (commercially made queen excluders are the only way to go) it will have ONLY honey. When it's nearly full, add another box atop it.

To get honey from all this without much investment, each September smoke the bees' entrance well, then lift the tight lid, and smoke from that direction. Wait a few moments, then start removing combs. My mom used to place 'em in a tight picnic cooler to keep the bees off 'em. You will probably have to brush bees off the combs as you take them, using a very soft-bristled bee brush.

Once you have your combs safely away from the bees, you can press them in a simple squeezer made by hinging two clean flat planks together. Just do this over a clean bucket, and let the bees clean up the sticky wax later. Then use the wax as you like. This isn't as efficient as a centrifugal extractor, because the bees spend time making wax and comb instead of honey, but it is very cheap.

One loses his initial fear of bees with experience, working them while protected by a suit. Also, witnessing the calming action of smoke a few times does much for a new beekeeper's morale. Keeping bees in improvised colonies is a very jolly thing indeed, especially if the deadly Varroa mite is not a problem in your region.

A number of good books exist for new beekeepers, but I think the best is *First Lessons in Beekeeping* by C. P. Dadant. It's also among the cheapest.

Bees tend to separate the honey within the hive according to what kind of flowers the nectar came from. The reason for this is they only work what's in season and most abundant that week or month. The most obvious distinguishing trait is color. Most folks prefer the lighter color honey, like basswood, locust, alfalfa, clover, sourwood, etc. Tulip poplar honey is a deep red color and has a distinctive cane molasses taste. Buckwheat is darker and though strong flavored is preferred by those who have acquired a taste for it. Chinkapin honey is dark brown and quite bitter. Not only that, but it makes any recipe bitter. By the way, honey can be substitute on a pound for pound basis with almost any recipe calling for sugar. It weighs about 50% more than granulated sugar, so by volume use a third less. Lubricate the inside of the measuring cup all around with vegetable oil, and the honey won't stick--more accurate measuring.

We always had folks beating a path to our door for ever more honey because we separated ours according to nectar source. It is a really well appreciated holiday gift, because it's so hard to find non-blended honey.

Maple Sugar

The northeastern portion of North America is famous for sugar maples *(acer saccharum)*. Whenever the night temperatures are less than 32°F and daytime temperatures are above, maple sap is running. Spring and fall. Many assume only springtime yields sugar.

It ordinarily takes 50 lbs. of sap (about 6 gallons) to make a pound of sugar or a pint of syrup. A very few trees sometimes yield up to 4 lbs. of sugar from 6 gallons, under ideal conditions. All maples *(acer spp.)* yield sugar, but you have to boil some longer. The sap sours easily, so boil it the day it is collected. Clear days make the water in the sap boil away much quicker. Special pans are made for the purpose, but even a washtub is o.k. I have heard of folks making maple syrup by partially freezing sap, removing the ice, and repeating until the sugar gets concentrated. Everyone I know boils it.

The trees are tapped by boring a ½ inch hole 1½ inches deep to fit the tap, called the *spile*. It's just a spout pounded into the tap hole. Hardware stores in sugaring country nearly always carry spiles. They have a little hook to hold the bale of the bucket. A horse drawn sled or wagon with a collecting tank mounted on it makes the work more fun. Some use mechanical power, while others run pipelines from the trees, downhill to the sugaring house. The latter method is preferred for many large commercial outfits. Even if no commercial operation is envisioned, it is best to put the cooking apparatus under cover with wind protection and a chimney to take away smoke. The team can be profitably employed accumulating wood for the fire once the day's sap has been retrieved.

Pour the sap through a strainer, a cloth bag will do, and just boil it until it's ready, without adding more sap. The old fashioned way to tell if it was done was when the syrup just barely would still pass through a tight woven cloth. It needs longer boiling if it is to be made into sugar. It tends to run over as it thickens. A piece of butter the size of a marble will cure this; sometimes 2 or 3 are needed. The best way is to monitor the fire better.

To tell when it is done enough to make sugar, some folks pour it on a bowl of snow. If it won't run into the snow, but hardens like candle-wax on the surface it can be placed in tubs to drain. Most prefer it caked; then it must be boiled still more, until it will break like glass or ice on the snow. When snow isn't available, stirring it in a bowl will make it granulate. If cakes are wanted, try making a loop of wire or a hickory twig, and dip it in the liquid sugar. Blow through it like blowing bubbles; when ready it will make feathers or ribbons of candy. Take it from the fire forthwith and pour into the molds of your choosing. Maple syrup and sugar are also trade goods. I know people who have made considerable returns working sugar during February and March, a 24-hour job. Then they took their ease during the remainder of the year. Large maples yield about 2 to 5 lbs. of sugar (or pints of syrup) apiece in a two-month season.

An Evaporator Pan

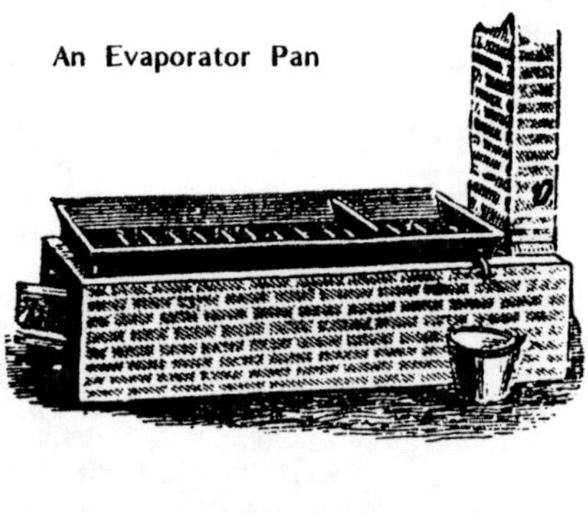

Sorghum

Sorghum is a plant greatly resembling field corn, but the stalks are more slender, and the seed heads are at the top of the plant. It is a close relative of the *kafir* or *milo* of the western states. Its true name is *sorgo*.

Grow it just like field corn, except that because the seeds are so much smaller than corn, the soil should be gotten to a finer tilth. Many old hands at growing it soak the seed until it's just ready to sprout, and then plant it by hand in "hills" about 3½ feet apart each way, about 25 seeds to a hill. When it comes up, they thin it down to about 7 stalks to a hill. This way it can be tilled both ways. It is much tenderer than a young corn plant, and more susceptible to frost, so its planting is delayed until the corn is in the ground. It is also slower growing, so weeds overtake it if not given perfect culture. It is also grown (in comparatively weed-free soil) in rows 3½ feet apart, standing at least 6 inches apart in the row, but up to a foot apart in dry regions.

Sorghum roots penetrate the soil more thoroughly than other crops, and some think it is more exhausting. It is more drought resistant than corn. The best growers remove the *suckers* or side shoots.

When the seeds are completely ripe, or nearly so, the sap becomes more saccharine. It is best to give it its full season, and most people who grow it around here make sure not to cut it until frost threatens.

Frost will ruin the sorghum, causing its juice to sour very quickly. To forestall this, go right down the rows with gloved hands ripping all leaves off each stalk. This is called *stripping*. The seed tops may also be cut off at this time. Just arch a stalk over, lop the top off and release. Once stripped, the stalks will stand light frosts only, but it better be harvested within a week or so, or it will begin to spoil.

Harvest by just lopping off the stalk with a corn knife, close to the ground. Load the stalks into a wagon and take them to your mill.

Oh yes, sorghum making requires one expensive piece of equipment, the mill. The only ones we have used were horse powered, but there are tractor powered ones. Tight rollers mash the stalks, which are 60% sap, and the juice runs into a bucket. When the bucket is full, pour it through a cloth into a barrel. When you collect enough to fill your boiling pan, you should stop pressing stalks until that batch is boiled. But it is more fun to have helpers (or be helpers like we always are, because we don't own a mill) and divide the work of boiling and pressing. Many call milling the stalks "grinding" because of the smashed-up look of the pressed remains.

I found making sorghum "molasses" much quicker than maple sugar. This is because good sorghum juice is about 10% sugar. It is too acid to granulate. Milk of lime, made by slaking burned lime in water and filtering it into the strained sap, will remedy this. Just get it alkaline enough to turn pink litmus paper blue. A small excess of lime isn't injurious.

We never bothered with the lime, because we were making syrup or "molasses." We use a pan about 3 feet by 6 feet by a foot deep. Ours was stainless steel but iron is fine. When it was reduced to between an eighth and a tenth of its original volume, we dipped it into canning jars, and sealed it.

If it's sugar you're after, and you've limed the juice, boil it for 15 minutes first, then filter again. Boil it hard until it has lost half its bulk. Then moderate the fire, so it won't boil over, stirring constantly to prevent it sticking and burning on the bottom of the pan. Take a bit out from time to time and test like maple syrup to see if it will crystalize. If it is ready, dip it into tubs or crocks. In 3 or 4 days it should granulate. Sometimes it granulates so completely as to leave very little molasses, which is that portion of rendered sugar which cannot be crystalized regardless of treatment.

Once solidified, scoop it out into clean conical bags made of coarse woven material, hang them in a very warm room (80°-90°F) over a tub to catch the dripping molasses. In a week or so the bags will contain brown sugar.

If you want your sugar more refined, dissolve it in hot water. Mix one fresh egg white in a cup of cold water and add it to up to 100 lbs. of dissolved sugar. Boil for an hour. Skim and filter out the coagulated egg white, which will take with it most of the impurities of the sugar. I have never refined sugar; it just looks too laborious to bother with. The slightly acid taste of the "molasses" is agreeable to many, and the final product sells well, if you find you don't like it. An acre of sorghum averages about a ton of sugar, but much higher quantities have been obtained by diligent growers.

34
DUNG

...When thou wilt ease thyself...thou shalt dig...and shalt turn back and cover that which cometh from thee.
Deuteronomy 23:13

Sanitation is essential to good health, which is essential to good living. Husbandmen are mindful of the fact that practically every offensive substance on the estate can be rapidly ameliorated through ingenuity. Things which most often must be dealt with are gray water, excrement, animal carcasses or parts thereof, spoiled food, and noxious discharges from the sick room.

Gray water is the runoff from everyday washing. Mostly it is just water, but is also partly such things as sweat, soap scum, liquid food residues, hair, etc. The time honored way of disposing of these things is to dig what is known as a dry well. This is just a large hole which has been refilled with boulders and coarse gravel. The soil, if it's not too poorly drained, absorbs the liquid matter issuing from the drain pipe. Water spreads itself out best in rock-filled pits, and is absorbed at a leisurely pace. Though it is best to place a dry well downhill from the house and well, it is a known fact that the majority of water borne matter will be acted upon by soil organisms long before it is able to find its way to an aquifer. This is especially true of large, relatively shallow installations. A dry well deep in the subsoil, or worse, in always-wet soil is nearly useless.

Aerobic Decomposition

Except for gray water, all the above hazards worsen with age, but share a common trait by which their dangers may be neutralized. That trait is that they are heavily infected with *aerobic* and *anaerobic* germs. The latter are organisms which thrive only in the absence of oxygen. If large amounts of oxygen are presented to them, they cannot compete with *aerobic* or oxygen-loving organisms. Indeed, they become *oxidized* themselves, and become food for their competition.

Oxidation is hastened if the offending matter's moisture can be rapidly reduced by absorbent, aerobic, germ-laden material. This principle was first satisfactorily explained by one Prof. Way, chemist to England's Royal Agricultural Society well over 100 years ago. The concept was incorporated in an invention made by the Rev. Henry Moule, Vicar of Fordington, Dorsetshire, England, also well over 100 years ago. His invention was something like all those modern so-called "composting toilets" advertized in primitive and survival type magazines.

Moule's invention was known as the *earth closet.* It consisted of a toilet seat over a box containing a bit of dry, powdery soil. After relieving one's self, the toilet seat cover was lowered, and an internal mechanism covered the wastes with a layer of more dry soil. The invention was remarkably sanitary, never foul smelling and hardly ever got out of order. Moreover, though the wastes had to be removed regularly, they could be stored safely in a dry shed, where their desiccation became complete in a few weeks, whereupon it could be reused safely. It could be thus reused in fact, up to 10 times! By that time, though still inoffensive, it had become extremely potent as a soil amendment, or *manure.*

But while Mr. Moule was inventing the earth closet, Sir Thomas Crapper was inventing his "Crapper toilet." It became so popular, especially in highly populated areas, that it was the source of several new English words. Though not actually a closet, like Moule's invention, the term *water closet* stuck. Today urban folk yet enjoy the convenience and simplicity of quickly sending wastes away to estuaries unknown.

The sewage treatment plants one sees just outside of town with their circular tanks, rotating arms and bubbling pumps are aerating the sodden sewage as best they can to render it less dangerous before they more or less get the water back out of it. But even modern flush toilets have numerous clogging problems. They're also a convenient way for urban/suburban dwellers to get rid of all kinds of chemicals which can't be broken down by natural means. The same always find their way into settled sewage sludge. This is one reason why "municipal wastes" are not recommended for fertilizing food crops. Of course, during the time it spends enveloped in the flush water, the anaerobic germs are multiplying fast, and so are the poisons they generate. This is why Asian "night soil" is the catalyst of digestive disorder when visiting Westerners forget to wash *and cook* all vegetables.

Septic Tanks

Most ruralites now have septic tanks for their flush toilets. They are made of concrete or plastic, about 500 to 1000 gallons capacity. In theory, the wastes break down to liquid and are dispersed in a drain field. In fact, their decomposition is anaerobic, which is rarely complete enough to avert having to be pumped out every few years. This is a revolting and health-hazardous job for which very substantial fees are collected. A man with the properly equipped truck can often make several hundred dollars daily doing it. Unlike with the earth closet, the effluent isn't safe or odorless by any standard.

Compost

A *myco-rhizal relation,* which means a mutually beneficial association between plant and fungus roots, was discovered in the late nineteenth century by Dr. M. C. Raynor of Wareham, England. She demonstrated it to have specific and unique curative powers on worn out soil.

Following in her footsteps was Dr. Albert Howard. This man acted upon Dr. Rayner's discoveries, experimenting with ways to improve soil fertility while modifying noxious waste into harmless humus. Wealthy, but less knowledgeable "activists" (notably J. I. Rodale in the U.S.) followed, crusading to alarm folks about the dangers of non-organic food.

While I do agree about the dangers of foods grown in soils lacking humus, and even more so about the chemical treatments which necessarily follow such growth, I don't agree with our modern tendency to make a crisis of everything. When folks get tired of one crisis and go looking for others, old problems still remain, but become stale in the eyes of fashion.

One problem is most soil has lost its organic basis--the humus. Compost is humus made under man's direction, rather than at random in the wild. A lack of humus is indicated by an ever growing need for poisons to kill bugs and diseases. Its presence doesn't insure against loss, but nasties have much less appetite for crops growing in truly healthy soil.

To make compost, pile up a 6-inch layer of highly carbonaceous material, such as straw, leaves, sawdust, etc., then top that with 2 inches of dung, then a very thin dusting of ground limestone, slack lime, or wood ashes, etc. Repeat until the pile is at least 3 feet tall, and preferably at least 6 feet diameter. Or all components can be mixed before making the pile. I collect litter daily from my pig sties; the pigs mix all the dung with the litter themselves. A central "chimney hole" is punched in the pile with a pointed steel bar, to assist aeration. The whole pile is turned and mixed at the end of about 3 weeks, or whenever the temperature has risen and begins to fall. When it again heats, and begins to decline, it is turned again. This is all very laborious, and no one except a fanatic is likely to keep it up for long, especially if one has many animals, and more than a few square yards of tilled ground. It does have one use though. I have seen with my own eyes compost piles which included an entire sheep carcass, left undisturbed for a year or two, and when dismantled, no trace of the sheep could be found in the humus! The carcass was laid on about 6" of vegetable matter first, then the dung/vegetable matter layers were built up to a height of at least 4 or 5 feet. No turning was ever done, no noxious fumes were emitted, and no varmints dug it up. I have seen it done several times, so I know it's repeatable. I think any noxious organic matter could be purged this way.

My Compost Tower

I invented a process which works much better than the above, even in the dead of winter. Even the expensive "compost tumblers" displayed in gardening magazines aren't as efficient.

I call it a *compost tower*. Four posts are set 2 feet into the ground. Then a 3 story, 3-sided chamber made of planks, is built on them. Its bottom is far enough from the ground to roll a wagon under. Removable slats or angle irons form slat floors for each story. The open side may be closed using removable boards. A roof is optional.

As dung-mixed-with-litter is removed daily from animal pens, it is added to the top story until it's full. By this time it has begun to heat very well, and may even require a little watering so it won't dry out. When full, the floor slats are removed, one at a time, causing the contents to fall into the box below. They receive a complete mixing by this action. Within a few hours of falling/turning the temperature will rise very high, often over 160°F. Kept well moistened, but not sodden, little nitrogen is lost.

Meanwhile the top compartment slats have been replaced, and it once again receives the daily accu-

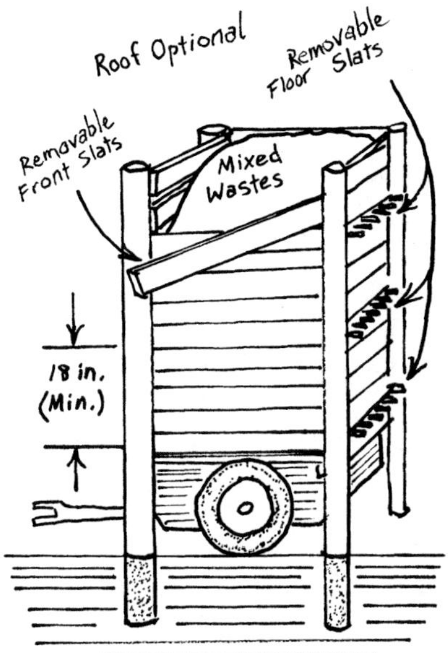

My Compost Tower

mulations from the barn or pig sties. When it's full, the contents of the box beneath it are tumbled into the bottom compartment. The top load is again tumbled into the middle box. The slats returned, it begins filling up again. When all 3 compartments are full, the bottom one is emptied into a wagon, and the others are tumbled, one at a time into the box below, and the process continues *ad infinitum*.

With the compost tower, large amounts of fertilizing matter are generated, and only the daily labor of lifting dunged litter need be endured. It is actually only slight labor compared to turning piles with a fork and loading a wagon to haul it afield. This plan would probably not work with more than a couple dozen hogs, or a few dairy cows, so probably isn't applicable to commercial farms unless they're very small. I once added human dung to a load from time to time, and found it a completely safe and unobjectionable process.

35
ENERGY

He heweth him down cedars, and taketh the cypress and the oak, which he strengtheneth for himself among the trees of the forest: he planteth an ash, and the rain doth nourish it. *Then shall it be for a man to burn: for he will take thereof, and warm himself; yea, he kindleth* it, *and baketh bread. Isaiah 44:14,15*

I think the only really good excuses for taking a chance on the power company's reliability are electric lights, refrigeration and laundry. I have used all sorts of alternatives to electric lighting, and some are really good, but none are as convenient as flipping a switch. Ice houses and ice boxes are in many ways superior to electric refrigeration, but I can buy a refrigerator this morning and preserve food in it this afternoon. If I build an ice house I have to wait until winter to stock it with ice. I have washed laundry by hand for months at a stretch, and know that the invention of the washing machine is one of the greatest advantages of living in modern times. All the other gadgets and doo-dads which run on electricity are just a waste of time for me--I rather look for low-tech ways to get my work done easier, ways in which I'm not dependent on others in a pinch.

Home Heating

Here in the Appalachian highlands, summer heat isn't unbearable. It is among the most pleasant of regions from late spring until late autumn. The hottest weather only brings discomfort on a few days out of the year, and even these have easily bearable temperatures for at least 18 hours of the day. Well built houses or shady woods agreeably get folks through the few hours when temperatures climb above 85°F.

But the very coldest winter days and nights have no such respite. We must have heat or freeze. Gullible folks spend their money on gas or oil heat. What they forget is rural electric lines are likeliest to fail when winter storms bring polar blasts. Most oil or gas heating systems depend on electricity to run their controls.

Yes, yes, I know there are those who place their faith in contraptions to generate the power from water, wind and photovoltaics. While these contrivances are admirable in many ways, nothing is as reliable as simple, old-fashioned wood and coal stoves.

There are many kinds of solid fuel burning stoves on the market. I know little about coal stoves, so I'll refrain from comment. But that most noble of fuels, wood, is renewable, and it only gives off as many pollutants when it is burnt as when it is allowed to rot in the woods. Tree tops left behind by loggers accumulate until there's as much wood rotting each year as would be burnt if it were collected. Researchers aren't paid to think of this when they burn wood in a laboratory and analyze the smoke. And urban bureaucrats don't take it into account when they over-regulate rural people's fuels. It's jolly indeed that it's commonly too costly for faraway officials to bother harassing them.

But to return to the stoves. They are way ahead of fireplaces for burning wood. Fireplaces very often suck more cold air into the house than they heat. The so-called Rumford fireplace is an exception; it's about as efficent as a good woodstove. If you use a stove, get one with a big enough firebox that you don't need to keep adding wood every hour. It is this constant adding of wood which fills a house with dust. I would prefer one which allows loading only once a day--or even better, once every other day! Furthermore, the bigger the firebox, the less work you have dealing with the wood. Why have to cut and split every stick as small as a chair leg?

Wood

Some people burn nothing but green wood. About ¾ of the energy given off is then used up boiling the sap out of it! Too, it is far heavier to handle than dry wood. Some innocents have been led to think they're saving time by not having to wait until it has *seasoned* (dried out). This also is mistaken. You not only have to handle 3 to 4 times as much of it, but you have to clean the creosote out of your chimney every 2 weeks.

If a tree is merely felled and

left for a few weeks, it will evaporate much of its sap through the leaves. Then it can be *bucked* into pieces short enough to fit in your stove. It's best stacked right there in the woods, off the ground, and covered against rain and snow. By next year it will have lost from 20-50% of its weight.

Firewood is measured in *cords.* A cord is a stack 8 feet long, 4 feet wide and 4 feet tall. Some species of wood produce more heat than others. For example, a cord of white oak has about twice the total heat output as a cord of white pine. The former weighs over 2 tons per cord; the latter, 1 ton. The heavier the wood, the hotter it burns. This assumes it's dry; all green wood is heavy. I only know three trees which burn well green: White ash *(Fraxinus americana)* Yellow birch *(Betula lutea)* and Holly (*ilex opaca).* Punky or waterlogged wood is about as bad as fresh cut green stuff.

Lumbermen call the coniferous (pine family) trees *softwoods.* The species which have leaves instead of needles are called *hardwoods.* This is misleading. Yellow pine is about as hard as some oaks. Willow, sassafras, basswood, and white birch are softer than most conifers.

Most hardwoods burn slowly and leave hot, long-lived coals. Most softwoods make a cheery blaze, but burn quickly to dead ash.

Some woods are practically nonflammable when green: Aspen *(Populus spp.),* Balsam fir *(Abies fraseri),* Basswood *(Tilia spp.),* Black ash *(Fraxinus nigra),* Box elder *(Acer negundo),* Buckeye *(Aesculus spp.),* Cucumber Tree *(Magnolia acuminata),* Pitch pine *(pinus rigida),* Sarvis, also known as "service berry" and "shad bush" *(Amelanchier spp.),* Sassafras *(sassafras variifolium),* Sour gum or tupelo *(Nyssa sylvatica),* Sourwood *(oxydendrum arboreum),* Sycamore *(Platanus occidentalis),* Tamarack *(Larix laricina),* Tulip tree *(Liriodendron tulipifera),* water oak *(Quercus nigra),* and White pine *(Pinus strobus).*

Some woods burn very slowly, giving off practically no heat when green: Butternut *(Juglans cinerea),* Chestnut *(Castanea dentata),* Dogwood *(Cornus spp.),* Persimmon *(Diospyros virginiana),* Red maple *(Acer rubrum),* Red Oak *(Quercus rubra).*

Certain woods, like fallen down pines and the cherries won't burn unless split. The outside catches fire but goes out.

Many woods spit and pop a lot, and are not to be trusted in an open fireplace: Arbor vitae *(Thuja occidentalis),* Balsam Fir, Box elder, Pines other than Yellow *(Pinus spp.),* Red cedar *(Juniperus virginiana)* Hemlock *(Tsuga canadensis),* Sassafras, Spruce *(Picea rubra),* Tulip tree, and Tamarack.

These woods need to be watched for popping until they are well started: Beech *(Fagus grandiflora),* Hickory *(Carya spp.),* Sugar maple *(Acer saccharum),* and White oak *(Quercus alba).* The coals these shoot are long lived and so are more dangerous, but in a stove they're hard to beat as fuel.

Certain woods are quite troublesome to split, even when they have no knots: Blue ash *(Fraxinus quadrangulata),* Birch, Box elder, Buckeye, seasoned Beech, Cherry *(Prunus serotina & virginiana),* Seasoned Dogwood, Elm *(Ulmus spp.),* most Hemlock, seasoned Hickory, Honey Locust *(Gleditsia tricantha),* seasoned Slippery Elm *(Ulmus fulva),* sour gum, seasoned Sugar Maple, Sweet gum *(Liquidambar styraciflua),* and sycamore (this can be split "bastard," that is, tangentally rather than radially).

Let me here insert that the best tool for splitting logs is a splitting maul, not an axe. The accuracy of the eye is more important than the the strength of the arms. And safety must always be considered. Too many people think "Oh well, I'm insured." Keep in mind that rural areas are far from hospitals, and not even an ambulance is going to get to you quickly. Often what is merely a major setback in or near the city is fatal in the country. Don't take chances when felling trees; keep an eye out for "widow maker" dead branches overhead. And if a tree gets lodged against another, don't place yourself in its path when getting it down--best use a long rope or cable attached to a truck, team or tractor. And one more thing about safety; I have actually seen, in print! illustrations of people holding a stick vertically in one hand, a hatchet poised atop, bringing it down on a log. DON'T DO THIS! I know 3 different people who are missing fingers because they learned it that way! The correct method of "contact splitting" for kindling is as follows: 1. Hold the stick by one end, like a hammer handle. 2. Gripping the axe handle in the other hand, set its bit against the free end of the stick. 3. Keeping the axe bit in contact, swing it and the far end of the stick so that only the stick's tip strikes the chopping block. The axe bit will be far from either hand as it sinks into the corner of the stick. Another such tap will get it imbedded into the cleft. 4. Pivot the stick so that the next blow will bring its unsplit butt squarely onto the chopping block. *Don't hold the stick with your hand!* A tap or two will rive this stick asunder. 5. Repeat this procedure on halves and quarters, as needed. Many sticks will split fully apart after step 3.

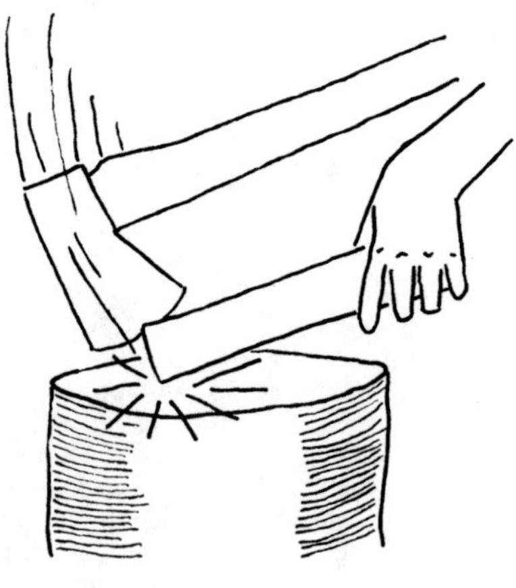

The Right Way to Start Splits

My favorite firewoods are, in order: 1. Seasoned Yellow (black) locust *(Robinia pseudoacacia).* This wood is often found dead on the ground in the woods of Appalachia. Even then it tends to be dry and combustible. It takes a long time to rot away, so who knows how long it has been seasoning there? 2. Seasoned, dry oak. It has a good amount of heat from flames, and has excellent long-lived coals. 3. Hickory. This wood has less flame and more coal. It is probably the hottest burning wood in our region, except maybe sarvis and persimmon. It burns with a blue flame, and makes the best blacksmith's coal.

My least favored firewoods are, in order: 1. Willow *(Salix spp.)* This wood is so non-flammable green, and such poor fuel dry, that it needs other wood in the fire to consume it. 2. Cherry. I have occasionally seen a fairly good fire of this, but most often its coals are so degenerate as to require other wood to keep it lit. These coals in a blacksmith's fire require so much air to get them hot that it oxidizes iron too much to weld with. It also burns away too fast. 3. Red oak, unless it has been kept dry and off the ground for a year or two. Its pores are so long that an 8 foot plank whose end is submerged will emit bubbles when blown against at the other end! These cavities positively retain water, which is also why this wood rots so much quicker than white oak.

Staying Warm

Down and wool comforters keep us warm at night, even when the fire dies down low toward morning. Keeping one's head covered, except for nose and mouth, does much to warm the rest of the body. Hot water bottles, or heated bricks wrapped in a towel, heat cold beds so the first hour of sleep isn't spent warming it. Sleeping warm makes even a cold morn much more bearable when one sets to work on the morning chores first thing.

36
SMITHING

The smith with the tongs both worketh in the coals, and fashioneth it with hammers, and worketh it with the strength of his arms. Isaiah 44:12

The only things about smithing an autonomous husbandman need know are: Sharpening, shaping, welding and heat treating metal. If one can't do these things, the cost of rural life goes way up. I have discovered it's easier to remake hand-powered and horse-drawn equipment than to locate replacement parts.

Some will dismiss smithing as too difficult to master. If readers purpose to master every skill useful to autonomous husbandry, self reliance may be unattainable. Mastery is the result of diligent and exclusive devotion to a narrow goal. Just because the jack of all trades is commonly master of none, a master of one needn't forswear becoming a jack of all the rest.

As a boy I made my first makeshift forge by digging a hole in the ground for the fire in the back yard with a post-hole digger. I then punched a 1-inch diagonal hole, starting about a foot away, meeting the bottom of the first hole. I slid a pipe down the diagonal hole, and attached the end to a vacuum cleaner exhaust. The blast from the vacuum cleaner exhaust fed a fire composed of common barbequeing charcoal briquets.

That first forge certainly worked, yet it made the work itself inconvenient because just about all of it was done on my knees. Later I made a few much better forges. The first was based on the same principle, built into a 55 gallon barrel filled with dirt. The fire hole, instead of being excavated into packed dirt, started as bricks laid dry to form a square hole, and the dirt was packed around it. The blast hole was a horizontal pipe which came out through the side of the barrel. Its blast entered the side of the fire hole instead of the bottom.

I have since owned commercially made forges, none of which worked as efficiently as those home made ones. The fire hole should be about 6 inches square and 8 inches deep. The vacuum cleaner exhaust was entirely too powerful a blast, I later substituted a brushless electric hair drier whose heating element I had disconnected. Later yet, I used a hand powered blacksmith's blower I got from a junk pile. A tall chimney on a small fire pot also pulls enough air through the coals to keep it hot. Recently I've done a bit of smithing with a carbon-arc torch attached to an electric welder, and for small items it's quite useful. Alas, it doesn't work when rural electricity goes off--as it does about once a month.

I mention all this to show that the heart of metalworking, the forge, need not be costly or hard to find. The important thing to remember is the fire must be contained safely, and must have an adjustable air blast to govern heat output.

Hardening Steel

Iron is the most common and versatile metal. Every junkyard is filled with old iron. Steel is iron which has higher carbon content. Steel with low carbon content is commonly just called iron, because that's basically all it is. If you heat it red hot in the fire and quench it in water, it won't be any harder or softer than before. Medium carbon steel hardens when quenched, preventing bending or deforming when battered, or wearing away in contact with the soil. High carbon steel can be made so hard that it will even cut other kinds of steel.

Quenching red-hot high carbon steel makes it so hard that it becomes brittle enough to crack when dropped on the floor. So it must be *tempered.* This is done after quenching the red hot metal, by reheating it (usually using a red hot iron plate as the heat source) until its shined-up surface changes colors due to oxidation. The first such surface change is a faint whitish sandy color, often not detectable without practice. Then the color of common straw appears. Then comes bronze, followed by purple, light blue, dark blue, then just black as the metal gets hotter and hotter. Each color indicates the hardness which will remain when it is quenched. Tempering at a lower heat leaves the steel very hard, but still a bit brittle. Faint straw is for cutting edges which will not be struck. Bronze is for metal-cutting hardness, but will sustain the impact of hammer blows. In fact, the surface of a hammer is normally tempered at this heat. Purple is for woodcutting tools, light blue for softer things, like food, and deep dark blue is mainly used for springs and catches.

Often one needs to soften the steel, for instance when it needs to be filed to shape while cold. To do this it can be either *normalized* or *annealed.* Normalizing means cooling slowly, by heating it red hot and letting it cool in a can of sand or over the fumes of the fire for a few minutes. Annealing is softer yet, and is done by burying the metal in the coals and letting the fire go out slowly.

Shaping Metal

At some point the black-hot metal begins to glow a faint dull red, only visible in a dark room.

From there it takes on the color some call "cherry," though I've never seen a real live cherry that color. It's more of an orange color. Then it glows yellow, getting lighter yellow until it is white, with sparks flying off because it is melting. Just before it gets to this heat, a careful observer sees the surface looking "wet" because it's just starting to melt. This is called *slippery heat* by smiths and is the point at which two pieces of metal can just be welded into one.

Most shaping of metal is done at yellow heat. The easiest shaping is simple bending. Hold long pieces by the cool end, short pieces with pliers or tongs. Then using a U-shaped piece of metal, a vise, a pair of pliers or tongs, bend the metal to the required shape.

Sometimes iron or steel must be made narrower, pointy, or just elongated. It's done by hammering yellow-hot metal against an anvil or other large chunk of steel. I used a discarded piece of train track as an anvil for years. I've seen old bulldozer parts used for it too. This pointing or elongating of metal is called *drawing out.* To make a good point, hammer first near the tip, otherwise metal will mushroom towards the tip, making weak shallow tubing out of the tip. This fault is called *piping.*

When metal must be thickened at the end, it is heated yellow hot and the end is first tapered slightly. Then it's reheated to yellow, and thrust downward like a spear against a hard steel floor plate. Don't keep pounding when it gets too cool, because it will deform crookedly. You may need to correct this with hammer work. Most importantly heat only the extreme tip when preparing to thicken a piece of metal. Thickening metal by pounding is called *upsetting.* Metal is sometimes also upset in the middle of a bar, but it's more fiddly.

Punching holes in yellow hot metal is usually a two man job. One holds the hot iron just above the anvil (so the latter won't suck all the heat out of it) while the other positions and then beats a tapered punch almost through the metal. The yellow-hot metal is then flipped, and the thin spot will cool to a black while the rest remains aglow. The dark spot is positioned over a small hole in the anvil (called the *pritchel hole*) and the cold slug is now driven out from the now visible back side. First quench the punch if it's too hot. A large hole is best made by first punching a small hole and enlarging by pounding a tapered *drift* through it until it's wide enough.

Welding

Two pieces of metal melted together to form one are said to be *welded.* Forge welding isn't easily learned. Many things deter one's efforts. First, the fire must be clean. Some kinds of coal contain impurities affecting the ability of iron to cleanly puddle on the surface. Hickory charcoal is best, or coke. The fire must also be deep enough that raw air from the blast doesn't reach the hot metal. It causes oxidation, which makes the molten surface flake off. Result? No weld. Keep copper or brass out of the fire. And allow no *clinkers,* which are glass-like puddled impurities at the base of coal fires.

Bring both pieces of metal to be welded to an orange heat. Sprinkle *flux* on them. Flux is a substance which melts and sticks to the hot iron. Most common fluxes are made of plain river sand, household borax, glass pounded to powder, or any of these mixed with iron filings.

Now get both pieces of iron up to white heat; sparks may just begin to fly off them. Immediately bring the hot metal swiftly to the anvil and pound them together. Long welds require welding a small portion at a time, reheating, and welding a bit more until it's one piece. Many smiths upset the metal first, because it gets pounded thinner when finally joining them.

The easiest welds to learn are called *faggot* welds. Bend the end of a bar over, and weld it to itself. This is a way to avoid having to upset too much. The next easiest (none are easy) is welding a loop or a link of chain. This also uses but one piece of metal, bent back to itself.

It's easier to weld iron (see above, under *Hardening Steel)* than steel. Steel must not be heated much above slippery heat, because too much sparking indicates the carbon in it is burning away. I always needed iron filings in the flux for success.

Brazing

No matter how you try, you may not succeed in welding, so brazing is worth learning. Brazing is the easiest way to strongly join metal. It's also excellent for joining dissimilar metals. And it prevents burning away thin pieces.

The metal is brought up to red heat, and borax sprinkled on it. Now the metal is heated to just between orange and yellow, and a piece of brass touched to it. The latter melts readily, spreading to every place covered by the melted borax. Small diameter brass rods are available for brazing thin parts. I have used any old junk brass for bigger parts. Cast iron can hardly be repaired by any other method. Simply hold together the orange-hot parts, touch a piece of brass to it so it runs everywhere it's needed, and let it cool. Best to cool cast iron slowly to avoid cracks. Brazing is 60% to 90% as strong as welding. Some use nickel-based rods to achieve brazed joints which are sometimes even stronger than a weld!

Practicing

Having built your forge, and assembled materials and a few tools, you can begin making more tools, especially simple ones like chisels, punches, hooks, latches, etc., as you need them. Flea markets and garage sales are among the best places to obtain old hammers, pliers, punches, and chisels. All these can be reforged to the shapes which suit your purposes. The only hard-to-improvise essential is the vise. There is an insatiable market for blacksmithed items, especially household wares.

Sharpening

Honing edged tools is probably the most commonly needed metalworking skill. First, learn a bit of theory about it. Cutting action is obtained by using a wedge narrow enough to pry asunder adhering particles. Certain metals, steel with a high carbon content foremost among them, can be *abraded* with a file or grindstone until a "feather" of metal appears at the edge. This feather-edge can then be folded foreward and back until it breaks off. The edge is then as sharp as the metal will allow.

The quickest way to accomplish this is to scrape the metal on a stone or other abrasive surface, carefully avoiding fluctuations in the angle of contact between blade and stone. The feather edge is most rapidly formed when the blade is drawn away from the edge. A tiny feather edge can be detected by the fingers by dragging them from the back of the blade towards the edge--never along the edge! A well formed feather edge has a sort of catching drag to it. I never oil a stone. After all, lubrication isn't the goal, and oil eventually clogs a stone, ruining it. Besides, water's free.

Once the feather edge (also called the "wire edge") has been established, the feather must be bent back and forth until it breaks off. This action is called *stropping*. Barbers who use straight razors keep a leather belt handy called the *strop*. Again, the edge is drawn back and forth--away from the edge--until the feather edge has worn off.

Understanding this principle, anyone with common sense can apply it to any sharpening situation. One item always requiring razor sharpness is the scythe. Another is hand sheep shears.

37
ROUGH CONSTRUCTION

Storehouses also for the increase of corn, and wine, and oil; and stalls for all manner of beasts, and cotes for flocks. 2 Chronicles 32:28

Let it not be imagined that one is going to learn how to build houses from a single chapter of a single book. I have read many books on construction, and few address more than a glimpse of once-common building methods which are now dubbed "alternative."

Conventional stud frame or block and brick building methods have considerable reference material in public libraries. There are even a fair number of books on pole building, log cabins, timber frames, rammed-earth (also called *pisé*, cobwall, etc.) stone masonry and even straw-bale houses. So why discuss any of that here?

Jenny Lind

In Southern Appalachia there is a method of building little-known in the "outside world." It is called *Jenny Lind* construction. How it came to be associated with the name of the great nineteenth century "Swedish Nightingale" no one seems to know. Her golden voice and great humanitarian works don't seem appropriately compared with rough, rustic building methods.

Characterized by having no skeletal frame, a Jenny Lind building is quickly set up and inexpensive. There are many variations, some excellent, some rather shoddy.

When we moved here we didn't know our house was of Jenny Lind construction. Unfortunately it is of the mediocre sort, and is rather cold in the winter. But we've lived in it for over a decade now, and are not the worse for it. We decided not to put money into major repairs or renovation. We're saving our money for a better house.

But the Jenny Lind has served well. It is about a hundred years old now, has weathered many a storm and flood. It has seen the birth of many children within its walls. And though it might be sneered at by many moderns, it provided a house and home for us and the families before us without a great financial outlay or some hideous mortgage. The remarkable thing is when property is sold with a livable house on it, the price is raised by many times the cost of a Jenny Lind.

Lumber

West Virginia is a sea of hardwood, with small local mills everywhere. At these places, "low-grade" hardwood lumber can be obtained very cheaply. I have built fences, livestock housing, boxes and bins, livestock feeders, gates, even a 40 foot bridge with low-grade lumber. I will not pretend I built a house with the stuff. I only mention it here because one can get get building experience using wood which cost about a third. Once confidence is cheaply acquired, a house could be built out of better grades of lumber.

Low grade wood is discounted by the mills because of some defect in the individual piece. For example, cracked, rotted or excessively knotty wood would be rejected. Interestingly, many times these defects occur at one end of say, a 10 foot board. If the bad end is cut off, there may still be a usable 8 feet! Some wood is rejected just because it was cut slightly too thin to plane down to the required ¾ inch final thickness. Or it may have been sawn a little too thick to plane in just one pass through the planing mill. Labor costs force mill operators to just discount it. These non-uniform pieces are perfect for learning Jenny Lind construction.

For a first project, I built a tool shed, which was first used as a chicken brooder house. It was about 5 feet square, and about 7½ feet high at the ridge; 6 feet high at the eaves. It had a dirt floor. Following is a bill of materials:

Bill of Materials

Braces: 8 pieces of 4 in. scantling, 5 ft. long*

Siding: 120 board feet of random width, 6 ft. planks (a board foot equals 144 cubic inches)

Gable Siding: 15 board ft., random 1½ or 3 ft. planks

Rafters: 3 pieces of 4 in. scantlings, 6 ft. long

-OR-

6 pieces of 4 in. scantlings, 3 ft. long

Roof Decking: 36 Bd. Ft. of random width, 6 ft. planks

-OR-

(cheapest option) 6 planks, 6 ft. long

Door Braces: 2 pieces of 2 ft. plank, plus 1 piece of 4 ft. plank (diagonal brace)

Roofing: 3 sheets of 7 ft. by 2 ft. galvanized roofing, cut in half to make 6 sheets of 3½ foot length, plus 6 ft. of galvanized ridge cap

Nails: 10 lbs. of 3 inch (12d) common or sinker nails & 1 lb. of 2½ in. (10d) common or sinker nails

Roofing Nails: 1 lb. Neoprene washer nails

Corner Nails: 1 lb. of 4 inch (20d) nails

Two Door Hinges

**Scantlings* are boards 2 inches thick; *planks* are boards 1 inch thick.

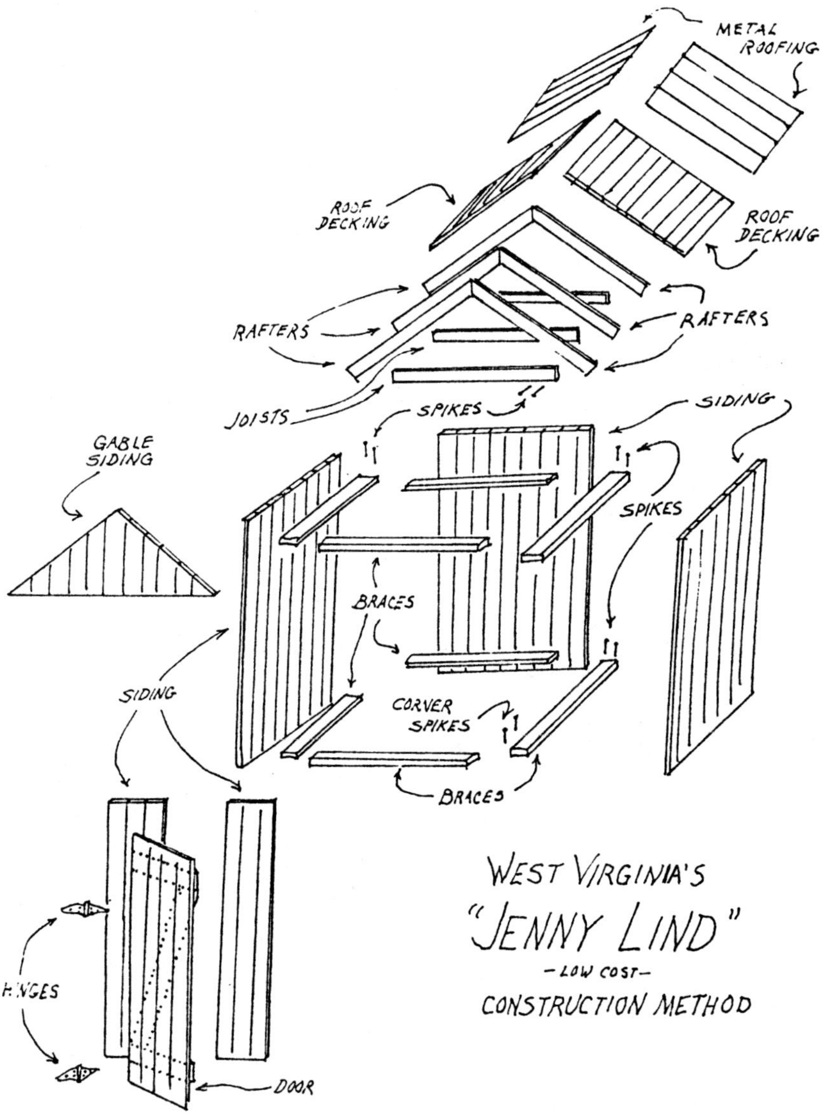
METAL ROOFING
ROOF DECKING
ROOF DECKING
RAFTERS
RAFTERS
JOISTS
SPIKES
SIDING
GABLE SIDING
SPIKES
BRACES
SIDING
CORNER SPIKES
BRACES
HINGES
DOOR
WEST VIRGINIA'S
"JENNY LIND"
-LOW COST-
CONSTRUCTION METHOD

The shed may be set upon flat corner rocks for a foundation. No more tools are needed than a hammer, saw, tin snips, framing square, and screwdriver (for the hinges). At current West Virginia prices for materials, $75 would be the cost, new. If roofing and hinges were salvaged materials, the cost would be under $45. If the lumber were salvaged too, the cost would be about $10, for nails. Sometimes even these can be salvaged! This size building was chosen because it could be used for a garden shed, small poultry house, grain storage building, horse tack room, privy, etc.

Putting It Together

The order of construction is straightforward: 1. Each side of the building is made (or partly made) on the ground. 2. One side is lifted vertical, and propped up with a spare board, a pole, or rope. 3. An adjacent side is raised to meet the corner of the first one. 4. The top and bottom brace portions are fitted (married) and 20d spikes are driven vertically at the corners, to hold the braces. 5. Overlapping siding edges are joined at the corners, using 10d nails. 6. Any remaining siding is nailed on, except door space. 7. Door is assembled and attached to siding by means of hinges. 8. Rafters and joists are measured, cut and joined. 9. Gable siding is attached, then sawn to shape using the rafter it is nailed to as a guide. 10. Roof decking is nailed on. 11. Roofing and ridge cap are attached. 12. Latch is made and attached, using scrap wood. 13. Rocks may be used to fill in between the corner foundation rocks.

Most builders use extra long planks for siding the gable ends from the ground level all the way to the top of the rafters, thus bracing the end rafters and saving time. The rafters then form a guide for sawing the siding planks off before attaching the roof decking.

No measurements are given in the accompanying illustrations, because random low-grade lumber often determines the actual dimensions of small buildings. I saved a great deal of time by first drawing some pictures of what I intended to do. It is a cheap way to encounter most of the problems before they develop.

Some Further Points

If the eaves of the building are higher than 6 feet, another set of horizontal braces must be used midway up between the ground and eaves. This middle brace may be 8 or more inches wide if you like, forming a handy shelf.

When making buildings whose sides are longer than the building's height at the eaves, it becomes necessary to nail diagonal supports to the inside of the walls. This keeps the structure from *racking* (collapsing like a parallelogram) when the wind gets up.

There are Jenny Lind buildings of all sorts: Houses, sheds, stables, pigsties, etc. I have even seen two-story Jenny Lind barns! More braces and supports are needed, and I can't say I recommend the idea.

Jenny Lind construction is nearly always built with green lumber. It was designed for it. But the siding planks will shrink, leaving drafty gaps. Battens of narrower planks (say, 2-3 inches wide) are simply nailed vertically, covering the cracks. This is called "batten on board" or just "board'n' batten" siding.

Jenny Lind buildings, off the

ground, built with long lasting wood (such as white oak, white pine, cucumber tree, chestnut, etc.) and using galvanized nails, ought to last AT LEAST a century. After that perhaps, we needn't care what becomes of them.

Bush Timber

The term *bush timber* is not in my dictionaries. It means wood not cut to dimension sizes like at the saw mill; in fact it isn't sawn at all, but *riven.* That is, it has been carefully split to size, like the old snake rail fences. There are two ways of doing this.

To split rails, slabs or puncheons from a log, one uses wedges and a sledge hammer, or gluts and a maul. (The former are of steel, the latter are home made out of wood, but look the same.) A log has the wedges driven into it from the end, and then wedges are driven into the resulting crack, widening it only a little at a time by leapfrogging the wedges all the way down the length of the log--one wedge loosens the crack enough to remove the other. This process is repeated on each half-log and then on each quarter and so on, to make rails. If slabs are wanted, one must be much more careful with the wedges. And the tree must be carefully chosen to avoid twisting grain, knots and crooked logs. All of these will eliminate straightness in the board. Rather than splitting the logs radially after first splitting them in half, the wedges are driven to form a rift parallel (or nearly so) with the flat surface of the half-log. When continuing the splitting process, one must at all costs avoid being over eager and driving the wedges in so far as to extend to rift too far. A gentle split is the rule, all the way down--far more leapfrogging of wedges will be necessary. Riven boards are called *puncheons*. They aren't quite as true as sawn timber but serve very well indeed for rustic cabin floors. I think they're best for floors in livestock buildings.

The other way to rive wood is using a *froe,* which is a tool which looks a little like a meat cleaver except the handle is set at right angles to the blade. The blade is tapped into the end of a log no bigger than about 5 inches diameter, then the handle is pressed, widening the rift. This extends the

RIVING WITH A FROE

rift a little bit farther down the wood. Now the froe is forced a bit farther into the rift and the process is repeated until the pole is halved. The pole is best supported in a device called a *riving brake.* There is more than one kind of riving brake, but the most common one is made by attaching 2 horizontal crossbars to 2 posts or trees about 4 feet apart. One crossbar is fastened onto the front side of both trees and the other is on the back a few inches higher. The pole to be riven is inserted so that the crossbars will suspend it off the ground by its own weight and leverage.

The reason for the riving brake is that the rift often runs off course. That is, it doesn't follow the center of the wood. Often it veers so badly that one piece winds up thick, and the other is a mere scrap. To remedy this, when you see the rift is riving untrue, turn the wood thick-side-down in the brake. Now using the free hand, press down fairly hard on the thick half while widening the rift with the froe. The strain on the thicker side's grain draws the rift back into it, but be careful not to overdo it. With practice, you can rive true 95% of the time or more, and the resulting timber is stronger than sawn, because it doesn't cross any grain. It makes good gate slats, fence rails and many other useful things where lightness and strength are required. It is a good way to use up trees too small for sawing and too big to just burn as brush.

Shingles & Clapboards

Shingles and clapboards used to all be riven. Sawing makes them rot sooner because the grain of the wood is crossed by the saw, and the exposed pores catch and hold moisture.

To make shingles, split a 1½ to 2 foot length of log at least 2 feet in diameter (3 or 4 feet is better) and at least 4 or 5 feet away from the butt of the tree (wood grain interlocks close to the roots). First split into quarters, and then eighths, using a common firewood maul and wedges. Then rive off the wedge-like center portion of each of these *bolts*. Now the bolts may be riven with a froe, using a wooden club to just tap it on through. Split a bolt in half, then each half into halves, etc., until you have shingles.

Clapboards are similar, but they are made at least 3 feet long. Because they are so long, they are apt to run out, like riving poles. The remedy is the same--rive them in a brake, and make the rift run true. Clapboards have an advantage over shingles in that they can be nailed directly onto rafters or vertical wall studs. Shingles have the advantage that they'll usually last longer on a roof. The best wood for either is white cedar or arborvitae *(Thuja occidentalis)* but it doesn't grow around here. The next best is shingle oak *(quercus imbricaria)* but most oaks, and also white and yellow pine are good if not too knotty. I have used tulip tree *(liriodendron tulipifera)* when rot resistance wasn't required.

A large true-riving tree will side a whole house with clapboards. Ditto for roofing a house with shingles. Livestock housing and utility buildings could thus be made without any investment except nails!

38
VERMIN

Thou makest the darkness, and it is night: wherein all the beasts of the forest do creep forth. *The sun ariseth, they gather themselves together, and lay them down in their dens. Psalm 104:20,22*

There are animal companions which have been at man's side throughout history. Their devotion to man and his habitations is unrivaled by any domesticated beast. Trouble is, these attentions are unwanted. I remember how I dealt with cockroaches when I lived on a farm in Maryland. I sprinkled the druggist's common eyewash, *boric acid powder,* on sheets of paper and left them in crevices where the bugs crawl. Our roach problem abated.

Other insect pests include ticks and chiggers, which though uncommon in the higher elevations of the Alleghanies, do find habitat in brushy woods and overgrown fields. I've never seen either in properly managed hay meadows and pastures, or in husbanded woodlands.

Mosquitoes only abound where stagnant water is allowed to stand. Clean up old tires, bottles, buckets, and such. If you're not in low lying areas, you won't have mosquitoes.

Clothes moths, just like in the city, mainly attack dark closets and trunks where woolens and furs are kept. Same for carpet beetles, though they also attack hides, hams and bacons stored in outbuildings. Grain eating moths can be avoided by keeping grains in sealed containers.

Fleas are the result of keeping cats or dogs indoors, it's that simple.

Poison

The first choice of most urbanites to rid themselves of pests seems to be poison. This is a shame for 2 reasons: They are very non-selective; in most cases damages cannot be reversed. Second, another animal can be poisoned by feeding on the carcass of a poisoned animal. Poisons have their place, but it is only as a last resort to careful husbandry.

Mice

No matter what precautions you take, mice seem to get in. Common snap-traps are very effective, but only when properly used. The first principle of trapping any animal is to put the trap where the animal is already traveling. Not observing this rule is why many mouse-trappers fail. They place a baited trap out in the middle of the floor somewhere; the wrong way.

Place the trigger side of the snap-trap against the baseboard of a wall--this is where mice travel most. It is part of the nature of such prey animals to travel where their predators would least be likely to detect them. Cats and birds of prey don't enter small tunnels formed by items leaning against a wall. Mice also like to pause in these places to review the invasion plan. Mice will scramble over the carcasses of trapped mice seemingly without noticing them. Set plenty of traps.

Rats

Rats are more insidious than mice. When a trapped rat dies, other rats in the vicinity will come to view the demise, instinctively avoiding the spot for some time. It may be weeks or even months before another rat will be caught at the same location. Moving the trap a few feet away may help. There are "repeater" rat traps, but I've never had much success with 'em. The really bad thing is that snap traps are not effective on many rats. They aren't big enough. Standard steel traps, both body gripping and limb gripping types are the only way to go, in my experience. The drawback to these is that they require considerable precaution in the vicinity of small

Trail-Set Deadfall

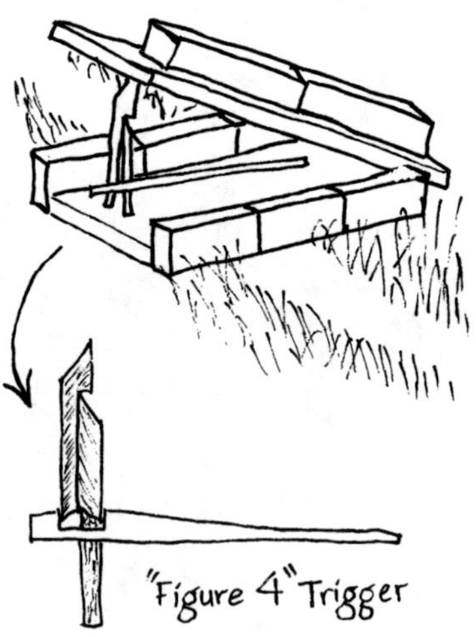

children and pets. I had good success trapping rats from a relative's back yard in Maryland, using a *deadfall.*

The deadfall is a contrivance which employs a weight, called the *fall;* propped up with an easily sprung trigger, called the *trip;* which drops towards a hard base, called the *bed.* Sometimes it is necessary to use *guides* of some sort to make certain the fall will drop straight down.

The deadfall I used on the rats was made as follows: The fall was an 8 inch wide plank, propped with a "figure 4" trip. The bed was another plank. I put 2 bricks on the fall plank to give it enough weight to kill a rat, but not enough to hurt a cat or playing children. I also hemmed in the bed area using bricks laid flat, to slow the escape of a rat. The deadfall was set squarely in the beaten path made by rats from a municipal drainage ditch to the food dish of a dog's kennel. I used no bait, and caught a rat every other day for two weeks. Apparently they weren't at all alarmed by the sight or smell of a rat which had died instantly, or else they just couldn't easily see it under the fall.

The real secret of abating rat infestations is twofold: 1. Don't tempt rats with constantly available food. 2. Dont provide nooks and crannies for the rats to nest in. Either fill such holes, or make them big enough for a cat to comfortably fit there. Cats are the best deterent to rats and mice, but not if the former is excluded because of tight access.

'Possums

The opossum is a fur bearing animal, but likes to live in and under old buildings. It likes to eat trash, but its main diet in the wild is carrion. Still, it is the number one predator of birds, including your chickens. Its favorite food is eggs, particularly those just about to hatch. It will endure the bites and pecks of even a sitting goose as it eats the nearly-hatched eggs underneath her.

Luckily, 'possums are not very bright, and will walk right into any baited trap without fear. A trap intended to kill its quarry, like a body-gripping steel trap, if it's big enough to take a 'possum, is dangerous to cats. I prefer box traps and steel traps for them, the former safest around children and pets. Bait it with any kind of meat, bones or eggs. Tainted baits are best.

Once you've caught the 'possum, disposal is the problem. Jelly-spined folks take the box trap a few miles away and let the problem animal loose where it will probably cause problems for someone else. In fact, the 'possum may have gotten to their place in that manner! Best to just dispatch it; the world has plenty of 'possums, because they can have up to 8 litters a year!

To kill a 'possum in the cleanest, most merciful manner, tap it sharply on the head with a tire iron, pipe or other blunt object. This will stun the 'possum. Then swiftly bring your heel down hard just behind its front leg, squarely on the rib cage. This will stop the heart, and the animal dies without much pain. Also, if you were fortunate enough to have to get rid of it during the late fall or early winter, the pelt won't be damaged this way, so you can skin it and get at least some recompense from those chickens it killed. Shooting is no good, because a 'possum's brain is so small you may shoot it's head many times before killing it, and this cruelty is not to be countenanced by kind hearted husbandmen.

Skunk

The next most common pest of the countryside has the same diet as 'possums, and an even better fur pelt. But it shoots back. The best way to trap a skunk is in a box trap. It need not be bigger than about 6 inches square and 18 inches long, and a skunk is just as fearless as a 'possum. Its favorite bait is anything with lots of fat to it--preferably tainted, just like with 'possums.

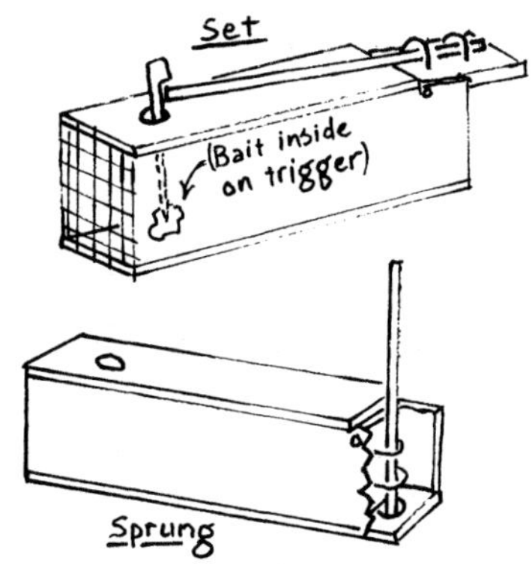

The hard problem with a skunk is not getting sprayed by its stink when you dispatch it. The best way is with a box trap with solid sides, so he can't see you approaching or carrying the trap away. Place the box into a woven feed sack, and attach its open end to the exhaust pipe of a motor vehicle. Vehicles with catalytic converters don't kill, but only torture the poor beast. You may need to rig up a tail pipe for a small engine exhaust. The skunk will fall asleep and not wake up--no spray. If you have a skunk in another kind of trap it can be killed by shooting it through the backbone, and then through the head, but this isn't as merciful as the previous way. Drowning makes them spray, as does shooting, most of the time.

To remove the stink from any object, smoke it. The best tool for this is a beekeeper's smoker. I have even smoked a friend's dog I was babysitting, and the next morning my friend never knew it had been sprayed. Neither tomato juice, strong detergent, coffee nor gasoline will completely remove the stink. Only smoke does this.

Weasels & Mink

These animals kill hens and rabbits on the small holding. Weasels do it by sucking blood from the hen's neck. Minks remove a chicken's head. Rabbits may have a small portion of the body eaten. The best way to rid these pests is to enclose the chickens or rabbits so that only one ground-level hole is left for entry. A weasel can get through a hole the size of a nickel. Place the trap in front of this. Steel traps are best, so if there are cats around, lean a board in front to make a tunnel to exclude them. A steel trap with a jaw spread of 4 to 5 inches will kill weasels and most minks. The best way to kill a trapped mink is with a good, hard rap on the head with a club. Be advised they can spray just like a skunk, though they aren't so cavalier about its use.

Groundhogs

Most folks get groundhogs in their rural gardens during the summer. These beasts can quickly make hash of all your work. There will always be an entry point, though it takes time to build up a beaten trail--by that time your garden may be half gone! Look for openings in the fence. Your garden's not fenced? Shame on you! A fence will make the beast at least have to work to get at your crops. And having dug under once, there will be an entry hole. Put your trap there, a killer trap is best, but snares work well without danger to pets or children. A snare is just a running loop of cable attached to the bottom of the fence. It looks like a honeysuckle vine to the varmint. He puts his head in the center of it instinctively, and his shoulders offer the resistance necessary to close the loop. The harder he fights it the tighter it gets. Make the loop only

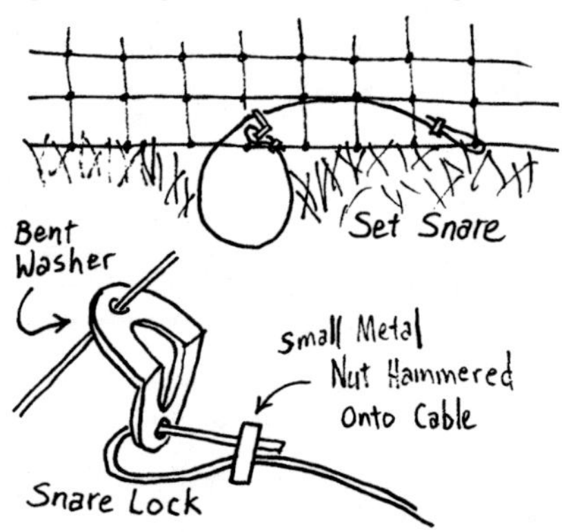

about 4 inches in diameter. Box traps are good too, if they are of the type which opens at both ends, because why should a groundhog fight to get to your bait when the fresh growing vegetables are all there to tempt him?

Rabbits

Rabbits are sometimes nearly as destructive to gardens as groundhogs. All the same methods apply in getting rid of them. A rabbit may work in either day or night, unlike groundhogs, which are mainly just diurnal.

Canines

Feral dogs, foxes and coyotes are the bane of stock keepers. The latter will kill everything from chickens all the way up to newborn calves and foals. Foxes kill anything up to the size of newborn lambs. I have seen where feral dogs had killed even stabled horses!

All these canines can be caught by one of three methods: You can just wait, day and night and shoot 'em. But they have all day and all night to hunt your livestock. Coyotes, at least here in the Alleghany moutains seem to work on a schedule which brings them around to your place once every 2 to 20 days.

All canines are susceptible to snares. The loop needs to be about 5 to 6 inches in diameter for foxes, and 6 to 8 inches for coyotes. The most common fault is making the loop too big. Believe me, the varmint will seek to put his nose right through it. For best results, use galvanized cable at least a sixteenth of an inch thick for foxes, and half again this for coyotes and feral dogs. There are a number of locking mechanisms for snare loops on the market but the easiest to make is a washer bent over at a right angle with holes drilled on the flat ends, slightly larger than the cable. I have never seen a canine, wild or domestic, killed by a snare. I know it's possible though, by providing a "tangle" stick for the enraged captive to choke itself on. But it's better if you just kill it like a 'possum or with a clean shot between the eyes.

Another way to catch canines, and the most versatile way, is with steel traps, the limb gripping kind. In short, a hole is dug and baited, imitating the way a dog hides a bone, and a trap is buried beside it. The dirt covering the trap must be sifted to prevent pebbles clogging the springs or jaws of the trap. It is important to remember that the smallest trap practical should be used. This means a size 1½ to 2 for foxes; number 3 for coyotes. Catching the animal around the main portion of the leg invites bone breakage when the canine jerks away. The more elastic portion of the leg near the foot is a better target. Two traps may be best.

Canines have exceedingly powerful senses of smell, and if they detect human odor, they will either shy away, or dig the trap up, expecting trash--men leave plenty of that lying around. When they dig up the trap, the scraping action of their foot simply springs the trap. If this happens, the "dirt hole" method is over for that individual. He's wise now. Then one may use a "urinating post" set. Wild canines have no fire hydrants, so must mark territory by peeing on trees, stumps, weed clumps, etc. Set up such an object yourself, and bury a trap beside it. Sprinkle some fox pee on the stump and "urine business." I have even caught foxes by using my own urine on the stump, but I don't know if it would always work. Deadfalls aren't commonly usable with canines, and it's just as well. A thoughtfully set steel trap rarely injures anyone's dog or cat, despite the horror stories promulgated by the animal rights protest industry. Those stories which are true are universally the grisly results of inept trappers. The calls to ban these age old tools are no more reasonable than banning all automobiles because of the grisly consequences of inept driving.

Raccoons

These animals cause considerable damage to cornfields. I lost many chickens to 'coons one year. I tried catching the coons in box traps, but they got wise to that. They will commonly trip such a trap if they can reach their paw in through wire mesh to the trigger. I finally cleaned them out using steel traps and snares. The snare should resemble a coyote snare because 'coons are very powerful animals for their size. In cornfields try to find their trails coming and going. Set your snares there. The best baits are fish and very sweet things like honey. Both these are attractive to bugs by day and other varmints by night. At first, 'coons are unafraid of traps, but if they trip a trap without getting caught they show a high degree of shyness. An "educated" 'coon is best taken in a snare. It is a rare animal which learns to avoid snares.

39
ODDS & ENDS

Whatsoever thy hand findeth to do, do it *with thy might. Ecclesiastes 9:10*

There are a few skills which also are useful for husbandmen to know, but I couldn't categorize them.

Charcoal Making

Charcoal has several uses. It is not just a fuel, it is a medicinally valuable absorbent, a pigment, a toothpaste substitute, and I don't know what all else!

I make charcoal when I burn my brush piles, by waiting for the pile to burn down to a big bed of coals, then I shovel it all into 55 gallon barrels, and cover it as well as I can with a sheet of tin. I used to make it in a pit, using mainly dismantled buildings and old wooden objects as fuel. But since brushy tree tops and prunings must be burned every now and then, why waste the by product?

In the old days, folks would stack up a cord or two of split wood around 3 thin poles stuck in the ground about 8 inches apart, in the shape of a teepee, and let it season for a year. Then it was covered with tamped earth at least 6 inches thick, and a fire was started with a long burning stick inserted to its center through an opening near the ground. A half-inch steel rod was used to poke holes in the covering every 2 feet in all directions, with a "chimney hole" at the top formed by the 3 poles. Once there was plenty of heat visible from the chimney hole, it was tamped full of dirt, and each of the other holes would be stopped as they quit smoking, and only clear heat came from them. The *collier's kill* or *kiln* as this covered stack was called, was left for several days or weeks, depending on its size, to extinguish itself for lack of air.

No matter how you make your charcoal, sift it later to separate ash, dirt and fine charcoal dust from the chunks, which are the only really pure charcoal you'll have. You get about 25 to 30, 20 lb. bushels of charcoal from a cord of wood.

Clay

Most creeks seem to have clay in them somewhere. It can be dug and then laid out on planks to dry in the sun in a half-inch layer. When dry, collect it and grind it in a common feed mill, so it's powder. Keep it dry in buckets. It can be used for making pottery, but also for encasing seeds into a clay ball for easy sowing (see small grains). Besides this, clay has many other uses, so it's nice to have some handy when you need it.

Lime

Limestone is commonly ground into "crushed stone" and pulverized into "sand" at quarries nowadays. Before the days of power crushers, it was turned into powder by burning it. The simplest way is to stack up sandstones, or other non-burning stone into a cylindrical chimney. You can use moist sand or clay for a mortar if you like. Make it about 5 feet in diameter and about 3 feet tall. Leave an opening at the bottom big enough to allow you to ignite the fire which will be contained in the ring of stones. Carefully lay kindling material at the center of the chimney. Surround this kindling with brush wood, then two fairly close layers of dry hardwood sticks on top, no bigger than your lower leg. Now cover this with a layer half as thick of limestones, broken into pieces no bigger than double your fist; the smaller the better. And keep alternating layers of wood and limestone into a conical form a couple of feet or so above the height of the chimney. Light the *lime kill* or *kiln,* as this apparatus is termed, with a burning pole inserted through the hole in the wall to the center of the pile. Once it seems to be going pretty well, heap about 6 to 8 inches of dirt atop, to slow the combustion. Try to keep it burning for 2 full days or more. You can substitute broken marble, or sea shells for limestone.

Once burned, the lime is no longer *calcium carbonate,* but is *calcium oxide,* or *quicklime,* and is highly caustic! Use rubber gloves or rakes and shovels to collect it. Sift out the rocks and ashes. Now it can be pulverized by *slaking* it.

Slack Lime

Quicklime is very unstable. It is a "thirsty" substance, and reacts fiercely with water. Thus, if quicklime is piled in a convenient place, and water sprinkled on it, it will crack, swell, and steam, falling apart at last into a powder finer than any mill could make. If you contemplate doing this, you ought not delay because though properly burned lime is easily reduced into a fine powder, this may not be the case if the *slaking* is postponed. Sprinkle it with water, and when the steaming abates, turn the whole mass with a shovel, to inspect for unslacked portions. Sprinkle those spots again. The result is called *slack lime,* and can be used as mortar for laying bricks or stones, especially when mixed with 2½ to 3 times its volume of clean sand. All the old castles of Europe were laid this way. It is also good for liming acid land, and purifying sour feed troughs, precipitating iron from water, soaking hides to remove hair before tanning into leather, preparing sorghum sap for sugar making, and on and on. It can be made back into quicklime by baking it in a slow oven for several hours until the water has been removed. The somewhat granular result used to be used for dissolving bodies without a trace.

Concrete

Portland Cement may be the single greatest boon to modern husbandmen in that it enables us to rapidly build long-lasting, very strong and weatherproof structures, including sanitary dairy buildings, pigsties, water troughs and tanks, fence posts, and other benefits too numerous to recount. It is the main ingredient in *concrete,* a combination of cement and *aggregates.* Commonly used aggregates include sand, gravel, crushed stone, broken brick, pumice, etc.

Portland cement is made in factories by burning a mixture of 12 to 13 parts limestone, 4 to 5 parts sand or other silicious material, and 1 to 2 parts ground shale or clay at 2000 to 3000°F in special kilns. The resulting clinkers are ground in a mill finer than flour. It is most commonly packed in 1 cubic foot bags which weigh about 92 lbs. When water is added to portland cement it slowly becomes very hard. It also adheres very strongly to aggregates mixed with it.

Sand is used in practically all concrete. It is defined as particles of stone small enough to pass through a ¼-inch screen. *Gravel* too, is practically always used in concrete, being particles which pass through a 1½-inch screen, but not a ¼-inch one. It is easy to accumulate these aggregates from most creeks by shoveling the *bank run* gravel (gravel mixed with sand, as it comes from the ground) through home made screens. The screens are simply wooden frames, made of "two-by-fours," covered with the appropriate meshed woven wire. They're about 2 to 3 feet wide and about 3 to 4 feet tall, propped up at a 60° angle. Shovel the bank run gravel against them at the top, and the oversized stuff will not pass through. The screened aggregates may need to be "washed" with water to float off organic matter. It isn't imperative that the water come off clear. If you buy aggregates from a quarry, it's best to get it washed and sized, not bank run or *crusher run.*

Sand and aggregates are mixed in a wheelbarrow for small batches, or on a smooth wooden or concrete floor for larger ones. The biggest jobs (those requiring more than a cubic yard of concrete) are best handled by ordering "ready-mixed" deliveries by truck. There are many places trucks can't come to. If the job is large, it must be planned so as to *pour* a portion at a time, the previous batch constituting part of the form for the next one, until the entire job is complete. Concrete won't bond well to other concrete which has *set* for more than a day or two.

Truly dry sand is exceedingly rare; it will become moist from sitting in the open air, even when protected from rain. This must be taken into account, because too much water *per sack of cement* makes weak concrete. Always pay attention to your water/cement ratio. A bottomless box is the usual way of proportioning aggregates, but many of us have just "eyeballed" it using a shovel of this and 2 shovels of that. The following information may help you plan your concrete mix:

Rough Concrete--for thick structure which need not be water tight, such as footings, retaining walls, etc.: 1 part portland cement--6¼ gallons water (max.) per sack of cement; 3 parts sand; 5 parts gravel--max. gravel size 1½ inches.
Medium Concrete--for most work; moderate wear and watertightness, such as walls above ground level, dams, hot beds and cold frames, cellar walls, citerns, septic tanks, watertight floors, sidewalks, patios, thick corner or gate posts, piers, chimney caps, and for tree surgery, etc.: 1 part portland cement--5½ gallons water (max.) per sack of cement; 2¼ parts sand; 3½ parts gravel--max. gravel size 1½ inches.
Fine Concrete--for thin structures or those subject to severe wear or weather, very watertight, such as fence posts, flower boxes, benches, bird baths, sun dials, feed troughs, etc.: 1 part portland cement--4½ gallons water (max.) per sack of cement; 1¾ parts sand; 3 parts gravel--max. gravel size ¾ inch.

First put the coarse aggregate (gravel) on your working surface. Then measure the appropriate volume of sand on top of it. Lastly, add the right amount of cement on top. Mixing in wheelbarrows is done with a hoe. On a platform it is mixed by moving it, one scoop at a time with a flat-bladed shovel, from one side of the platform to the other, and then back again. Water is then added gradually, stirring it in until there's just enough to make the mix about like fairly thick oatmeal. If it's too soupy, much of the cement will be carried away by the draining water, and of course, unless you add more cement, will be weak.

Concrete slabs are formed in frames of "two-by-fours" laid on edge (if your slab will be four inches thick) braced with stakes driven into the ground beside them. The concrete mixture is poured or shoveled into the frame, and *struck* or *screeded* by scraping a rigid board across the top to flatten and spread everything which rises higher than the frame itself. Sometimes the mix is *tamped* to remove any air pockets, using a pole with a short bit of plank attached to the bottom.

After screeding, the top may be *floated* using flat wooden or metal devices called *floats*. These can be as simple as a 2 foot long flat board with a drawer handle attached. Lay the flat side on the concrete and wiggle lightly over the surface, to encourage the finer particles to accumulate at the surface.

Once the slab is floated, it needs to stand undisturbed, and be kept moist, for at least 2 days; 7 days would be much preferred. Concrete continues to harden for at least 28 days, so it's best to keep it wet as long as possible. Special types of portland cement harden more rapidly.

In hot weather, concrete *sets,* or hardens, much more rapidly than in cool weather. Once it starts setting, any movement will drastically weaken the final result. Keep it moist with a layer of sand or straw, hosed down at least daily.

In cold weather, concrete must be kept from freezing, or it will crack on the surface, known as *crazing.* Crazy concrete is as annoying as crazy people. Crazy concrete can be remedied by pouring a *sand coat* which is a 1 inch thick layer of a rich mix of sand and cement. Do it right the first time. If you must pour concrete in winter, use special quick-setting portland cement, or special ingredients to hasten hardening. Then, once it begins to set, cover it all with wet straw, then dry straw. The heat produced by the concrete will help keep it from freezing. Check on it every day, and keep it moist and thawed. By the way, I don't know any remedies for crazy people.

Cyclopean concrete is the cheapest of all. It is made by filling your concrete with stones no bigger than about two-thirds the concrete's total thickness. The stones must not protrude above the forms, interfering with screeding, nor take up more than about two-thirds to three-fourths of the total volume of concrete. If they overlap, with plenty of concrete between them, the resulting structure will probably be improved in strength, since most stones are stronger than concrete.

Concrete for fence posts ought to be richer and finer than common mixes, and needs steel reinforcement of some sort. Rusty metal, as long as it's not flaky rust, sticks to concrete best. Barbed wire, chicken wire, old cables, and store-bought *re-bar* are most often used. Whatever's cheapest. Most concrete posts either have bolts protruding to attach wooden or wire fences, or have properly spaced holes cast into them. Either way, a single rod of re-bar, or 2 evenly spaced strands of fairly tight barbed wire will serve as reinforcement. Always tamp the forms full when making fence posts, and let them set well before being tempted to move them.

Lye

If your house is heated with wood, or you burn brush or charcoal, you will be blessed with fair quantities of ashes. Hardwood ashes are particularly useful for making *lye*. Lye is an ingredient in soapmaking, is used to clear clogged drains and even soft pretzels must be first dunked in a weak solution of boiling lye water before baking! Lye solution is the first step in preparing *potash,* or *black salts,* both of which were once a common fertilizer.

To leach lye, get a large, non-metallic vessel with a drain hole at the bottom on one side. Prop this up and set a basin under the drain hole. Put a 3 inch layer of coarse (sift the sand out) gravel in it, and mash in some straw or hay so that there is a fairly tight, 1 inch mat of it above the gravel. Now tamp the rest of the container full of dry hardwood ashes. The best lye-producing wood for ashes is water elm *(planera aquatica)* but any of the lowland elms are excellent. After them, in order, are: ash *(Fraxinus spp.)* maple *(Acer spp.)* basswood *(Tilia spp.)* hickory *(Carya spp.)* buckeye *(Aesculus spp.)* and beech *(Fagus spp.)*. Coniferous woods are no good for lye ashes.

Scoop out a funnel shaped depression in the ashes, and pour in a bucket of rainwater. If the ashes were well tamped, the water will take from 6 to 18 hours to filter through and start dripping out the drain hole. You may get 2 leachings from 1 batch of ashes. The first "run" is strong enough to cut grease. The second run should be leached through twice. Never add water until the funnel shaped depression is emptied. If any water in the catch basin evaporates, it'll just strengthen the lye. The resulting lye can be bottled as is in glass bottles, but no metal for the top. This stuff is quite caustic, and consequently somewhat hazardous in the hands of the careless or vicious. It can be boiled down to a cake, called black salts, a crude form of the pure lye one can buy so cheaply in the store. I give the recipe here only because it is so useful for so many things, and it may not forever be so easily bought.

Whitewash

The cheapest "paint" for fences, trees and outbuildings consists of plain slack (hydrated) lime and salt mixed with water. Mix 15 lbs. of salt to 6 gallons of water, then stir in 50 lbs. of lime and it's whitewash. Add 6 oz. of alum right after the salt, but before the lime, and the finished product won't be so prone to rub off easily.

Apply whitewash thinly first, so that it looks somewhat transparent. When it dries, it will be more opaque, and you coat it again. It is very cheap, not poisonous, and easily made.

Soaps

There are some naturally soapy plants, such as Soapwort, also called Bouncing Bette *(Saponaria officinalis)*. The poisonous kernels of the Buckeye *(Aesculus spp.)* are soapy, and have been used to clean fine fabrics, but I never tried them.

Leftover fat that might ordinarily go to waste can be turned into very good soap. Fat from the rendering plants is the main ingredient in most commercially made soaps. By altering the basic recipe, you can make all sorts of household soaps. Don't use rancid fat. Rendered fat and drippings are o.k.

First clean the rendered fat if it has bits of meat or rind in it, by adding an equal amount of water and bring it to a boil. Remove from the heat, stir, and add a quart of cold water for every gallon of the hot mix. Then let it sit until cool; the impurities sink, and clean fat hardens on the top.

The lye you got from leached ashes needs to be boiled until it is strong enough to just float a fresh laid egg. This is the strength used for soap making.

If you don't keep hens you won't have fresh laid eggs. So the correct strength of the lye will have to be determined as follows: Mix salt into some water until no more will dissolve. Then take a clean stick, and weight one end with a small rock or a fishing sinker so it floats vertically. Mark the place where the brine comes to when it floats. Now the stick can be floated similarly in the lye water. When the mark you made on it is poised at the surface of the water, the lye is strong enough. If it is too high, add water; too low and you'll need to boil it longer to concentrate it.

Mix six pounds of melted fat with about 1½ pints of the hot home made lye. Don't forget lye is caustic: Don't splash it on yourself. Use an iron or enamel pot, not aluminum. Boil this mix until it gets about the consistency of thick porridge. You can test if it has boiled enough by putting a few drops of it in boiling water; if it dissolves fast, it's ready. Take it from the heat and mix in about a pint of salt. This will make the soap harder when it has cooled. Pour the hot liquid soap into any flat container lined with cloth. Traditionally it is a wooden box or trough which has been well wetted. Discard any brine formed by the salt you added to the pot, when pouring the soap into molds. Let the soap stand in the mold for a day and cut it by wrapping a fine wire around it to choke off each cake. Keep it away from drafts and frost, and it will be ready to use in 2 weeks. It keeps improving until it's 6 months old.

It's easier to make soap just following the recipe on the back of the can of store bought lye.

Tanning

I have had occasion to tan hides once in a while. The way I did it wasn't strictly "tanning" but it served my purpose. The easiest hides to learn on are road-killed cats, because they are so common and their hides have so little fat in the pores. The common advice is to use groundhog, 'possum or other vermin, but 'possums are awfully fat and groundhogs are awfully tough and have no fur value. Skin the carcass and scrape off any adhering flesh and/or fat.

If the hide has been dried, soak it for an hour or two to soften it. Then rinse and partly dry it so it is still flexible. Sprinkle the flesh side with salt, fold it flesh side in, and leave it to cure for 2 days. While it is curing, prepare a tanning liquor of 1 gallon of distilled or rain water, 1 lb. of salt, and 1 oz. (by measure; approx. 4 tsp.) of concentrated sulfuric acid. Plumber's suppliers have this. If you can't find it, use 2 to 2½ oz. of automotive battery acid, which is dilute sulfuric acid. These proportions can be reduced if the hide is small. Use a plastic, wooden or ceramic container, and slowly add the acid to the water, not *vice versa*. You can store this stuff in a **WELL MARKED** plastic pop bottle until needed; it is very poisonous!

When the hide is ready, scrape again, getting all the adhering salt, flesh and fat off. There is often a film of membrane like strong tissue paper, get this off too. But mind also that bits of sliced skin tend to look like strong tissue paper too, just get the hide clean or the tanning liquor won't do its job.

Now thumb tack the hide out on a wooden floor or large board, flesh side up. It needn't be real tight, but all the flesh side of the skin should be exposed. Pour a bit of the tan liquor in the middle and swab it all around with a piece of wool or cloth stuck in the end of a cleft stick. Get it right to the edges all around, but try to avoid the hair, if you've left that on. Very heavy hides need a second application after that one has soaked in, say after a half hour or so.

Leave the hide stretched out for a half hour, then roll it up with the flesh sides together. Lay it away for about 12 hours, then check it for whiteness. Do this by pulling and stretching; if it gets white all over as you do this, it's done. If not, roll it back up as before and leave it for a few days.

When the acid has done its work, mix a heaping teaspoon of baking soda in a pint of water and swab the flesh side of the skin with it. This kills the action of the acid. Drain this off, and apply a coat of neat's foot oil to the flesh side. Hang it up overnight then wash off any excess oil with soap and water. Just before the skin gets dry, "break up" tightening fibers by pulling and stretching in every direction. You may find it easier to do this over a chair back or other improvised beam.

The above is not actually true leather, and I wouldn't advise it for heavy hides to make harness or shoe soles. These need to be tanned by first de-hairing in a strong solution of slack lime and water, then soaked in a 60:1 dilution of strong vinegar for 24 hours, shifting the hide occasionally to rinse out the lime.

Now immerse the hides in a cooled mix of 30 lbs. finely ground oak bark, ½ gal. of vinegar, and 20 gallons of boiling water. Minimize folds and wrinkles, and shift the hides daily. As soon as they've begun to soak, start another batch of the same mix a-steeping. After 2 weeks move the hide to the new batch, discard the old batch and start another, but with no vinegar. Be sure to shift the hide daily in the brew. At the end of 3 more weeks, move the hide to the new mix, discard the last batch, and make another; no vinegar. If the edge of the skin is cut through with a sharp knife at this time, the outer portions will appear dark and the center will still be light colored. In 6 more weeks move the hide again. Soak it for another 4 to 6 weeks, and check the hide for the raw streak in the center. If it is gone the hide is done. Thick hides like sole leather need up to 8 weeks more.

When the hides are finally done, scrub them well, let them dry a bit, rub 'em full of neat's foot oil as before, and then work them back and forth (2 men are needed) over a wooden beam nailed between two posts or trees until softened. This leather can be used for harness and anything else leather is good for, but I confess I've never done it this way.

Weather Lore

Here are some old sayings I will pass on, which I have found even more reliable than the government's weather reports.

No dew at sun-up; rain will soon come up.

Flying fowls perch, when foul weather lurks.

Northwest winds a-blowing, bad weather is a-going.

Buildings dry the best, with no trees to their west.

Stagnant air, don't build there.

Chimney soot falls down, rain is coming 'round.

Red sky in morning, sailors take warning;

Red sky at night, sailor's delight.

Halo round the moon, uncertain weather soon.

When sounds get clearer, rain is nearer.

To cricket's chirps in 14 seconds, two score more, degrees are reckon'd

Rain doth ride the turning tide

South or east the lightning flies,

Thunderstorm will pass you by;

North or west, the flash of lightning

Soon will give you cause for fright'ning

Five seconds from the flash to thunder: Lightning struck one mile yonder.

Wooly worm is mostly red;
mild winter lies ahead.

40
EPILOGUE

Behold that *which I have seen:* it is *good and comely* for one *to eat and to drink, and to enjoy the good of all his labor that he taketh under the sun all the days of his life, which God giveth him: for it* is *his portion.*
Ecclesiastes 5:18

Becoming a husbandman involved no change in my vocation as a musician, nor did I intend such. The hope at the outset was to better my lot in life. Did I succeed? Yes, most emphatically! Let me count the ways.

No Need to Rely on Commerce

Near the city, monetary costs were high, so paying work was vital to survival. Or I should say *pay* was vital. Today I'd still miss a lack of income, but no more than the failure of any other crop. Then again, there are trifling cash returns from every surplus produced here, and where one money-harvest falls short, another might buoy finances enough to get by. We satisfy our human needs directly, trying to avoid foolishly converting our commodities to money, and then wasting leisure time shopping to buy it all back.

Though money is no longer all-encompassing to us, it is important. It may even be central, I suppose. Life just feels more secure now as I remember the robberies, high taxes, and overall hindrances to prosperity near the city.

We don't spend much time selling our goods; if we did, who would spend their time growing them? Besides, when you have a good year, so does most of your competition, and prices drop; when you have less to sell, the prices may go up, or foreign producers may have had a good enough year to keep everyone's prices low. The city man's buying habits display no loyalty or kinship to tillers of the soil. City-made farm tools and machines will not drop in price when the price city folks pay for food goes down. If one opts to live by commerce, the city is definitely the place to do it. But we don't, and so we enjoy not only the serenity and health urbanites lack, but also almost all of the tax advantages of the money-farmers.

Freedom From Crime & Violence

It hardly needs to be said that crime rates are lower in rural areas than urban. If someone stole from me, the thief couldn't enjoy his new possessions for long, because everyone would know the legitimate owner. Where fewer folks live, a man's property is recognized. Moreover, folks recognize each other. Half my county could identify any local ne'er-do-well seen committing crime. City vandals are often not recognized by people dwelling within the same apartment house! Violence is also limited in the country because most rural homes are armed. The fear of fierce defense deters most heinous acts; such few rural barbarians as may periodically arise find they soon must flea to the city or face an uncertain future.

Cultural Opportunities

I was appalled at first by the ignorance of school-age rural young people. But on reflection I came to realize the learning of the present generation of students in the city is even more dismaying. The main difference is that in our hinterlands classrooms are not beset with the violence, depravity, and intoxication tolerated in city areas of equal or greater affluence. I venture to say the rural youngsters learn more, at least as regards knowledge essential to their future success.

Though I was warned by my family and friends before I left urbanity that there was no culture in the provinces, I must say quite the contrary is true. Yes, "country music," or perhaps rather its urban-generated surrogate, is common fare here, but is this banality any less ubiquitous or more boring than the throbbing ditties of our urban culture? In our remote highlands there are greater numbers *per capita* of persons of refined taste and sophistication--where great art and music are concerned--than I ever witnessed in the city. Naturally, the sparse population makes it impossible for their overall numbers to be able to support strong artistic institutions. Festive pilgrimages to the shrines of Parnassus edify more than a wearied lifetime at its very doorstep. What a special treat fine arts are when they haven't become mundane by their endless presence!

Better Health

The fresh air and sunshine of the countryside are proverbial in their healthfulness. Contact with fewer persons in the day slows or stops the spread of contagion. Degenerative diseases have long been linked with diet and exercise; both are beyond reproach in the country. Cosmopolitans must pay for the privilege to work their muscles at unproductive weight lifting in sweat-reeking gymnasiums. Impure food is really beyond their control, which frightens and angers them into lashing out with myriad new regulations and edicts meaningless to ruralites. They at least pacify townsmen until the next disaster. When husbandmen contract degenerative diseases, they at least can rule out poor diet and exercise as the causes.

Contentment

I think the boredom of modern life is the reason so many hours of the day are spent seeking entertainment. Our old television set has been packed away in a closet for over a decade. When I mention this to acquaintances they say "You haven't missed much." But they still daily spend hours of their humdrum lives before it. If they were contented with their lives they wouldn't seek entertainment so voraciously. I know this sounds strange coming from one whose career is in the entertainment industry, but let the facts speak for themselves. If entertainment produced contentment, then why must it be repeatedly resorted to? I think mere diversion from an otherwise dreary existence is the real effect. I speak of one who remembers this sort of life all too well.

Discontent is the basis of all mass movements to better the lot of humanity. Hidden in every mob are individual souls, hoping, praying, that the grand cause they are championing will ease the torment of their own discontent. Each has been led to believe his own world will brighten if some villain can be vanquished, or some situation changed.

Let me counsel any reader who thinks mass alliances will improve this world to rather work personally to better his own lot in life. You can do more good for the world by throwing off the tether which swings you inexorably back to bleak reality when either the righteous cause or the entertainment stops. The only successful mass movements are those which were never organized by anyone.

The best reason for being a husbandman then, is because it improves the condition of your own life, not somebody else's. It is based on the principle that your life is your responsibility. Therefore its pleasures should be your reward. For one who has never known it, the joy of returning home to husbandry after travels abroad cannot be imagined. The security, goodwill, love, and fulfillment are never tinged with the fear that somehow, someone may take it all away; once learned, it can be resumed almost anywhere away from the city. Contentment arises not from dependence, but true autonomy; in the end, only husbandry gives us that.

SOLITUDE

Happy the man, whose wish and care
A few paternal acres bound,
Content to breathe his native air
In his own ground.

Whose herds with milk, whose fields with bread,
Whose flocks supply him with attire;
Whose trees in summer yield him shade,
In winter, fire.

Blest, who can unconcernedly find
Hours, days, and years slide soft away
In health of body, peace of mind;
Quiet by day.

Sound sleep by night; study and ease
Together mixed, sweet recreation,
And innocence, which most does please
With meditation.

Thus let me live, unseen, unknown;
Thus unlamented let me die,
Steal from the world, and not a stone
Tell where I lie.

Alexander Pope, circa 1700

APPENDIX 1
HUNGRY GAP

All grazing animals prefer green grass to hay or other stored fodder. It's doubly true during the *hungry gap*. What this term means is that portion of mid-spring during which the grass suddenly takes on a vivid green, and begins growth again. Pastures are not just alluring because the alternative is mostly dry fodder. That first herbage is sweeter and more nutritious than later growth. The problem is there's not much of it, and its early growth is very slow. Livestock, especially sheep and horses, will spend so much time seeking it that they neglect the more abundant fodder you supply. It is amazing to see how rapidly such animals lose flesh. It's even more amazing to see how rapidly they regain it if during these early weeks the animals are shut up in a small lot where they have access to abundant stored fodder, but no grass.

It is always bad husbandry to turn the animals onto pasture during the hungry gap. Take sheep for example. Ewes will wax thin, eventually losing even wool. Their lambs get stunted for lack of milk. Even creep fed the lambs will never look as good as those whose dams are kept in dry lot during the hungry gap. A further peril of the hungry gap is the short herbage is far more likely to harbor sheep parasites, and they are far more likely to ingest them. Lambs are even more susceptible to worms than adults, and so are yet more stunted unless constantly fed worm medicine. Even then they're never the equal of those whose mothers are fit and full.

Horses are afflicted in the same way as sheep during the hungry gap, but suffer the further risk of *grass founder*. The grass, if allowed to grow to just flowering, will not be so over-nutritious, and the horse's system can adjust.

Cattle and goats will lose condition somewhat also during the hungry gap. The eternal excuse made for discontinuing hay too early in the spring is: "My stock won't eat it, so I'm not going to waste it on 'em."

Claptrap!

Confine them during the hungry gap and the profits are twofold: First, the animals will do much better. Second, the grass will in the end yield much more (even subtracting the extra hay consumed) if you let it grow unhampered during the hungry gap.

When turning the animals onto grass after hungry gap confinement, feed them in the morning as usual. Only after they have consumed their usual portion should you admit them to grass. Even then, wait until the dew is off it. Dew on new grass is the number one cause of bloat and colic.

APPENDIX 2
DONKEYS AND DOGS

I hesitate to expose myself to the ire of the proponents of herd-protecting animals by cautioning that it's not for everyone.

There is currently something of a fad in the use of donkeys to protect sheep from canine intruders. I have seen with my own eyes the results of keeping donkeys with sheep. Canine attacks do not abate, but they are greatly reduced, and their severity is also lessened. A single donkey grazing amid a flock of sheep will learn to intimidate intruders, especially canines.

But donkeys often protect newborn lambs from their mothers. When it's feeding time, donkeys frequently kick pregnant ewes who get too close. I have watched them bite off, or bite and then wrench off sheeps' and lamb's ears. Not all donkeys do this. But the fact is no one knows when buying a donkey whether it will do these things, or worse, to his flock. When one does obtain a peaceful guardian, he's very unlikely to sell as long as he's got sheep. Thus, certainty in the donkey's merits is too rare to take the chance with a valuable flock, at least for those of us with humble means. And the first several donkeys one goes through in finding that perfect example will probably be hard to get rid of, since they are proven hazards to a flock.

As for stock dogs, there are two sorts. One kind has a special herding instinct in certain bloodlines. Border collies are the best known breed for this, but there are others. Again, if one buys a mere puppy, even if it has inborn instinct, it must be trained. This requires a faithful commitment. So far, I have yet to see people who want to use a dog to cut the time and work of herding adhere to the daily training of that dog.

Good herding dogs are best looked on as the wise but substantial investment they are. One should only buy such a dog already trained and consider himself fortunate that the hundreds or even thousands of dollars they cost are yet far cheaper than hiring human help. The cheapest is seldom best, but the best seems to always be the cheapest in the end.

The other kind of stock dog functions solely as a guardian. There are numerous breeds, but one stands alone, the *Great Pyrenees.* This dog is big enough to ward of practically any predator short of a big bear or a lion. It doesn't "play" with sheep like other guard dogs. It kills an invading predator by knocking it off its feet through its own massive assault, and then simply ripping the belly,and guts, open. I have heard amazing stories of one dog having thus thwarted whole packs of coyotes, leaving a ripped open individual as soon as it's down and then attacking the next. The coyotes eventually die, but in the meantime they're no threat to a sheep. It isn't pretty and such dogs probably aren't safe to human intruders either. Sadly, the majority of the breed seems to have no guardian instinct at all, so it becomes needful to buy only from bloodlines of true working animals. The price is higher, but a non-working dog is worthless to a husbandman.

BIBLIOGRAPHY

Rather than follow the usual custom of listing author first, I have instead begun each entry by title, in the hope that average readers may more easily find further reading in subject areas of interest to them. Many of these books are now rare, some simply out of print; it is possible they are still available in extensive libraries and through used and antiquarian booksellers:

1. *ABC & XYZ Of Bee Culture.* Root, A. I. & E. R.; A. I. Root, Medina, Ohio, periodically published book, *A good dictionary of apiculture & related subjects.*

2. *Agricultural Testament.* Howard, Albert; Oxford U., 1943 (Rodale Reprint 1972 Emmaus, PA) *Howard, the 1st "organic" farmer, rational, no mystical mumbo-jumbo*

3. *Agriculture.* Benson & Betts; Bobbs Merrill, Indianapolis, 1927, *School text, good non-mechanized beginner's book*

4. *Back To Basics.* Mack, Norman, ed.; Reader's Digest, Pleasantville, NY, 1981, *Great deal of useful info, much unnecessary info, too*

5. *Barns, Sheds & Outbuildings.* Halsted, Byron D., ed.; Stephen Greene Press, Lexington, MA, (1973 reprint) 1881, *The best book on all types of small-farm buildings*

6. *The Basic Book of Organic Gardening.* Rodale, Robert, ed.; Rodale, ©1971, Ballantine reprint, *Excellent guide & reference for all gardeners.*

7. *The Butchering, Processing & Preservation of Meat.* Ashbrook, Frank; Van Nostrand Reinhold, NY, 1955, *Best home meat processing book available.*

8. *Camping & Woodcraft.* Kephart, Horace; MacMillan, NY, 1922, *An essential book for all rural residents.*

9. *Carla Emory's Old-Fashioned Recipe Book.* Emory, Carla; Bantam, NY, 1977, *Self-sufficiency handbook, warm, inspiring.*

10. *Charcuterie.* Grigson, Jane; Knopf, NY, 1968, *Excellent, French sausage & curing, with rare recipes.*

11. *Compost.* Billings, F. H.; Faber, London, 1942, *Simple approach to humus-making.*

12. *Country Woodcraft.* Langsner, Drew; Rodale, Emmaus, PA, 1978, *"Treenwork" & old-time implement plans.*

13. *Driving The Horse In Harness.* Kellog, Charles; Stephen Greene, Lexington, MA, 1986, *Good book on carriage use of horses.*

14. *The Essential Cookbook.* Camrass, Zoë; Bonanza Books, NY, 1985, *Reference book on the nature of food & cookery.*

15. *The Farmer's Handbook.* White, John M.; U. Of Oklahoma, Norman, OK., 1948, *Good beginner's book of commercial farming, useful tables.*

16. *Farmers Of Forty Centuries.* King, F. H.; Rodale Reprint, Emmaus, PA, ©1911, *Intriguing memoir of Far-East husbandry.*

17. *Farming For Self-Sufficiency.* Seymour, John; Schocken, NY, 1973, *Superb, by 20th Century's greatest self-reliance writer.*

18. *The Fat Of The Land.* Seymour, John; Taplinger, NY, 1962, *Warm, rich story of Seymour's first farm years.*

19. *Feeds And Feeding. 21st Ed.* Morrison, Frank B.; Morrison, Ithaca, NY, 1948, *Greatest pre-antibiotic era livestock/poultry book.*

20. *Five Acres & Independence.* Kains, M. G.; Dover, NY, (1973 reprint of 1940 ed.) *Reads like a collection of U.S.D.A. pamphlets, but much useful, if dry info.*

21. *Flight From The City.* Borsodi, Ralph, School of Living, Suffern, NY, 1947, *A founder of America's self-reliance movement.*

22. *Forage Crops.* Vorhees, Edward, B.; MacMillan, NY, 1913, *Superb book on roughage growing & feeding.*

23. *The Forgotten Crafts.* Seymour, John, Knopf, NY, 1984, *Useful, once-common rural skills.*

24. *The Gardener's ABC of Pest and Disease.* Dimock, A. W.; M.Barrows & Co., NY, ©1953. *Useful as a beginner's book of symptoms, but heavily reliant on dangerous chemicals for remedies.*

25. *The Gardener's Bug Book, 4th Ed.* Westcott, Cynthia; Doubleday, Garden City, NY, 1973. *Many color plates & descriptions of horticultural pests.*

26. *The Golden Age Of Homespun.* Van Wagenen, Jared; Hill & Wang, NY, 1953, *Unique, informative memoir of U.S. pioneer life.*

27. *Growing For Market.* Yepsen, Roger B., ed.; Rodale, Emmaus, PA, 1978, *Small-scale commercial horticulture.*

28. *The Guide To Self-Sufficiency.* Seymour, John; Hearst, NY, 1976, *Greatest British self-reliance book written to date.*

29. *Heating With Wood.* Gay, Larry; Garden Way, Charlotte, VT, 1974,

Still the best book on stoves & firewood.

30. *The Herb Book.* Lust, John; Bantam, NY, ©1974, *Most complete low-cost herbal medicine book.*

31. *The Home And Farm Manual.* Periam, Jonathan; Greenwich House, NY, (1984 reprint of 1884 ed.) *Large book of useful rural info.*

32. *How to Build & Furnish A Log Cabin.* Hunt, W. Ben; Collier Books, NY, NY, ©1974. *Useful information on non-mechanized rustic construction.*

33. *How To Have a Green Thumb Without an Aching Back.* Stout, Ruth; Cornerstone Library Publications, NY, NY ©1973, *Mulch gardening methods.*

34. *The How-To-Do-It Book Of Beekeeping.* Taylor, Richard; Linden, Interlaken, NY, 1977, *My favorite beekeeping book.*

35. *Humus And The Farmer.* Sykes, Friend; Rodale, Emmaus, PA, ©1946, *A treatise on the benefits of true soil fertility.*

36. *I'm A Stranger Here Myself.* Seymour, John; Faber, London, 1978, *A sad memoir of a happy family's demise thru the commercialization of their family farm.*

37. *The Illustrated Book Of Herbs.* Bunney, Sarah, ed.; Gallery Books, NY, 1984, *Best identifier , European, wild & garden herbs.*

38. *Keeping A Family Cow.* Grohman, Joann Sills; Ballantine, NY, 1981, *Most understandable book on house-cow keeping.*

39. *Kerr Home Canning & Freezing Book.* Kerr Glass Mfg. Corp., Sand Springs, KS, *A great inexpensive resource on food-preserving.*

40. *Laboratory Manual For Meat Science.* Smith, Savell, King, Carpenter; American Press, Boston, MA, 1988, *Modern Meatpacking procedures.*

41. *The Living Soil.* Balfour, Eve; Faber, London, 1945, *Updated versions exist; soil/health connection.*

42. *Living The Good Life.* Nearing, Helen & Scott; Schocken, NY, ©1954, *Smug, authoritarian, inconsistent, but considered a "classic" by so-called "back-to-the-landers."*

43. *The Lore Of The Land.* Seymour, John; Schocken, NY, 1983, *property management, very good, Britain-oriented.*

44. *Lost Country Life.* Hartley, Dorothy; Pantheon, NY, 1974, *Super book on pre-industrialized rural life.*

45. *The Making Of Tools.* Weygers, Alexander G.; Van Nostrand

Reinhold, NY, ©1973, *Great book for novice metalworkers.*

46. *Merck Veterinary Manual.* Merck & Co., Rathway, NJ, 1979 (5th ed.) *Top book on livestock & poultry ailments, etc.*

47. *Metalworking.* Almeida, Oscar; Drake, NY, 1976, *Best beginner's manual for smithing & repairs. Shop class text.*

48. *Milk And Its Products.* Wing, Henry H.; MacMillan, NY, 1916, *Somewhat technical, but good cheese info.*

49. *The Milk Of Human Kindness.* Douglass, William Campbell; Last Laugh, Marietta, GA, 1985, *Devastating case against pasteurization of milk (very fun)*

50. *The Modern Blacksmith.* Weygers, Alexander G.; Van Nostrand Reinhold, NY, 1974, *Very good, 1 flaw--no welding advice.*

51. *The Modern Farmer's Cyclopedia of Agriculture.* Wilcox, Earley Vernon; Orange Judd, NY, 1944, *Nearly Complete Compendium of Commercial Farm Lore.*

52. *Nuts For The Food Gardener.* Riotte, Louise; Garden Way, Charlotte, VT, 1975, *A good beginner's book for non-commercial and useful nut growing on modern small holdings.*

53. *The One-Straw Revolution.* Fukuoka, Masanobu; Rodale, Emmaus, PA, 1978, (Bantam Reprint 1985) *Unique & useful information once you get past the mysticism.*

54. *The Operation, Care & Repair of Farm Machinery.* John Deere & Co., Moline, Il., *14th ed., & before has much horse equipment info.*

55. *Our Farm of Four Acres.* (Miss Coulton); Orange Judd, NY, 1865, *Two British city ladies' successful move to the country.*

56. *Pig Ailments.* Straiton, Eddie; Farming Press, Ipswich, UK., 1988, *Most understandable book on pig-raising problems.*

57. *Plant Propagation in Pictures.* Free, Montague; Doubleday, NY, 1957, *Superb practical manual on all forms of sexual & asexual plant propagation.*

58. *Plowman's Folly.* Faulkner, Edward, U. Of Oklahoma, Norman, OK, 1943, *A rational case against plowing.*

59. *Practical Beekeeping.* Tompkins & Griffiths; Garden Way, Charlotte, Vt, 1977, *A good start-up manual.*

60. *Practical Poultry Production.* Lamon & Kinghorne; Webb, St.Paul, MN, 1922, *Excellent book on pre-factory-farming poultry.*

61. *Production of Field Crops.* Hutcheson, Wolfe, & Kipps; McGraw-Hill, NY, 1948, *Milestone book on commercial farm crops.*

62. *Productive Sheep Husbandry.* Coffey, Walter C.; Lippincott, Phila., PA, 1923, *Excellent pre-mechanized manual.*

63. *Productive Vegetable Growing. 4th ed.,* Lloyd, J. W.; Lippincott, Phila., PA, 1923, *Excellent pre-mechanized manual.*

64. *Raising Poultry The Modern Way.* Mercia, Leonard S.; Garden Way, Charlotte, VT., 1976, *Very good, especially for novices.*

65. *Raising Sheep The Modern Way.* Simmons, Paula; Garden Way, Charlotte, VT, 1976, *Excellent small-scale shepherd's book.*

66. *Sheep & Wool Science.* Ensminger, M. E.; The Interstate, Danville, IL., 1970, *Expensive, general commercial sheep farming book.*

67. *Sheep Shearing.* Bartlett, E. S.; Breeder Publications, Chicago, 1938, *Contains good hand shearing and historical info.*

68. *Starting Early Vegetables & Flowering Plants Under Glass.* Nissley, Charles; Orange Judd, NY, 1935, *Exc. greenhouse, hotbed, and coldframe management book.*

69. *Success With Small Fruits.* Roe, E. P.; Dodd, Mead, NY, 1898, *Delightful, unique look at many small fruits.*

70. *Swine Production.* Bundy & Diggins; Prentice Hall, Englewood Cliffs, NJ, 1970, *The last swine textbook before factory-farming took over.*

71. *Tan Your Hide!* Hobson, Phyllis; Garden Way, Charlotte, VT., 1983, *Many tanning recipes & leather work info.*

72. *Ten Acres Enough.* Roberts, Isaac Phillips; James Miller, NY, 1864 (3d ed.) *First U.S. author on small farm self-reliance.*

73. *Three Acres And Liberty.* Hall, Bolton; MacMillan, NY, 1922, *A founder of American self-reliance movement.*

74. *The Training & Breaking Of Horses.* Harper, Merritt W.; MacMillan, NY, 1912, *Best All-around horse training book.*

75. *Types And Market Classes Of Livestock.* Vaughan, H. W.; R. G. Adams, Columbus, OH, 1916, *One of the finest stock-judging treatises.*

76. *Vegetable Gardening.* Watts, Ralph; Orange Judd, NY, 1926, *Excellent home/market gardening reference.*

77. *The Weed Problem.* King, F. C.; Faber, London, 1951, *First successful no-till, no-chemical gardener.*

78. *The Wise Encyclopedia of Cookery.* Wm. H. Wise & Co., NY, 1949, *Much hard-to-find food prep. knowledge.*

INDEX

A

Aerobic decomposition 236
Air 9
Alfalfa 88, 154, 158
Alsike clover 159, 160
Anemia 60, 64, 66
Annealing steel 244
Apple pie filling 227
Apples 212
Apricots & peaches 213
Artichoke dibble 75
Artichokes, Jerusalem 71
Ashes, firewood 27, 29, 61, 70, 152
Assembling Jenny Lind Buildings 250

B

Bacon, dry curing 145
Bacon, smoked 142
Barley 177, 178
Barn 15
Barrow 58, 70
Bean curd 170
Beans, field 173-174
Beans, horse, or fava 173
Beans, Mexican beetle 174, 197
Beans, pole 197
Beans, soy 168-170
Bee hunting 231-232
Beef jerky 225
Beehive 229-231
Beestings 51
Beets 198
Beets, fodder 190-191
Beets, see root crops 190, 191
Bermuda grass 151, 156, 157
Black currant 216
Black Raspberries 217
Black salts 262
Black walnut 214
Blacksmithing 243
Blight, potato 165
Bloat 56, 91
Blueberries 217
Bluegrass 155
Bluegrass, Canada 156
Boar 60, 70
Bonemeal 152
Bot flies 93
Brazing 246
Bread, recipe 178
Brine curing 143-144
Brine curing ingredient table 145
Bromegrass 151, 153, 154
Brush clearing 28, 46
Buckwheat 180
Budding 211-212
Burr clover 156, 157
Bush timber 251
Butchering mutton 149
Butchering, chickens 128
Butter 122-123
Buttermilk 122

C

Cabbage 198
Calcium 27, 152
Calf 48, 49
Calf, castration 49
Calf, enterotoxemia 48
Calf, pulpy kidney 48
Calf, scours 48
Calving 50
Canine pests, control 257
Canning 226-227
Carrots 189
Castration 49, 67, 68, 69
Cattle, feeding 49
Cattle, feeding & watering 50, 52
Cattle, gestation 50
Cellar 221, 222
Charcoal 245
Charcoal making 258
Charcuterie 136-146
Check reins 104
Cheese 123-125
Cheese molds 125
Cherries 213
Chicken feed 39, 42
Chicken mash 39
Chickens, battery raising 42, 43
Chickens, predators 40
Chickens, scalding & plucking 128
Chicks 38
Chicks, watering 38
Chitting potatoes 164, 165
Churning (butter) 122
Churns 122-123
Clamping roots 190
Clapboards & Shingles 252
Clay 259
Clearing land 61
Clothing 11
Clover, alsike 159, 16
Clover, burr 156, 15
Clover, crimson 16
Clover, ladino 15
Clover, mammoth 15
Clover, red 15
Clover, sweet 15
Clover, white 15
Colic 9
Colorado potato beetle 16
Colostrum 5
Commerce
Compost 15, 23
Compost tower 23
Concrete 260-26
Construction, rough 247-25
Coppicing 20
Corn 5, 181-18
Corn crib 187, 18
Corn fodder 18
Corn growing 184
Corn stover 18
Corn-shock binder 187
Corn, harvest & storage 186
Corn, saving seed 185, 186
Corn, types of 181
Cossets (bottle lambs) 79
Courthouse 13
Cow 47-58
Cow, bedding 52
Cow, bloat 56
Cow, fodder 52
Cow, hardware disease 58
Cow, mastitis 56
Cow, milking 55, 56
Cow, pumpkin feeding 199
Cow, shed 52
Cow, soiling 53
Cow, watering & feeding 50, 52

Cracklings 142
Creep feeder 88
Crimson clover 161
Crupper 103
Cryptorchid 79
Cucumbers 196, 197
Cultivation 5
Curds & whey 123-124
Curing & larding 226
Currant, Black 216
Currant, Red 216
Currants & gooseberries 215

D

Dairy utensils 121
Dairying 121-126
Deadfall trap 254
Decomposition, aerobic 236
Dent corn 181
Dibble, artichoke 75
Dog rose 220
Dogs, working 270
Donkeys 270
Draft beasts 94-117
Dry curing 145
Dry sausage, farmer's 145
Drying meat 225
Drying vegetables 225
Ducks 44
Dung 152, 236-239
Dynamite 28

E

Earth closet 236
Eggs 38, 41
Eggs, candling 38
Eggs, hatching 38, 41
Eggs, incubation 38
Eggs, to boil fresh 41
Elastrator 78
Electricity 12
Energy 11, 239-243
English walnut 215
Enterotoxemia 48, 78
Ewe, feeding 88
Ewes, milk 78

F

Farrowing 60-61, 63-65
Fava beans 173
Feed, calculating protein 42
Fences 6, 11, 29-33, 40, 46
Fences, swine 61
Fermenting 227-228
Fertility 10
Fertility of land 152
Fertilizer, artificial 153
Fescue, meadow 153
Field beans 173-174
Financing 13
Firewood 11, 240-242
Fleece worms 80, 91, 92
Flint corn 182
Flour 178
Flour corn 182
Flux 245
Fodder beets 190-191
Food trees 206
Food, storage 221-225
Foot rot 92
Forge 243, 244
Founder 99

Fowl, exotic 45
Fruit, soft 215
Fruits & nuts 206-220

G

Garden 11
Garden seeds, longevity 205
Garden tables 200-205
Gardening 193-200
Garlic 195
Geese 43
Gestation 50, 60
Gilts 63
Goats 45-47, 53
Goats, and fruit trees 46
Goats, buck 47
Goats, gestation 46
Goobers 171
Gooseberries 216
Gooseberries & currants 215
Grafting 210-212
Grape propagation 219
Grape pruning 219
Grapes 218
Grapes, Peret's Mixture 166
Grapes, spraying 220
Grass 151-152
Grass or grain founder 99
Grazing 152
Grinding flour 178
Groundhog 256

H

Hairy vetch 160, 161
Ham, dry curing 145
Ham, smoked 142
Hames & traces 108-109
Hand shears 80, 82, 86
Hard sausage 145
Hardening steel 244, 246
Hardware disease 58
Harness, abrasions 112
Harness, backstrap 109
Harness, belly band 109
Harness, breeching 107, 109
Harness, crupper 109
Harness, hip drop 109
Harness, holdbacks 110
Harness, reins 115
Harness, saddle 109
Harness, shoulder strap 110
Harrow 114
Harvest, storage of 221-225
Harvesting wheat 176
Hay 15, 154
Hay cock 162
Hay fence 162
Hay making 162
Hay meadows 151-156
Hay plants 158
Hay, moisture content 163
Hay, storing 163
Headcheese 141
Headstall 103
Heat (estrus) 50, 60, 67
Heat, standing 60, 70
Hedges 31, 32
Hens 33-45
Hens, coops 34-36
Hens, crossbreeding 33
Hens, culling 41
Hens, dung 36, 46
Hens, egg laying 34, 41
Hens, limestone and 34

Hens, oyster shells and 34
Hens, parasites 34, 42
Hens, sex-links 33
Hickories & Pecans 215
Hip rose 220
Hog, butchering tripod 132
Hogs 6, 58-75
Hogs' heads 141
Hogs, butchering
in hot weather 136
Hogs, cleaning the carcass 131
Hogs, eviscerating 134-135
Hogs, intestines 135
Hogs, killing 59
Hogs, pumpkin feeding 199
Hogs, scalding 131-134
Hogs, scorching 131
Hogs, shooting 131
Hogs, slaughtering 131-132
Hogs, sticking 132-133
Home heating 240
Honey 229, 233
Horse 96-98
Horse beans 173
Horse, "lugger" 115
Horse, "poling" 107
Horse, age to teach 101
Horse, ailments 98-100
Horse, bit luggers 106
Horse, bitting 103
Horse, bitting harness 103, 104
Horse, butt-rope 102, 118
Horse, colic 161
Horse, colic 174
Horse, collars 106, 108, 110, 111
Horse, determining age 117
Horse, driving 105
Horse, feeding 98
Horse, foot rope 114-115
Horse, grooming 116
Horse, harness 104, 106-112
Horse, harnessing 111-112
Horse, harrowing 114
Horse, hobbling 104
Horse, parasites 99-100
Horse, picketing 116-117
Horse, plow 113-114
Horse, plowing 113
Horse, Quarters hitch 102
Horse, reins 115
Horse, runaway 114
Horse, safety 101-102
Horse, teaching "back" 106
Horse, teaching "get up" 106
Horse, teaching "whoa" 105
Horse, third hand 102, 107
Horse, training 100-117
Horse, training bits 103
Horse, training to lead 102
Horsepower 16
Humus 11, 152-153
Hungry gap 269
Hydraulic rams 26, 27

I

Ice cream 125-126
Ice house 223-225
Indian wells 19, 20
Iron, shaping 244, 245

J

Jam, rose hip 220
Jassid or leafhopper 165
Jenny Lind assembly 250

Jenny Lind building 247
Jerky, beef 225
Jerusalem artichoke dibble 75
Jerusalem artichokes 71

K

Kale 192
Kentucky bluegrass 155, 156
Kiln, charcoal or collier's 258
Kiln, lime 259
Kilning barley 178
Kohlrabi 192
Kosher dill pickles 228

L

Labor 14-19
Ladino clover 155
Lamb Chart 150
Lamb feed 88
Lamb, "bottle" 79
Lamb, bottle rams 79
Lamb, cryptorchid 79
Lamb, tail docking 79, 85
Lambing pens 79
Lambing problems 89
Lambs 77, 90
Lambs, bloat 91
Lambs, castration 78
Lambs, creep feeding 77
Lambs, parasites 92
Lambs, pneumonia 78
Lambs, pulpy kidney 78
Lambs, scours 90
Lambs, slaughtering 146
Lambs, soremouth 90
Lambs, weaning 78
Land 10, 13, 14, 28
Land's poverty 152
Land, fertility 10
Land, grazing 10
Land, hazards 12
Lard 142
Layering 210, 214, 216
Leafhopper or jassid 165
Leeks 198
Legumes 151
Lespedeza 151, 154, 156
Lettuce 198
Lime 152, 153, 259
Lime, slaking 259
Limestone 160
Livestock 15, 18
Longevity of garden seeds 205
Lucerne 158
Lumber 248
Lye 262

M

Maize 181-188
Maize fodder 186
Maize stover 186
Maize, harvest & storage 186
Maize, types of 181
Malting barley 178
Mammoth clover 159
Mangels 190-191
Manger, sheep 87
Manure 152
Maple sugar & syrup 233,234
Maslin 173
Mast 71, 72
Mastitis 56, 121-122

Materials for outbuilding 248
Meadow 151, 157-163
Meadow fescue 153
Meat curing 142-145
Meat curing, dry 145
Meat curing, overhauling 143
Medlar 214
Melons 199, 200
Metal, shaping 244, 245
Mexican bean beetle 174, 197
Mice, controlling 253
Milk 121-122
Milk fever 51
Milk goats 45-48, 52, 55
Milk, and tuberculosis 56
Milk, good & bad germs 123-124
Milk, pasteurizing 56, 122
Milking 55, 56
Mineral supplements 27, 70
Minerals, salt 50, 61
Minks & Weasels 256
Molasses, sorghum 235
Mowing with scythe 162
Mulberry 213, 214, 216
Mulching 5
Muskmelons 199, 200
Mutton 146-150
Mutton & lamb, legging out 147
Mutton, chart 150
Mutton, cutting up 149
Mutton, eviscerating 149
Mutton, fell 148
Myco-rhizal relation 237
Mycoplasmal pneumonia 60, 66

N

Nitrogen, soil 153, 154
Normalizing steel 244
Nose bots 93
Nut trees 214

O

Oats 179, 180
Odor removal, skunk 256
Onion seeds 196
Onion sets 196
Opossums 254
Orchardgrass 151, 154, 155
Orcharding 206-220
Outbuilding 248
Ox yokes 118-119
Ox, team 119
Oxen 96, 97, 117-121
Oxen & horses 94, 97
Oxen, calf yoke 118
Oxen, training 117-121

P

Paddocks 78
Pantry 222-223
Pasteurizing 56, 122
Pasture 78, 151-157
Pasturing aftermath 151
Paw-paw 214
Peaches & apricots 213
Peanut crop, shocking 172, 173
Peanuts 171-173
Pears 212
Peas 173, 197
Pecans & hickories 215
Pelletizing seeds 174
Peppers 198

Peret's mixture 166, 213
Persimmon 214
Pests 253-258
Phosphorus, soil 153
Pickling meat 145
Pie filling, apple 227
Pig heads 141
Pig skins, fried 142
Pig, butchering tripod 132
Piglets 60, 62, 63, 65, 66
Piglets, white scours 73
Pigs 58-75
Pigs' feet, pickled 138
Pigs, anemia 60, 64, 66
Pigs, cleaning the carcass 131
Pigs, congenital abnormalities 74
Pigs, creep feeding 65, 66
Pigs, eviscerating 134-135
Pigs, intestines 135
Pigs, mashed 73
Pigs, pumpkin feeding 199
Pigs, scalding 131-134
Pigs, scorching 131
Pigs, slaughtering 131-133
Pigs, weaning 60, 66
Pigs, young 58, 62, 67, 70
Planting table, vegetable 203-204
Plow 113-114
Plums 214
Pole beans 197
Popcorn 182
Pork 70, 130-136
Pork chart 137
Pork cuts 139-140
Pork skins, fried 142
Pork, cracklings 142
Pork, curing 142-145
Pork, cutting the carcass 136-146
Pork, dry curing 145
Pork, pig heads 141
Pork, sausage 141
Pork, wholesale cuts 136-137
Posts, driving 30
Posts, gate 30
Posts, wood for 31
Potash 262
Potash, soil 153
Potassium 27
Potato blight 165
Potato leafhopper 165
Potato Pests 165
Potato, Colorado beetle 165
Potatoes 164-168
Potatoes, chitting 164, 165
Potatoes, growing methods 164
Potatoes, straw 164, 166-168
Poultry 4, 33, 127-129
Poultry, butchering cone 128
Poultry, drawing 129
Poultry, scalding & plucking 128
Poultry, vermin 40
Preparing land 27-33
Preserving food 221-225
Prolapse 90
Propagating grapes 219
Propagating trees 210
Pruning grapes 219
Pruning trees 207-209, 213
Pulpy kidney 48
Pulses 151, 168-174
Pumpkins 199

Q

Quince 214

R

Rabbits 93-94
Raccoons 258
Rails & slabs, splitting 251
Rainwater 23, 25
Ram 79
Raspberries 217
Rats, controlling 254
Real estate 13
Red clover 159
Red clover,
effect on lamb crop 159
Red currant 215
Red Raspberries 217
Redtop 151, 154, 155
Reins 115
Rennet 123-124
Roosters 37
Root cellar 221, 222
Root crops 188-192
Rose hips 220
Roto-tiller 94
Rough construction 247-252
Rutabagas 190
Rye 158, 180
Ryegrass 153

S

Saltpeter 143
Sanitation 11
Sauerkraut 228
Sausage 141
Sausage casings 135
Sausage, dry 145
Saving seed 185, 186
Scalding hogs 131-134
Scion 210
Scours' 48, 90
Scythe 162
Seeds, longevity of 205
Seeds, pelletizing 174
Septic tanks 237
Shaping metal 244, 245
Sharpening tools 246, 247
Sheaf, wheat 176
Shearing 79-86
Shears, electric 82, 86
Shears, hand 80, 82, 86
Sheep 76-93
Sheep feeder 87
Sheep, bots 93
Sheep, breeding 79
Sheep, Cotswold 91
Sheep, crutching 79
Sheep, dry lot feeding 89
Sheep, feeding 87
Sheep, feeds 78
Sheep, foot rot 92
Sheep, gestation 77, 79
Sheep, meat 76
Sheep, minerals & salt 77-78
Sheep, parasites 77, 78, 92
Sheep, pasturing 78
Sheep, prolapse 90
Sheep, ram 79
Sheep, shearing 79-86
Sheep, slaughtering 146
Sheep, Troubleshooting 89
Sheep, vaccines 78
Sheep, wool 77
Shelter 11
Shepherds 76, 77
Shingles & Clapboards 252
Shoats 62
Shocking corn 187
Shocking wheat 176-177

Singletree 111, 112
Skunk 255
Skunk odor removal 256
Slack lime 259
Slaughtering sheep 146
Slaughtering, chickens 128
Slaughtering, hogs 131-132
Small grain, tillering 175
Small grain, winnowing 177
Small grains 174-180
Smithing 243
Smoke house 143
Smoking bacon 142
Smoking ham 142
Snares 256-257
Soaps 263-264
Soft fruit 215
Soil 10
Soil, flocculation 61
Soiling 53
Soremouth 90
Sorghum 234
Sorghum "molasses" 235
Sorgo 234
Sow 60, 64, 66
Sow & boar, foot problems 73
Sowing grain without tillage 174
Sows, agalactia 73
Sows, constipation 73
Sows, farrowing 60-61, 63-65
Sows, nervous 73
Soybean curd 170
Soybeans 168-170
Spinach 199
Splitting rails & slabs 251
Splitting wood 242
Spring house 223
Springs 9, 10, 20
Squash 199
Standing heat 60, 70
Starting tomato seeds 194
Steel, annealing 244
Steel, hardening 244, 246
Steel, normalizing 244
Steel, shaping 244, 245
Steel, tempering 244
Steers, training 119-120
Stones 28, 33
Storing the harvest 221-225
Stoves 240
Straw 15
Straw potatoes 164, 166-168
Straw, rye 180
Strawberries 218
Strength, horses 17
Strength, human 16, 17
Stump puller 28
Stump removal 27, 28
Sugar & syrup, maple 233, 234
Swarm 230
Swedes or rutabagas 190
Sweet clover 159-160
Sweet corn 182, 195
Sweets 229-235
Swill 70
Swine 58-75
Swine housing 61, 62, 64-66, 70
Swine, and wood ashes 61, 70
Swine, breeding 67
Swine, brood hogs 70
Swine, castration 67, 68, 69
Swine, clearing land 61
Swine, coliform infection 72
Swine, eviscerating 134-135
Swine, feed 62, 66, 67, 70, 71
Swine, fences 61
Swine, hypoglycemia 72
Swine, litters 59, 67

Swine, milk 60, 61, 64, 66
Swine, mycoplasmal pneumonia 60, 66, 75
Swine, poor growth 74
Swine, protein supplement 65, 71
Swine, pumpkin feeding 199
Swine, raising and fattening 70
Swine, rear paralysis 75
Swine, salmonellosis 73
Swine, skin parasites 74
Swine, slaughter 63
Swine, slaughtering & cleaning 131-134
Swine, streptococcal meningitis 75
Swine, swollen ear 74
Swine, teats 59, 61
Swine, trichinosis 74
Swine, troubleshooting 72

T

Taxes 13
Tegs 79
Tempering steel 244
Tetanus 99
Tillage 5, 10
Timber 31
Timber, bush 251
Timothy 53, 159-161
Timothy 53, 159
Tofu 170
Tomatoes 194
Tomatoes, blossom end rot 195
Tomatoes, Peret's mixture 166
Tools, sharpening 246, 247
Tower, compost 238
Tractor 96
Tractor, walking 95
Transplanting trees 208-209
Trap, deadfall 254
Trees 10
Trees, coppicing 206
Trees, fruit & nut 206
Trees, layering 210, 214, 216
Trees, propagating 210
Trees, pruning 207-209, 213
Trees, transplanting 208-209
Trefoils 151
Tripod, hog butchering 132
Tuberculosis 56
Turkeys 44
Turnips, rutabaga, or Swede 190

U

Urea 153

V

Varmints 40
Vegetable gardening 193-200
Vegetable planting table 203-204
Vegetable seeds, longevity 205
Vegetables, cool season 201-202
Vegetables, storage 221-225
Vegetables, warm season 202
Vermin 253-258
Vetch, hairy 160, 161
Vinegar 228

W

Walking tractor 95
Walnut, black 214

Walnut, English 215
Water 9, 10, 19-27
Water pipes 26
Watermelons 199, 200
Weaning 60
Weasel & mink 256
Weather lore 265
Weeds 10
Welding 245
Wells 10, 20-22, 24
Wethers 79
Wheat 175
Wheat flour 178
Wheat grass 153
Wheat sheaf 176
Wheat, harvesting 176
Wheat, shocking 176-177
White clover 155
Whitewash 263
Winnowing small grains 177
Wood ashes 152
Wood stoves 240
Wood, splitting 242
Woodchuck 256
Woodland 27
Woodlot 11
Wool 77, 79, 80, 82, 83, 86, 87
Wool, maggots 80, 91, 92
Wool, second cuts 86
Wool, shearing 79-86
Working dogs 270
Worms 92

Y

Yeast 178
Yogurt 125

USEFUL ADDRESSES

The following addresses are suppliers carrying products useful to husbandmen. I offer it with reservation, knowing there are others of which I'm unaware. Also, I would encourage anyone writing these companies to please include a dollar or two to cover the cost of sending you their literature. Most of these concerns are small, family-run businesses trying to make ends meet.

Bees

A. I. Root Co.
Box 706
Medina, Ohio 44256
216-725-6677
Beekeeping mfg. & supplies

Brushy Mtn. Bee Farm
610 Bethany Church Rd.
Moravian Falls, NC 28654
1-800-233-7929
FAX 1-336-921-2681
Low cost beekeepers' supplies

Dadant & Sons, Inc.
51 South 2nd St.
Hamilton, IL 62341-1399
217-847-3324, FAX 217-847-3660
Beekeeping mfg. and supplies

Cheesemaking Supplies

Erie Cheesemaking Supply Co.
7700 West Bargain Rd.
Erie, PA 16509
814-838-2841

Hoegger Supply Co.
P.O. Box 331
Fayetteville, GA 30214
1-800-221-4628

New England Cheesemaking Supply Co.
85 Main St.
Ashfield, MA 01330
413-628-3808, FAX 413-828-4061

USEFUL ADDRESSES

Cows

American Guernsey Cattle Club
P.O. Box 666
7614 State Ridge Blvd.
Reynoldsburg, OH 43068

American Jersey Cattle Club
6486 Main St.
Reynoldsburg, OH 43068-2362

Goats

American Dairy Goat Assn.
P.O. Box 865
Spindale, NC 28160

The American Goat Society, Inc.
RR 1 Box 56
Esperance, NY 12066
518-875-6708

Caprine Supply
P.O. Box Y
33001 W. 83rd St.
De Soto, KS 66018
913-585-1191
Goatkeeper's equipment

International Dairy Goat Registry
RR1 Box 265A
Rossville, GA 30741
706-375-4326

Work Horses & Supplies

American Morgan Horse Association
P.O. Box 960
Shelburne, VT 05482-0960

American Shire Horse Assn.
P.O. Box 739
New Castle, CO 81647
970-876-5980
Large Draft Horses

American Suffolk Horse Assn.
Rt. 1 Box 212
Ledbetter, TX 78946
Large breed specially bred for husbandry

American White & American Cream Horse Registry
RR1 Box 20
Naper, NE 68755
American Draft Horses

Haflinger Assn. of America
14570 Gratiot Rd.
Hemlock MI 48626
517-642-5307

Haflinger Registry of North America
14640 State Rt. 83
Coshocton, OH 43812
614-829-2790, FAX 614-829-2322
Small, docile draft horses

Norwegian Fjord Horse Assn. of North America
24570 W. Chardon Rd.
Grayslake, IL 60030
Small, docile draft horses

Welsh Pony Society of America, Inc.
P.O. Box 2977
Winchester, VA 22601
Small, docile draft horses (types C & D)

Magazines

Back Home
P.O. Box 70
Hendersonville, NC 28793
704-696-3838
Rural/Small town oriented self-reliance

Backwoods Home Magazine
P.O. Box 40
Montague, CA 96064-9496
530-459-3500
Rural living w/emphasis on self-reliance

Countryside & Small Stock Journal
W11564 Hwy 64
Withee, WI 54498
The Original Self-Sufficiency Magazine

Mother Earth News
P.O. Box 56302
Boulder, CO 80322-6302
Urban/Suburban oriented self-reliance

Mules & More
P.O. Box 460
Bland, MO 65014
573-646-3934
Draft mule farming

Rural Heritage
218 Dean Ridge Lane
Gainesboro, TN 38562-5039
615-268-0655
Animal-powered farming

Small Farmer's Journal
P.O. Box 1627
Sisters, OR 97759
541-549-2064, FAX 541-549-4403
Animal & human powered farming & self reliance

Rabbits

American Rabbit Breeders Assn.
1925 S. Main
Bloomington, IL 61701

Seed & Nursery Stock

J. W. Jung Seed Co.
335 S. High St.
Randolph, WI 53957-0001
1-800-247-5864, FAX 1-800-692-5864
Many heirloom varieties

Johnny's Selected Seeds
1 Foss Hill Rd.
RR1 Box 2580
Albion, ME 04910-9731
207-437-4301, FAX 800-437-4290
Many heirloom varieties

Mellinger's Inc.
2310 W. South Range Rd.
North Lima, OH 44452
330-549-9861, FAX 330-549-3716
Seeds, stock & supplies

Nichols Garden Nursery
1190 N. Pacific Hwy NE
Albany, OR 97321
541-928-9280, FAX 541-967-8406
Large variety of herbs & heirloom seeds

Northern Nut Grower's Assn.
4518 Holston Hills Rd.
Knoxville, TN 37914
Information & sources for nut growers

Pinetree Garden Seeds
Box 300
New Gloucester, ME 04260
207-926-3400, FAX 888-52-SEEDS
Very cheap heirloom seeds

R. H. Shumway's
P.O. Box 1
Graniteville, SC 29829-0001
1-803-663-9771, FAX 1-888-437-2733
Huge selection of heirloom seeds

Raintree Nursery
319 Butts R.
Morton, WA 98356
360-496-6400
Large variety of planting stocks; esp. for Pacific states

Ronniger's Seed Potatoes
Star Rt.
Moyie Springs, ID 83845
Large variety of heirloom spuds

Southmeadow Fruit Gardens
Box SM
Lake Side, Michigan 49116
616-469-2865
Huge selection of fruit stocks

Territorial Seed Co.
P.O. Box 157
Cottage Grove, OR 97424-0061
541-942-9547, FAX 880-657-3131
Supershmelz Kohlrabi & other Heirloom seeds

Thompson & Morgan
P.O. Box 1308
Jackson, NJ 18527-0308
800-274-7333, FAX 888-466-4769
Large Selection of pricey seeds

Totally Tomatoes
P.O. Box 1626
Augusta, GA 30903-1626
1-803-663-0016, FAX 1-888-477-7333
Huge selection of tomatoes & peppers

Sheep

American Cotswold Record Assn.
Vicki Rigel, Secretary
18 Elm St.
P.O. Box 59
Plympton, MA 02367
617-585-2026
Source list for Cotswold sheep & info.

Tools & Supplies

A. M. Leonard
241 Fox Dr.
P.O. Box 816
Piqua, OH 45356
1-800-543-8955, FAX 1-800-433-1633
Horticultural supplies

Aladdin Industries
Lamp Division
P.O. Box 100255
Nashville, TN 37224
1-800-456-1233, FAX 1-615-748-3220
Non-electric lighting

Crosscut Saw Co.
Seneca Falls, NY 13148
People powered tools

Cumberland General Store
Rt. 3
Crossville, TN 38555
615-484-8481
Hard-to-find tools for small farms

Gempler's
P.O. Box 270
211 Blue Mounds Rd.
Mt. Horeb, WI 53572
Work clothing & Safety equip.

Lehman's Hardware
One Lehman Circle
P.O. Box 41
Kidron, OH 44636-0041
216-856-5757, FAX 216-857-5785
Rare husbandry tools & non-electric appliances

The Marugg Co.
P.O. Box 1418
Tracy City, TN 37387
Austrian Scythes & fine husbandry tools

NASCO
901 Janesville Ave.
P.O. Box 901
Fort Atkinson, WI 53538-0901
920-563-2446, FAX 920-563-8296
Very large selection of mail order farm equip.

Animal Health

Jeffers Veterinary Supply
P.O. BOX 100
Dothan, AL 36302
1-800-533-3377, FAX 205-793-5179
General Livestock, Equine & Pet Supplies

PBS
P.O. Box 9101
Canton, OH 44711-9101
1-800-321-0235, FAX 216-492-1516
Livestock veterinary supplies

S. B. Wallace
P.O. Box 87
Marlinton, WV 24954
1-800-233-6914, FAX 1-304-700-7380
Livestock health products, wholesale to public

NOTES

NOTES

NOTES

NOTES

PUBLISHER'S NOTE

To order additional copies of this book please send $18.00 per book, plus $3.00 shipping and handling cost, together with your name and mailing address, to the address below.

For a current list of publications, prices, and information concerning volume discounts please send $1.00 together with your name and mailing address to:

COBBLEMEAD PUBLICATIONS
HCR 68 Box 185
Trout, WV 24991